Sustainable Development Goals Series

The Sustainable Development Goals Series is Springer Nature's inaugural cross-imprint book series that addresses and supports the United Nations' seventeen Sustainable Development Goals. The series fosters comprehensive research focused on these global targets and endeavors to address some of society's grand challenges. The SDGs are inherently multidisciplinary, and they bring people working across different fields together toward a common goal. In this spirit, the Sustainable Development Goals series is the first at Springer Nature to publish books under both the Springer and Palgrave Macmillan imprints, bringing the strengths of our imprints together.

The Sustainable Development Goals Series is organized into eighteen subseries: one subseries based around each of the seventeen respective Sustainable Development Goals, and an eighteenth subseries, "Connecting the Goals," which serves as a home for volumes addressing multiple goals or studying the SDGs as a whole. Each subseries is guided by an expert Subseries Advisor with years or decades of experience studying and addressing core components of their respective SDG.

The SDG Series has a remit as broad as the SDGs themselves, and contributions are welcome from scientists, academics, policymakers, and researchers working in fields related to any of the seventeen goals. If you are interested in contributing a monograph or curated volume to the series, please contact the Publishers: Zachary Romano [Springer; zachary.romano@springer.com] and Rachael Ballard [Palgrave Macmillan; rachael.ballard@palgrave.com].

More information about this series at http://www.springer.com/series/15486

Sinan Küfeoğlu

The Home of the Future

Digitalization and Resource Management

Springer

Sinan Küfeoğlu
Department of Engineering
University of Cambridge
Cambridge, UK

ISSN 2523-3084 ISSN 2523-3092 (electronic)
Sustainable Development Goals Series
ISBN 978-3-030-75092-3 ISBN 978-3-030-75093-0 (eBook)
https://doi.org/10.1007/978-3-030-75093-0

This Springer imprint is published by the registered company Springer Nature Switzerland AG
The registered company address is: Gewerbestrasse 11, 6330 Cham, Switzerland

Preface

What awaits us in future is a tempting question for many. Climate change and its possible impacts and depletion and contamination of the world's limited resources are alarming headlines that keep academics, authorities, and industry occupied. Tackling climate change, mitigating the adverse effects, reducing CO_2 emissions in all sectors, primarily in power, transport, buildings, and agriculture, and most importantly, building a sustainable future for future generations are some of the challenges.

To build a safer and more sustainable future, as humankind, we would like to use more renewable energy, increase energy efficiency, and reduce our carbon and water footprints in all economic sectors. The increasing population and humans' ever-increasing demand for consumption pose another question whether the world's resources are sufficient for all now and future generations. Fair access to water, energy, and food is the objective for all. In line with the United Nations Sustainable Development Goals, scientists, researchers, engineers, and policymakers worldwide are working hard to achieve these objectives. Rising sea levels will be another menace for us. A significant portion of the human population resides seaside. The growing urban population is another factor that planners should consider. These lead us to think of the possibility of sustaining life on the sea. In that case, energy and water supply seem straightforward, but what about the food?

Digital technologies and increasing digitalisation in daily life, society, and business bring hope and convenience to optimise resource production and consumption. Expanding digitalisation will be a vital tool to achieve Sustainable Development Goals. On the other hand, income generation and economic activities are needed for a sustainable life as well. COVID-19 pandemic has shown us that home offices or working from home is a prominent alternative from traditional working arrangements. Designing an environment that will be both a living space and a proper and comfortable working space is another tempting subject.

To answer all these challenges, we would like to introduce the core of smart cities of the future, the building block of the future's urban life: Open Digital Innovation Hub (ODIH). ODIH will serve as the 'Home of the Future', a fully digitalised and smart, self-sustaining building that answers all the motivation we highlight here. In ODIH, we introduce a living space that produces its water, energy, and food by minimising carbon and water footprints thanks to the Internet of things, artificial intelligence, and blockchain technologies. It will also serve as an open innovation environment for start-ups and entrepreneurs who wish to integrate their solutions into the infrastructure of ODIH and test those in real time. We believe this will be a true open innovation test bed for new business models.

Cambridge, UK
December 2020

Sinan Küfeoğlu

Contents

1 Sustainable Living Spaces and Open Digital Innovation Hub

Abstract

We dream of a self-sustaining 'Home of the Future' that will be a home and a workplace for two persons. This home should be utilising digital technologies to optimise its resource consumption as well as vital resource production. We envisage of a home capable of producing energy, water and food and an open office that will be a hosting bed for start-ups and entrepreneurs to test and monitor their daring ideas and solutions in real-life. The objective of this book is to investigate the self-sustaining concept under the water-energy-food nexus perspective. To apply the theory into practice, we will design a self-sustaining house that will serve as an Open Digital Innovation Hub (ODIH) for entrepreneurs, start-ups and researchers. ODIH, as a platform, will serve like a catalyst for exploring the boundaries of Internet of Things, Artificial Intelligence, Smart Charging and Blockchain that can be used in resource management.

The author would like to acknowledge the help and contributions of Zehra Çonguroğlu, Sefa Güngör Özkan, Umut Gergin and Hasan Ecehan Bayır in completing of this chapter.

1.1 Introduction

We dream of a 'Home of the Future' that will be a dwelling and a workplace for two persons. This home should be utilising digital technologies to optimise its resource consumption as well as vital resource production. We dream of a home capable of producing energy, water and food and an open office that will be a hosting bed for start-ups and entrepreneurs to test and monitor their daring ideas and solutions in real-life. The objective of this book is to investigate the self-sustaining concept under the water-energy-food nexus perspective. To apply the theory into practice, we will design a self-sustaining house that will serve as an Open Digital Innovation Hub (ODIH) for entrepreneurs, start-ups and researchers. ODIH will be a floating platform providing residence for two persons and can produce its own water, energy and food. The concept Self-Sustaining is defined by the Merriam-Webster Dictionary and by the Cambridge Dictionary as:

> maintaining or able to maintain oneself or itself by independent effort (Merriam-Webster 2020).
> used to describe an activity that can continue without more investment (Cambridge Dictionary 2020).

Water is a finite and essential resource for life. The growing population and inefficient use of freshwater can cause a water crisis. Over 80% of the wastewater generated in developing countries is discharged without treatment into surface water bodies, and globally, two million tons of

S. Küfeoğlu, *The Home of the Future*, Sustainable Development Goals Series,
https://doi.org/10.1007/978-3-030-75093-0_1

sewage, industrial and agricultural waste is discharged into the world's waterways (UN-Habitat 2010). Therefore, it is essential to use water in a controlled manner and manage it properly after use. In ODIH, clean water will be supplied through reverse osmosis (seawater) and humidity to water systems. It will also harvest rainwater and purify greywater to be used in cleaning and soilless agriculture.

The energy need is increasing day by day, and it becomes prevalent which resources we prefer to meet this need. In 2017, it was measured that only 17% of total energy consumption came from renewable energy sources (United Nations 2017). The three most used energy sources are oil, coal and natural gas (IEA 2020). Replacing them with renewable energy sources such as solar and wind energy has become a critical requirement to reduce carbon emissions. ODIH will essentially be a prosumer that produces its own energy and sell the excess power over Blockchain. It will have a rooftop Photo-Voltaic (PV), auxiliary Wind Turbine and a small Biogas plant. To store energy, we will employ batteries and heat pumps.

Food production is a complex process that requires many steps and natural resources such as water and land. Also, harmful chemicals used in industrial production methods for both herbal and animal food and the transportation of food after production to urban regions of the world by vehicles using non-renewable energy affect climate change. Moreover, a doubling of global food demand is expected in the next fifty years (Moll et al. 2003). Therefore, people should move towards a safe and sustainable food production that will reduce the food supply chain and harmful chemicals. The smart agriculture phase aims to provide a certain amount of food products annually. Each household will have its own food supply. The compound will also harness rainwater through smart agriculture.

In addition to these facts about water, energy and food ODIH's all household appliances will be the Internet of Things (IoT) compatible and smart. We will also have a smart charging infrastructure for Electric Vehicle use. We aim to design a Home Management System (HMS) to run and oversee the operations in the compound. HMS will make use of Artificial Intelligence, Machine Learning, Big Data, and it will be an evolved version of traditional smart meters. Also, ODIH aims to host all innovators and entrepreneurs who have ideas on digitalisation in smart homes and self-sustaining buildings. HMS will be open source, and innovators will integrate their solutions and test those in real life.

As defined by the World Commission on Environment and Development, sustainable development is "a development that meets the needs of the present without compromising the ability of future generations to meet their own needs" (World Commission on Environment and Development 1987). When 17 sustainable development goals of the United Nations' are examined, it is seen that 11 of them are related to the ODIH project. These goals and their relationship with ODIH are:

Goal 1 No Poverty: we introduce the self-sustaining concept to produce enough water, energy and food for two grown-ups over a year. We anticipate that the self-sustaining idea will reduce poverty for households.
Goal 2 Zero Hunger: Smart agriculture phase aims to provide a certain amount of food products annually. Each household will have its own food supply.
Goal 3 Good Health and Well-being: Home Management System will incorporate health applications thanks to its open-source software and suitable hardware (sensors and IoT devices) of ODIH. This way the residents' health condition will be monitored regularly.
Goal 6 Clean Water and Sanitation: Clean water will be supplied through reverse osmosis (seawater) and humidity to water systems. We will also harvest rainwater and purify greywater to be used in cleaning and soilless agriculture.
Goal 7 Affordable and Clean Energy: ODIH generates its own energy primarily from rooftop solar power (PV) systems. Auxiliary renewable energy supply will come from wind and biomass.
Goal 9 Industry, Innovation and Infrastructure: by harnessing digital technologies and building an open-source HMS, our aim is to foster innovation in smart cities. New digital solutions will be easily integrated and tested in ODIH.
Goal 11 Sustainable Cities and Communities: ODIH is designed to be the very core of smart cities of the future. By being genuinely self-sustaining and employing digital connection to the outer world, similar units will be easily connected to each other to form new sustainable communities.

Goal 12 Responsible Consumption and Production: HMS will monitor and reduce resource consumption within ODIH. Furthermore, liquid waste (black and grey water) will be completely recycled, whereas solid waste will be recycled to a certain extent.
Goal 13 Climate Action: By consuming 100% renewable energy over a year, ODIH will prevent 10 tons of CO_2 emissions per year, saving roughly 425 trees from cutting annually.
Goal 14 Life Below Water: ODIH is designed to be a floating platform. Similar floating objects such as ships and boats dump their liquid waste into the sea, thus contaminating the ecosystem and harming the life below water. ODIH recycles its liquid waste and therefore protects life below the water by not polluting the ocean.
Goal 15 Life on Land: Increasing human population on earth will bring an increased demand for food and, therefore, farmlands. Deforestation for farming is a significant concern for life on land. ODIH promotes soilless agriculture by using the Nutrient Film Technique. We aim to reduce land use for agriculture, thus protect life on land.

In the literature, there are numerous indices to assess the overall concept of sustainability in the built environment. To better understand the characteristics of ODIH, some prevalent concepts, which are self-sustaining, sustainability, net-zero, green building, zero waste, self-sufficient, passive house and smart home, are defined in Table 1.1. After definitions, the relationship of these concepts with ODIH will be evaluated.

Sustainability is quite a generic concept, so that there are sub-concepts like sustainable housing, sustainable design and sustainable construction. These concepts generally aim to use resources as efficiently as possible from production to consumption in the houses or systems. ODIH, on the other hand, uses digital technologies for the efficient use of water, food and energy but also produces its own resources. Nevertheless, we should make one primary remark that we do not aim to use sustainable materials in construction. We deliberately omitted material science and building technologies and solely concentrated on the integration of digitalisation with resource management.

Net-zero is generally associated with energy and includes concepts such as net-zero energy and net-zero energy building. Over a year time, ODIH produces its total energy need in net terms so that it can be evaluated under the definition of net-zero energy building.

The concept of Green Buildings has been defined by standards set by many institutions in the literature. Although each institution uses different criteria and definitions, five common issues emerged together. These five titles are listed as follows: Water, Energy, Pollution, Innovation, and Resources and Materials. Although ODIH might be included in the scope for Water and Energy, it is not fully covered by the definitions of Green Buildings. However, the metrics of Green Buildings are hazy, and there are no certain numerical standards. For instance, speaking of 'saving water in buildings', we are not sure how much water should be saved or recycled. Nevertheless, in ODIH we aim to produce all water demand over a year within the compound. If we briefly touch on Resources and Materials, Green Buildings are expected to be made of recyclable or sustainable materials. However, such a situation is not expected within the scope of the ODIH project.

Zero-waste is a very difficult concept to achieve. It aims not to generate any waste that will disturb the ecosystem from the production to the use of any commodity or product. Although ODIH has similar aspirations in terms of waste treatment, it does not promise to be 'zero-waste'. On the other hand, the concept of self-sufficiency seems to be fitting the very nature of ODIH. However, we believe the concept of self-sustaining is a better match since ODIH also aims to be engaging in economic activities via selling energy within the Smart Charging scheme. Our scope also covers harnessing digital technologies for resource optimisation. Thus, income generation and digitalisation make our perspective larger than self-sufficiency.

Passive houses use the sun effectively to heat themselves and store and reuse heat together with a highly efficient heat recovery unit. However, ODIH directly stores the energy it generates in its batteries. Unlike ODIH, passive houses use resources such as oil and gas, albeit sparingly.

In addition to ODIH's self-sustaining character, the smart home concept should also be explained.

Table 1.1 Sustainable built environment definitions

Terminology	Citation	Definition
Net zero	Hernandez and Kenny (2010a)	"The term 'net-zero energy' being applied when the balance is zero"
	U.S Department of Commerce (2019)	"A net-zero energy house produces as much energy as it consumes over the course of a year"
	The U.S Green Building Council (2020)	"Net-Zero Energy recognizes buildings or spaces that achieve a source energy use balance of zero over a period of 12 months"
	Tucker (2012)	"Net Zero Energy Building is one which generates as much power and energy as it consumes"
	Hernandez and Kenny (2010b)	"The overall annual primary energy consumption is equal to or less than the energy production from renewable energy sources on site"
Self-sufficient	Moench (2000)	"Self-sufficient homes make their own energy, produce their own food and do their part to save the world's environment"
	Open Access Government (2019)	"Self-sufficient homes supply all their own energy, water, sewer needs, and food"
Green buildings	Real Estate Investment Blog (2020)	"A high performing home that is energy and water efficient has good indoor air quality, uses environmentally sustainable materials and also uses the building site in a sustainable manner"
	Desjardins (2020)	"Green homes are healthy, energy-saving, affordable, accessible and sustainable homes"
	Heindenry (2016)	"A green home is with environmentally friendly materials and sustainably built, with a focus on the efficient use of water and energy"
Smart homes	Aldrich (2003)	"A residence equipped with computing and information technology which anticipates and responds to the needs of the occupants, working to promote their comfort, convenience, security and entertainment through the management of technology within the home and connections to the world beyond"
	Spigel (2005)	"The smart home is a sentient space where human subjects and domestic objects speak to one another via intelligent agents and internet connections"
	Ricquebourg et al. (2007)	"Houses that have user interfaces such as white goods, sensors, motors, control devices or graphics in the door environment, have a residential gateway that makes it possible to connect to the outside world and have a network that enables the communication between service providers and residents in the external environment"
Zero waste	Lehmann and Crocker (2012)	"It is shorthand for the better management of resources in an increasing number of corporations and governments around the world. Zero waste is a way of thinking and doing that will become even more commonplace and important as we attempt to deal with environmental, social and economic issues facing all of us"
	Kahn and Islam (2017)	"In an ideal, zero-waste scheme, the products and by-products of one process are used for another process. The overall goal of this approach is to generate new solutions that convergent over long periods of time, so the natural ecosystem is not disturbed"

(continued)

Table 1.1 (continued)

Terminology	Citation	Definition
Sustainable housing	Edwards and Turrent (2000)	"Housing that meets the perceived and real needs of the present in a resource-efficient fashion whilst providing attractive, safe and ecologically rich neighbourhoods"
	Saidu and Yeom (2020)	"Sustainable and affordable housing is housing that manages and coexists within the limitations of available scarce resources, at the same time preserving and conserving these resources for future needs"
Sustainable construction	Edwards and Turrent (2000)	"Sustainable construction "The creation and management of healthy buildings based upon resource-efficient and ecological principles"
Sustainable design	Edwards and Turrent (2000)	"Sustainable design Creating buildings which are energy-efficient, healthy, comfortable, flexible in use and designed for long life"
Passive houses	Passive House Institute (2015)	The Passive House is a building standard that at the same time, is genuinely energy-efficient, convenient and inexpensive
Self-sustaining	Oxford Learner's Dictionaries (2020)	"Able to continue in a healthy or successful state without help from anyone or anything else"

Table 1.2 Certificates

Certifying body	LEED	BREEAM	Green globes	Green Buildings Council Australia	Green Buildings Council Australia
Criteria Of Certification	Water efficiency	Energy	Indoor environment	Energy	Ecological quality
	Energy and atmosphere	Water	Emissions, effluents and other impacts	Transport	Economic quality
	Materials and resources	Waste	Integrated design	Land use and ecology	Sociocultural and functional quality
	Indoor environmental equality	Transport	Site	Health and well-being	Technical quality
	Innovation in design	Land use and ecology	Energy	Water	Process quality
	Regional priority	Health and well-being	Water	Materials	Site quality
	Sustainable sites	Management	Resources-systems options, analysis and materials selection	Pollution	
		Innovation		Management	
		Materials		Innovation	
		Pollution			

Smart homes can fulfil many functions such as comfort and security, and residents can communicate with interior objects. Nevertheless, the level of intelligent communication is hazy, and the whole concept lacks standardisation.

There are various organisations that issue numerous certificates for the buildings. These certificates and the assessment criteria are summarised in Table 1.2.

As can be seen in Table 1.2, in terms of sustainability, the common criteria that organisations emphasise are:

- Water
- Energy
- The Pollution
- Innovation
- Resources and Materials

ODIH will introduce its own criteria. ODIH will be a testbed for a novel sustainability index that we name 'self-sustaining' criteria. The assessment is done in three dimensions, namely: water, energy and food.

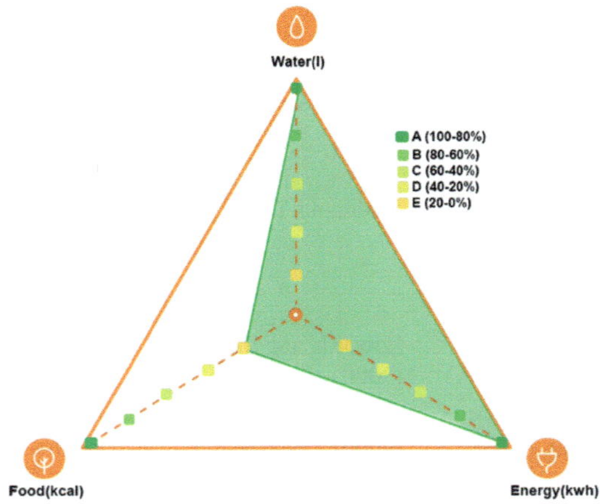

Fig. 1.1 ODIH self-sustaining assessme

1.1.1 The Self-sustaining Concept

In addition to covering resource optimisation through digitalisation, ODIH defines the concept of self-sustaining as supplying its water-energy-food resources and sustaining it over a year without the need for external investment. We focus on creating a self-sustaining system in terms of food, water and energy. While being self-sustaining in these three fields, the main objective is to utilise digital technologies to minimise resource consumption and optimise resource production with the aid of our HMS. We propose a new standard, self-sustaining, with three dimensions: water, energy and food. The assessment is done in terms of necessary resource generation over a year. We will have 5-levels: A: 100–80% B: 80–60% C: 60–40% D: 40–20% E: < 20%. Let us clarify the self-sustaining concept over an example. If a building generates 65% of its yearly energy consumption, its self-sustaining credit will be B (energy). Similarly, water and food will have credit notes as well.

In ODIH, we managed to produce 100% of yearly water and energy demand and only 20% of annual food demand. This means it will have the self-sustaining credit of AAE. Figure 1.1 illustrates the self-sustaining grading system.

We would like to extend the metric of self-sustaining to large public buildings, schools, and university campuses or even to cities or countries. Self-sustaining implies the amount of external investment or expenditure a closed system needs over a year. Through this study, we aim to demonstrate the significance of digitalisation in achieving true sustainability, especially in the built environment.

1.1.2 The Design of ODIH

To provide both a building block for the smart cities of the future and to show that life above the water is possible, we designed ODIH as a floating platform. A fully digital living space with communication capabilities with the outer world at the same time producing vital resources for the residents' survival in the long run. Figure 1.2 illustrates the fundamental relationship between ODIH and the smart cities of the future, and Fig. 1.3 shows the exterior design of ODIH.

ODIH will be a two-story floating platform. Downstairs is designed as an open office for researchers and entrepreneurs. There will be water storage tanks here. In the upstairs, there will be a one-bedroom flat, and two persons will be living here. Upstairs will also have four soilless farming rooms. The main purpose will be to observe and collect resource consumption data of these residents. All resource production data of ODIH, water, energy, and food and consumption data will be shared online with open access for research purposes. Figure 1.4 illustrates the components of ODIH and their locations.

Figures 1.5, 1.6, 1.7, 1.8, 1.9 present renders from interior spaces of ODIH (Fig. 1.10).

Fig. 1.2 Smart charging and data flow between the smart city and self-sustainable compound

Fig. 1.3 External design of ODIH

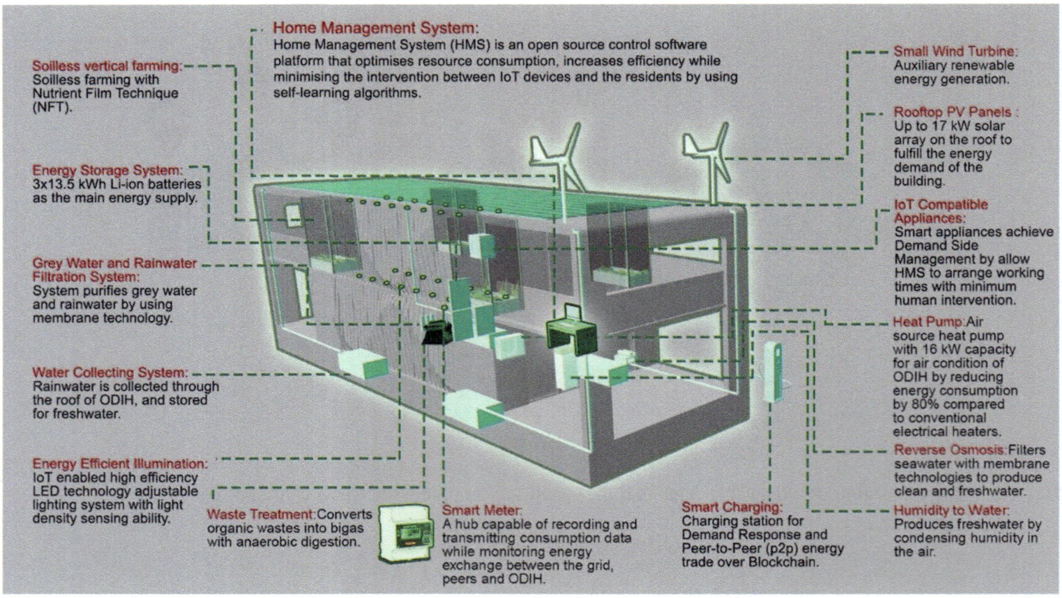

Fig. 1.4 Components of ODIH

Fig. 1.5 The effect of shades in the interior

Fig. 1.6 The effect of shades in the interior

Fig. 1.7 Smart farming lanterns

1.1.3 A Breakdown of the Detailed Content

There will be eight chapters in this book. The names and contents of the following chapters are as follow:

Chapter 2: Water

- Clean water supply
- Wastewater treatment and waste management
- Rainwater management
- Heat pumps

Chapter 3: Energy

- Energy generation
- Photo-Voltaic (PV) installation
- Small wind turbine installation
- Storage (Batteries)

Chapter 4: Food

- Smart Agriculture
- Soilless farming

Chapter 5: Enabling Technology: Internet of Things (IoT)

- Smart household appliances
- Other smart appliances and equipment
- Hardware research to build up the Home Management System (HMS)

Chapter 6: Home Management System: Artificial Intelligence

- Artificial Intelligence (AI) & Machine Learning (ML)
- Software research work to build up the Home Management System (HMS)

Chapter 7: Demand Response and Smart Charging

- Demand Response

Fig. 1.8 Working area for the researchers

Fig. 1.9 Landscape of ODIH and working area

- Smart charging
- E-mobility integration

Chapter 8: Blockchain and Peer-to-Peer Trading

- Blockchain applications
- Peer-to-Peer energy trade and smart contracts
- Establishing ODIH as a financial entity

Fig. 1.10 Living room

ODIH will be a test-bed for open innovation and novel business models. All electric devices will be IoT compatible, and the HMS will be open source. That means the start-ups and entrepreneurs will be able to integrate their products to ODIH and use its infrastructure. Following a Living-Lab concept, new business models will flourish here. Nevertheless, we deliberately omitted the business model development dimension of ODIH to keep the focus of this book on digitalisation and resource management. Therefore, the book covers science (water-energy-food nexus) and digitalisation in detail. We leave the business development dimension for a follow-up study.

References

Aldrich FK (2003) p 17

Cambridge Dictionary (2020) [Çevrimiçi]. Available at. https://dictionary.cambridge.org/dictionary/english/self-sustaining. Erişildi 30 Dec 2020

Desjardins (2020) Desjardins [Çevrimiçi]. Available at: https://www.desjardins.com/ca/personal/loans-credit/mortgages/green-homes-program/choosing-green-home/what-are-green-homes/index.jsp

Edwards B, Turrent D (2000) Sustainable housing : principles & practice. E and FN Spon, London, New York

Heindenry M (2016) Official website of realtor [Çevrimiçi]. Available at: https://www.realtor.com/advice/home-improvement/what-are-green-homes/

Hernandez P, Kenny P (2010a) From net energy to zero energy buildings: defining life cycle zero energy buildings, p 817

Hernandez P, Kenny P (2010b) From net zero energy buildings: defining life cycle zero energy buildings (LC-ZEB), p 817

IEA (2020) Total final consumption (TFC) by source, World 1990–2018 [Online]. Available at: https://www.iea.org/data-and-statistics/?country=WORLD&fuel=Energy%20consumption&indicator=TFCbySource. Accessed 20 Aug 2020

Khan MM, Islam R (2017) Zero waste engineering: a new era of sustainable technology development, 2 dü. Scrivener Publishing, Wiley, Hoboken, NJ; Beverly, MA

Lehmann S, Crocker R (2012) Designing for zero waste : consumption, technologies and the built environment. EarthScan, Abingdon, Oxon; New York

Merriam-Webster (2020) [Çevrimiçi]. Available at: https://www.merriam-webster.com/dictionary/self-sustaining. Erişildi 02 Jan 2021

Moench M (2000) Encyclopedia for self-sufficient homes: a complete reference book for new self-sustaining home designs. Osprey Press, Oxford

Moll H, Uiterkamp AS, Gerbens-Leenes W (2003) Design and development of a measuring method for

environmental sustainability in food production systems. Ecol Econ 2:231–248
Open Access Government (2019) Open access government [Çevrimiçi]. Available at: https://www.openaccessgovernment.org/self-sufficient-home/59276/
Oxford Learner's Dictionaries (2020) Oxford learner's dictionaries [Çevrimiçi]. Available at: https://www.oxfordlearnersdictionaries.com/definition/english/self-sustaining. Erişildi 10 Sept 2020
Passive House Institute (2015) Passive House Institute [Çevrimiçi]. Available at: https://passivehouse.com/02_informations/01_whatisapassivehouse/01_whatisapassivehouse.htm. Erişildi 2020
Real Estate Investment Blog (2020) Real estate investment blog [Çevrimiçi]. Available at: https://sdjsa.org/definition-green-home.htm
Ricquebourg V et al (2007) The smart home concept: our immediate future, p 24
Saidu AI, Yeom C (2020) Success criteria evaluation for a sustainable and affordable housing model: a case for improving household welfare in Nigeria Cities. Sustainability (switzerland) 12(2):4
Spigel L (2005) Designing the smart house, p 1
The U.S Green Building Council (2020) Official website of The U.S Green Building Council [Çevrimiçi]. Available at: https://www.usgbc.org/programs/leed-zero
Tucker LM (2012) Net zero housing: the architects' small house service bureau and contemporary sustainable single-family house design methods for the United States. J Interior Des 37:3
U.S Depertment of Commerce (2019) National Institute of Standarts and Technology [Çevrimiçi]. Available at: https://www.nist.gov/industry-impacts/net-zero-energy-house
UN-Habitat (2010) Sick water: the central role of wastewater management in sustainable [Çevrimiçi]. Available at: https://www.un.org/waterforlifedecade/waterandsustainabledevelopment2015/images/water_quality_eng.pdf. Erişildi 20 Aug 2020
United Nations (2017) Ensure access to affordable, reliable, sustainable and modern energy [Çevrimiçi]. Available at: https://www.un.org/sustainabledevelopment/energy/. Erişildi: 20 Aug 2020
World Commission on Environment and Development (1987) Report of the World Commission on Environment and. Oxford University Presss, New York

2 Water

Abstract

The world's freshwater sources are already limited and existed human activities constantly pollute sources. This chapter aims to present a comprehensive technology selection for freshwater generation, waste treatment, heating, ventilation, and air conditioning (HVAC) systems. Smart cities, which we believe should be built upon the concept of self-sustaining, cannot be considered independently of activities such as water, waste, and air conditioning management. We selected specific technologies in the Open Digital Innovation Hub (ODIH) to provide air conditioning, water generation and consumption, and waste disposal needs with optimum resource consumption. Greywater and rainwater will be purified by membrane technology to reproduce usable water. The salty seawater located beloved ODIH will be desalinated by the Sea Water Reverse Osmosis (SWRO) system, and lastly, the system which produces high-quality water from moisture is the main water production systems. Toilet and organic wastes are treated by a small-scale bioreactor that works with the anaerobic digestion principle. Exhausted gas will be used in a particular stove. For the HVAC, we will use an air source heat pump with a 16-kW capacity. According to the results of our calculations, ODIH can produce its required water without a water supply network connection.

The author would like to acknowledge the help and contributions of Ali Barışcan Kaya, Ali Kemal Dikmecli, Doğu Mert Özkan, Efe Durmazkul, Güney Yurtsever, Yasemin Beril Kılıç in completing of this chapter.

2.1 Introduction

Water occupies a prominent place in every aspect of our life. Continuity of life is directly connected to water. Water also takes part in enabling transportation, stabilizing temperature, cushioning, which provides protection during an earthquake, cleaning and breaking down wastes, enabling production, providing home, and being a key point for agriculture (Horspool 2019). Also, according to the research carried out, the human body can endure for a few weeks without food, while only a few days without water (European Federation of Bottled Waters 2020). Despite all these benefits, freshwater sources, lakes, and even the atmosphere are being polluted every minute. Water in the world, as a source, is limited. Rainfall regimes differ according to the location (Mullen 2020). Around 97% of the hydrosphere is saltwater, whilst only 3% is freshwater which is critical for terrestrial and freshwater species (Advancing Global Change Science and Solutions 2020). The water distribution in the world is shown in Fig. 2.1.

S. Küfeoğlu, *The Home of the Future*, Sustainable Development Goals Series,
https://doi.org/10.1007/978-3-030-75093-0_2

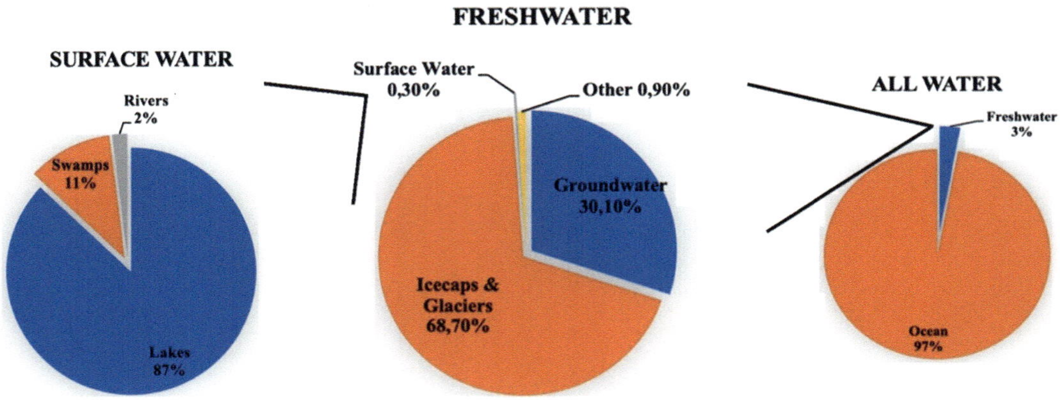

Fig. 2.1 Water distribution on earth (Solutions 2020)

Table 2.1 Water consumption per capita in the world, 2017 (Statista 2017)

Countries	Water consumption per capita (m^3)
USA	1206.8
Canada	883.3
Belgium	883.9
Turkey	746.8
Mexico	704

2.1.1 Current State of Water

3% of the water in the world is freshwater, and it might sound a fair amount for fair use. However, the increasing population limits access to this limited water supply. According to access to freshwater, countries might be classified as "water-rich" or "water-poor" based on the amount of water per capita in the country. Countries with an annual water amount of more than 8000 per capita are among the water-rich countries, countries with less than 2000 m^3 per capita fall in water scarcity group and countries with less than 1000 m^3 per capita are defined as water poverty group (Uyar 2018) However, given the population estimate of 100 million for the year 2030, the freshwater consumption will be 1120 m^3/year per capita when the current amount and consumption of water remains constant (Uyar 2018). Table 2.1 gives the top 5 countries water consumption per capita in the world in 2017.

Looking at the distribution of water in the continents of the world, North America comes first. The top 5 countries rich in water are Brazil, Russia, Canada, Indonesia, and China. Besides, the top 5 countries that are poor in terms of water are Israel, Jordan, Libya, Mauritania, Cape Verde (Shiklomanov and Rodda 2003).

As explained above, population and water consumption are increasing in parallel. Unfortunately, this consumption is not distributed equally to everyone in the world. 844 million people cannot reach potable water (WHO 2019), and almost 2 billion people are obliged to use contaminated water (Unicef Data 2019). As a result of low-quality drinking water services, food production, gender equality, education, health subjects crucially affected (Abundant Water 2020).

The world's atmosphere contains 37.5 billion gallons of water (Watergen 2020). When the atmospheric humidity increases to a limit, it falls as rain. By replicating this process, humidity can play as a secondary water source for ODIH (Open Digital Innovation Hub).

2.1.1.1 The Future of Water in the World

Scarce water resources are a risk for everyone and every generation. The number of countries affected by the water crisis will increase due to the growing population and economies. At this point, the risk perception regarding water becomes important. Water scarcity causes problems such as health problems (inadequate hygiene conditions even with access to water), climate change, and loss of biodiversity (Aksoy and Öktem 2014).

The gap between worldwide water supply and demand is projected to arrive at 40% by 2030. Local people, farmers, industries, and governments

Fig. 2.2 Shanghai (Central 2020)

are affected in climates with more water demand around the world (World Economic Forum 2017). If no action is taken between 2025 and 2050, potable water demand will be increased by 3,435 million litres (Environment Agency 2020).

Sea level will be increased around 11–16 cm in the 20th due to climate change (Kulp and Strauss 2019). It could rise another 0.5 m this century (Hay et al. 2015). If carbon emissions levels continue in the twenty-first century, the water level worldwide could exceed the 2-m threshold because of the melting of the Antarctic ice sheets (Wong et al. 2017). Therefore, the people in oceanfront settlements worldwide should be alerted to the increase in ocean water level, and coastal planning should be done (Kulp and Strauss 2019). On the other hand, rising oceans could influence three times greater than the number of individuals by 2050 than recently suspected, as per new exploration, threatening to everything, including eradicating a portion of the world's great coastal cities. Figure 2.2 shows maps of Shanghai, China. The map on the left is the current map of 2020; the map on the right is for 2050, according to research. Areas expected to be below the current water level are shown in blue.

In Shanghai, which is one of the biggest megacities in the world, water takes steps to expend the core of the city and numerous different urban communities around it. The new information indicates that 110 million individuals living arrangement will be submerged (Lu and Flavelle 2019).

2.1.1.2 The Future of Water in Turkey

Turkey is in the medium to the high-risk group. According to the water hazard index to be experienced in agriculture, Turkey is among the first 15 countries with the highest risk in the 2024–2050 period (OECD 2017).

2.1.1.3 What is Water-Energy-Food Nexus?

In 2011, more than 550 people representing stakeholder groups gathered for a conference in Bonn to declare a game-changer term: The Water, Energy and Food (WEF) Security Nexus Solutions for the Green Economy (Nexus-The Water, Energy & Food Security Platform 2020). WEF Nexus is targeting to improve basic services and economic development by tackling the issues in water, energy, and food holistically. The Nexus advance underscores the importance of commitment between water, energy, and food security. It is based on understanding the connectivity between those three (Nexus-The Water, Energy & Food Security Platform 2011). The three "supply securities" (water, energy, and food) depend on an ecosystem. The ecosystem involves the three resources land, water, and energy (atmosphere), and there is an interaction between supply securities and the ecosystem (land, water, and atmosphere). The ecosystem must be stable and protected.

2.1.2 Water Perspective

United Nations, World Bank, World Economic Forum, and other institutions are seeking ways to prevent water scarcity which can affect mostly the poorest regions in the world. In 2015, the United Nations launched a series of goals (Sustainable Development Goals) which are designed to make better conditions for the people who are living in poor conditions until 2030. The sixth Sustainable Development Goal (SDG) is about water and sanitation. According to World Bank, 2.2 billion people do not have a potable water supply, whereas 4.2 billion people lack reliable sanitation services, and 3 billion can not reach proper standard handwashing facilities (The World Bank 2020a, b, c). Clean water supply will be more critical for the Covid-19 pandemic and other expected pandemics. The amount of water that a person needs is 7.5 L per day in an emergency, and this amount of water includes drinking, basic hygiene practices, and basic cooking needs (World Health Organization 2013). According to UNICEF, 600 million children will experience water stress by 2040 (UNICEF 2017a, b). Lack of water also affects food and energy production. 70% of the total global freshwater used by the agriculture sector (The World Bank 2020a, b, c). By considering 26% of total employment in the agriculture sector, a water crisis would have a severe impact on some economies (The World Bank 2020a, b, c). Climate change poses another threat to water supplies. Research shows every 1 °C increase in temperature, 7% moisture can be held in the atmosphere, and this leads to more natural disasters. Because of this trend, 9 out of 10 natural disasters are water-related (UNICEF 2017a, b).

From the energy perspective, water is a vital element in energy-producing. In China energy sector occupies 10% of all freshwater usage (Pradeep and Lijin 2017). Water is a crucial tool of the energy industry not only in power plants but also in mining and energy storage.

When it comes to building a self-sustaining compound, Water-Energy-Food Nexus must be considered to approach prominent problems like salination, soil pollution, rising water demand, etc. The connectivity between the three supply securities will provide a broad understanding of the design process.

2.1.3 What is a Sustainable Compound?

The lexical meaning of sustainability is "causing little or no damage to the environment and therefore able to continue for a long time" (Cambridge Dictionary 2020). From this definition, "Sustainable Compound" means and aims recycling, resource and water preservation, waste and energy management, water production, and environmental sensitivity (Onubi 2019). In green building certification programs like BREEAM (Building Research Establishment Environmental Assessment Methodology) and LEED (Leadership in Energy and Environmental Design), sustainable houses have six main evaluands and subtitles (Somalı and Ilıcalı 2020). These are water efficiency, environmental impact, material, human prosperity and health, electro-mechanic systems, and general issues like energy management. Major subtitles of these evaluands are monitoring of energy and water consumption, waste treatment, energy efficiency, recyclable equipment usage in building, and thermal comfort (Somalı and Ilıcalı 2020). In a word, a sustainable compound produces all requirements by itself, monitoring and use them recursively.

2.1.3.1 Needs of a Sustainable Compound

ODIH aims to provide the requirements of two adult persons such as clean water, management of waste, and HVAC (Heating Ventilation and Air Conditioning).

A. Clean Water

   ODIH decided to build in Istanbul, Turkey. The daily water requirement is 189 L per capita in Istanbul for 2017 (TEMA 2020a, b). According to estimation in research, this value becomes 195 L per capita for 2020 (Ertem and Doğan 2016). 26% of this water use for water closet, 22% for washing machine, 17% for a shower, 16% for valve, 14% for leakage, and 5% for others (TEMA 2020a, b).

B. Waste Management
Average wastewater production was 183 L per capita in Turkey in 2016 (Turkey Ministry of Environment and Urbanisation 2016). On the domestic consumption side, 60–90% of usable water became wastewater after used in the house (Demir et al. 2017). Wastewater conversion ratio except as 75% (Alsulaili et al. 2017). According to this value, ODIH's estimated daily wastewater production is approximately 105 L per capita. Wastewater of toilets called black water. Remaining wastewater that can be reuse entitled greywater. Greywater is 80% of the total wastewater (Mughalles et al. 2012). In terms of solid waste, buildings cover 40% of the total solid waste generation (Xhensila and Vedat 2020). Daily 1.3 kg domestic solid waste produced per capita in Istanbul (TMMOB 2019).

C. Heat Ventilation and Air Conditioning
Heating and cooling are essential requirements for human life. Heating can be provided directly by fossil fuels or indirectly via electricity. However, those direct and indirect sources have environmental impacts. The heating and cooling sector takes up close to 40% of the final energy demand in the world (WBCSD 2009). This value shows an increase to 52% in Europe (Nowak 2018).

2.1.3.2 Sustainable Compound Versus Traditional House

Traditional houses annually spend $2200 for energy requirements, whereas in a sustainable house case, this value becomes $1100 (Tolson 2011). As mentioned in the previous part, the biggest portion of these requirements covered by the heating and cooling processes. In a typical sustainable house where air conditioning with heat pumps integrated with renewable energy sources, natural gas consumption on average is decreased by 21,623 m^3 (Taler et al. 2019). A sustainable house that is suitable for the LEED certification program consumes 18–39% less energy than a traditional house (Xhensila and Vedat 2020). This leads to a reduction in bills and dependence on fossil fuels. From a water perspective, the innovative technologies that overlap with LEED certification, 50% for wastewater and 80% for drinking water efficiency, can be achieved (Xhensila and Vedat 2020). Pipelines cost 91.5% of clean water distribution system costs (Tıryak et al. 2020). Taking into consideration the cost of clean water distribution, providing water in a house, and consuming it efficiently can decrease the water distribution system cost. Energy and water sources are running out gradually, and the deficiency will influence everyone deeply in the future. Therefore, sustainable compounds have a critical role in building a sustainable and green future.

2.2 Aim of the Study

The purpose of this research is to investigate all aspects related to water, such as clean water supply, wastewater treatment, and rainwater collection for the ODIH. We also aim to provide energy-efficient solutions for heat ventilation and air conditioning. Waste management is also of our concerns. Those main titles of the sustainable house are milestones to build a green future. Each house can be converted into a sustainable house by taking into consideration certain functions. The conclusive goal of this research is to build a sustainable house that can:

- Provide drinking water to at least two adult persons
- Reuse of greywater after filtration process, possibly for cleaning or hygiene purposes.
- Provide HVAC systems with a renewable energy source
- Manage wastes in a sustainable manner.

This research work comprises an integral part of the ODIH project via dealing with the water aspect of a self-sustaining compound. Digital technologies and smart applications will further foster our sustainable solutions.

2.3 Methodology

There have been many methods recommended to provide water, reuse greywater, and HVAC application. The most common ones will be discussed in the Methodology, and the most feasible methods will be selected for each subtitle. The subtitles of methods could be summarised as follows:

I. Sea Water Reverse Osmosis filters will be used if freshwater is required. Also, a system for converting air humidity into drinking water will be installed as a supplementary source.
II. The reuse of the greywater process and purification of rainwater will be met with MBR (Membrane Bioreactor).
III. HVAC requirements are supplied by heat pump systems powered by photovoltaic panels (PV) and wind turbines.
IV. Waste treatment is done by a biodigester system.

2.3.1 Providing Freshwater

Supplying a sufficient amount of water for two adult persons is one of the most important requirements of the self-sustaining concept. An adult person consumes 195 L of clean water per day in Istanbul (Ertem and Doğan 2016). When we consider common usage like a washing machine or dishwasher, water usage for two adults will be approximately 355 L per day, and 40 L of water is required to start Hydroponic agriculture. Also, four litres of water needed for weekly maintain continuity (Vahaa 2020). A study shows that seasonal water demand is not altered by bath, dishwasher, toilets, washing machine, and washbasin usage (Rathnayaka et al. 2015). Therefore, in this study, we assume that the seasonal water demand is constant. Irrigation can be a change in the conventional house; however, ODIH has a hydroponic agriculture system, so it has a stable water requirement. In conclusion, ODIH will have a steady water consumption value for each month.

ODIH aims to provide freshwater 24 h continuously and independently from the urban water supply. The primary sources for water production are seawater, greywater, rainwater, and water from humidity. The relationship between these four sources and the blackwater is shown in Fig. 2.3.

Greywater might seem a mid-product, but the primary source of water requirement is greywater. We prefer not to store greywater since it might cause an odour problem. To prevent odour and bacterial growth, greywater will be purified frequently stored as freshwater. The remaining requirement will be provided by harvested rainwater. Rainwater can be considered greywater, and it will follow the same filtration process as greywater. Depending on the seasonal rainfall change, in some months, the rainwater cannot supply the remaining water requirements. In these months reverse osmosis system will be used to generate freshwater from seawater. In this case, a reverse osmosis system is used as a dispatchable source since it does not depend on external factors such as temperature, seasons, etc.

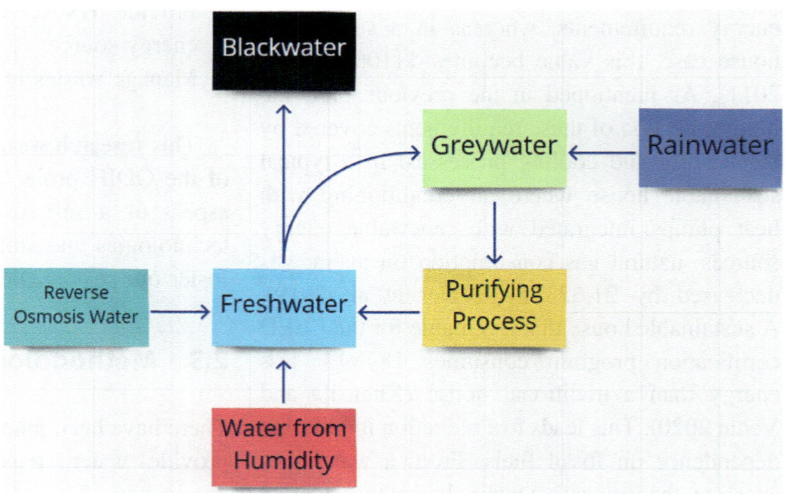

Fig. 2.3 Relationship between freshwater sources

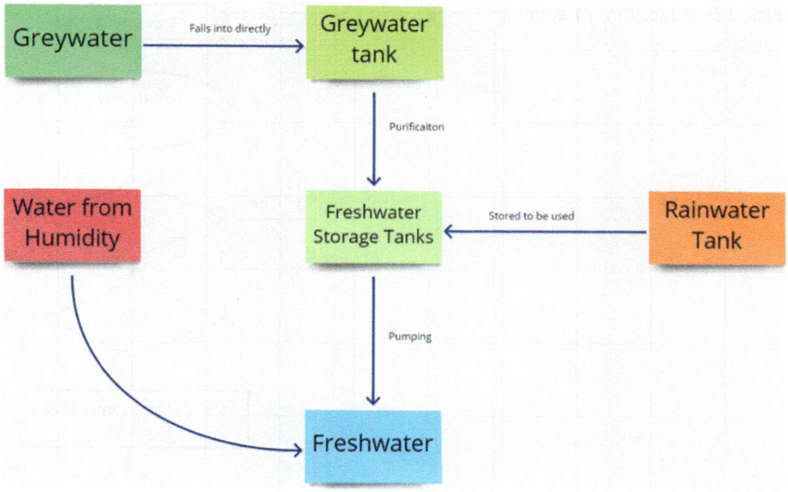

Fig. 2.4 Relations between water installation and water from humidity system

ODIH will provide freshwater from four primary sources, and the order of precedence aims to decrease energy consumption. The most energy-intensive method is humidity to the water system. Using humidity in the water system increases energy consumption and contributes very little water. However, the system offers an independent way to provide water that has no connection with the internal water installation of ODIH. Also, the system provides high-quality water, which is suitable for cooking and drinking (AKVO 2020). Relations between water installation and water from the humidity system can be seen in Fig. 2.4.

Other water sources such as rainwater, greywater, and seawater are used in a specific manner, and this is illustrated in Fig. 2.5. F_c and G_c refer to the current amount of freshwater and greywater. Also, F_n refers to the needed amount of freshwater. The differences between these three sources are energy consumptions and storage life. Since greywater and rainwater consume less energy than seawater, they will be used before the reverse osmosis system. Having regard to the greywater has less storage life than the rainwater, the greywater will be used before rainwater.

The freshwater algorithm starts with the steady-state condition, which means there is neither production nor usage. When freshwater usage begins freshwater tank is controlled by Home Management System (HMS) sensors. If the required freshwater is more than the amount of water in the freshwater tank, the grey water tank is controlled. There should be 60 L of water to use in flush. If the existent water in the grey water tank is more than 60 L, excess water will be filtered to use. If the filtration process of greywater counterbalances the requirement of freshwater, the system goes to the initial position (steady-state). Rainwater tank will be controlled in case the water amount in the grey water tank is less than 60 L. If there is enough water in the rainwater tank to supply the demand, this water will be purified, and the system goes back to the steady-state condition. Reverse osmosis system will be met the requirements of water in case of water deficiency in the rainwater tank. After discharging greywater, even if the freshwater requirement is met, to prevent both odour and pathogen problems, greywater is treated until its volume drops to 60 L. When the rainwater tank is full, depending on specific conditions, the rainwater gets filtered. To harvest more rainwater, the treatment process is put into action according to weather conditions. For instance, in a situation in which rain is continuing even though the rainwater tank is full, to keep more rainwater already stored, rainwater is filtered, and treated water will be kept in the freshwater tank as freshwater.

Since greywater is stored after the filtration process, small scale tanks can be sufficient for grey water storage. On the rainwater side, a larger storage tank will be more appropriate due to the uniform precipitation regime and utilise rainwater

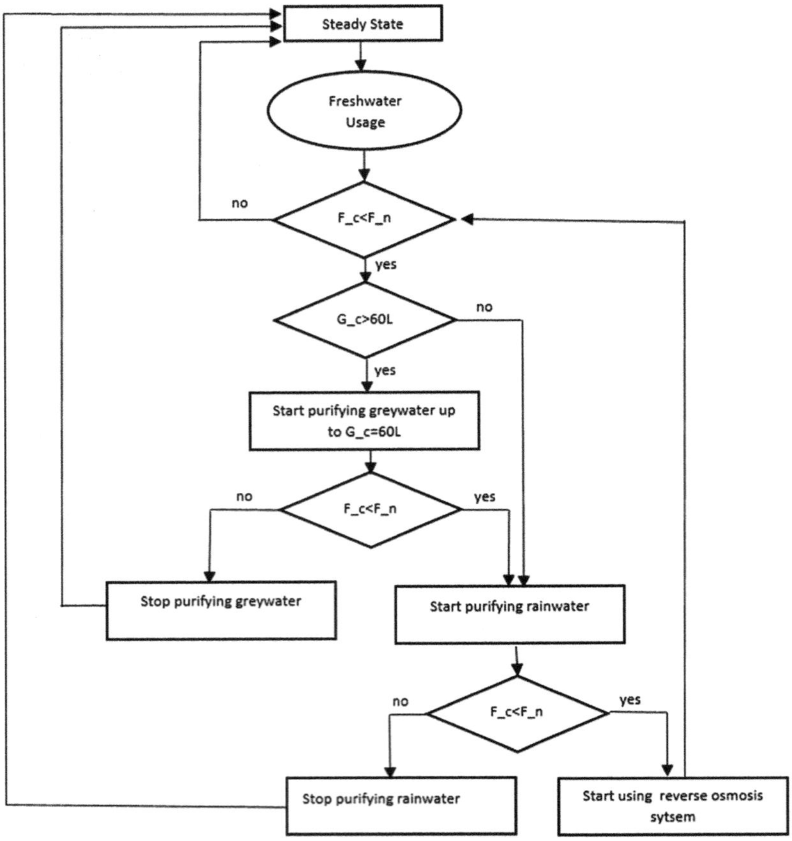

Fig. 2.5 Algorithm of water system

more efficiently on stormy days. Rainfall is not distributed equally for each day and each season. Articles indicate that rainwater can be stored for up to 30 days without any deterioration since it includes pollutant (Joanna et al. 2020). Thus, a large capacity rainwater tank can store a surplus of rainwater in winter for use in the summer months. A clean water storage tank also requires a large scale since greywater store clean water.

2.3.1.1 Technologies and Tools in Providing Freshwater

Salty or brackish water is not fit the purpose of drinking, agricultural irrigation, or industrial usage. However, the enormous water amount in oceans and seas has immense potential. Desalination will play a crucial role to make use of the potential in seas and provide freshwater in ODIH. Separation process of saline from salty water named desalination. There are many technologies for desalination. Two major technologies are thermal technologies and membrane technologies (Khan et al. 2018). The concept of thermal technology is based on heating salty water under high pressure and condensing the vapour. The membrane technology uses pumps powered by electrical energy. These pumps pressurise water to desalinate it. Different methods for Thermal and Membrane Technologies can be seen in Table 2.2.

Table 2.2 Desalination technologies (Khan et al. 2018)

Thermal technologies	Multi-Stage Flash (MSF)
	Multi-Effect Distillation (MED)
	Vapor Compression (VC)
Membrane technologies	Reverse Osmosis (RO)
	Electrodialysis (ED)

Table 2.3 Advantages and disadvantages of technologies (Antonyan 2019)

Technologies	Advantages	Disadvantages	Globally installed capacity
Thermal technologies	Highly efficient in desalination of briny water	The thermal and electrical energy required Phase separation of water is required	33% of total installed capacity
Membrane technologies	Do not need thermal energy; the system uses only electrical energy	Small water production capacity	67% of total installed capacity

Depending on the usage purpose, each technology has several advantages and disadvantages. Thermal technologies are more efficient than membrane technologies in the desalination process of briny water. Besides, membrane technologies have lower energy consumption than thermal technologies (Antonyan 2019). Also, thermal technologies have higher operation and maintenance costs since they cause corrosion problems (Antonyan 2019). Table 2.3 includes detailed information about thermal and membrane technologies.

Energy consumption of desalination hinges on parameters such as production capacity, salinity rate, and desalination method. In Fig. 2.6. five leading desalination technologies are compared in terms of energy consumption. According to pieces of information, membrane technologies are the most suitable option in ODIH since they are appropriate for small-scale usage and consume less energy. If membrane technologies are compared with each other, reverse osmosis is more suitable than electrodialysis. ED technologies consume less energy than reverse osmosis, but they are not ideal for desalinating water with a high salinity rate (Antonyan 2019).

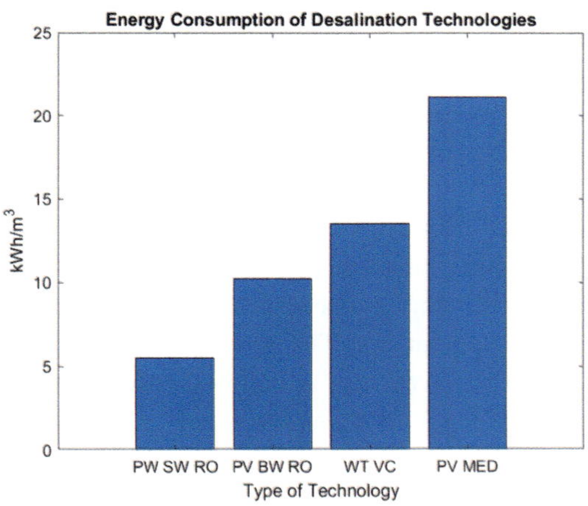

Fig. 2.6 Energy consumption of desalination technologies (Antonyan 2019)

Capturing water from humidity can be used as a supplementary source. The working principle of the system is decreasing the temperature of the air to condense it. Pumps force the air towards the filter system and condenser that cooled air. Water droplets create in the condenser when the temperature decrease below dew points (RainMaker 2020). These droplets can be collected in the water tank, and after the water filtration process, it becomes drinking water. This system consumes a significant amount of electric energy. With this system, 1 L of drinking water can be provided by using 0.35 kWh electricity (Watergen 2020). Therefore, it will be used for high-quality water requirements, like drinking and cooking. Also, external factors do not affect the efficiency of the system since it has a wide operating temperature and humidity range. It can operate between 18–45 °C and 35–75% humidity of air (AKVO 2020). The working principle of humidity to the water system can be clearly seen in Fig. 2.7.

The secondary water source of ODIH will be rainwater. Contrary to popular belief, rainwater is not comparatively clean. Test results indicate that rainwater is physically, chemically and microbiologically polluted (Monika et al. 2020). Major chemicals air pollutants are carbon dioxide (CO_2), carbon monoxide (CO), sulphur dioxide (SO_2), nitrogen oxide (NO_2), hydrocarbons, hydrocarbons, particulates, and heavy metals

Fig. 2.7 Working principle of water capturing system (AKVO 2020)

(Izzati et al. 2016). Rainwater contents depend on some factors such as local microclimate, roof material, atmospheric pollutions, etc. According to articles, rainwater impurity reaches the maximum value in urban areas (Monika et al. 2020). Therefore, harvested water from ODIH requires a comprehensive purifying system to reuse. The annual average precipitation for Istanbul is 56.4 mm. The monthly distribution of rainfall is indicated in Table 2.4 (MGM 2019).

Since ODIH has 120 m^2 of a roof surface, a monthly average of 4391 L of water can be collected, and it can satisfy almost half a month's water requirement. There are some cost-efficient systems for the treatment of harvested rainwater. Pasteurisation includes a combination of ultraviolet radiation and heat from solar energy (Jefferson et al. 2000). This technique is cost-efficient and reliable. The most efficient case for this treatment is fully oxygenated water at 50 °C (Jefferson et al. 2000). Also, this treatment is influential in diminishing E. coli and other pathogenic bacteria. Secondly, the disinfection process is applied to increasing the microbiological quality of water. Chlorine tablets or solutions are used for the inactivation of microorganisms (Li et al. 2010). After the harvesting process, water will be pumped and stored in the tank. Chlorine added to this stored water with a level of 0.4–0.5 mg/L (Li et al. 2010).

These two methods are cost-efficient systems; however, ODIH does not require them since it has a comprehensive filtration system for greywater. Rainwater and greywater can be filtered in the same network. Harvestable rainwater is calculated in Eq. 2.1.

$$\begin{aligned} Rainwater\ Capacity = {} & (Catchment\ Area) \\ & \times (Precipitaiton) \\ & \times (0.8) \times (0.9) \end{aligned}$$

The calculation for the Marmara Region:

$$Rainwater\ Capacity = \left(120\ \mathrm{m}^2\right) \times (56.4\ \ \mathrm{mm}) \times (0.8) \times (0.9) = 4872\ \mathrm{mm}^3 \left[\begin{array}{c} Catchable\ Water\ Constant = \\ 0.8, Filter\ Efficiency\ Constant = 0.9 \end{array}\right] \tag{2.1}$$

Harvestable Rainwater (TEMA 2020a, b).

The water generation and purification system will be integrated into other systems such as IoT (Internet of Things), smart agriculture, HMS, and energy generation system. Since producing freshwater is an energy-intensive process and daily water consumption can be estimated, HMS regulates the time interval for water production to prevent the peak load of ODIH. Also, HMS organises the water source sequence. The first source of water is greywater. After the filtration process of greywater, if there is still water demand, HMS switch on the rainwater tank to purify it. If it is not enough, the reverse osmosis system will be kicked-off, and it will meet the water shortfall. Total energy consumption of water production and purification system can be served by PV panels and wind turbines of ODIH.

2.3.1.2 Reuse of Greywater

Greywater, unlike the blackwater, is the type of wastewater that contains a low level of illness-causing organisms. The greywater produces in laundries, washing machines, dishwashing,

Table 2.4 Monthly rainfall in Istanbul (MGM 2019)

Months	Jan	Feb	Mar	Apr	May	Jun	Jul	Aug	Sep	Oct	Nov	Dec
Raining (mm)	86.8	72.1	62.3	44.1	31.3	24.8	22.6	27.7	44.1	70.0	86.1	105.3

bathrooms, and kitchen sinks (Oron et al. 2014). The domestic greywater reuse ratio is almost below 1% in Turkey (Rosegrant et al. 2002). ODIH's estimated daily water consumption is 355 L. 75% of this water, which accounts for 262 L, becomes wastewater. Greywater is 80% of the total wastewater or sewage. Therefore, ODIH produces 213 L of greywater per day (Mughalles et al. 2012). Greywater is mainly made in conventional houses by the washing machine 62.5 L, the shower 85.7 L, and kitchen and hand wash basin sink 60.9 L per day (Loh and Coghlan 2001). Greywater would be an environmentally hazardous type of water and should not be exhausted into seawater without treating it. However, it can be a stable water source for ODIH by using it as reused greywater after a purifying process. Greywater can be used directly in siphon or watering plants (The Reenage 2020). ODIH cannot use greywater without filtration in watering plants since it has hydroponic agriculture, even greywater will be used in siphon without filtration. Also, depending on the filtration process, it can be reused in houses.

The main greywater filtration processes are sequencing batch reactor (SBR), rotating biological contactor (RBC), constructed wetland (CW), and MBR system. Each method has advantages and disadvantages, depending on the purpose.

To begin with, the first technology is the CW system. It purifies grey water with sand, gravel, and plants. In this technique, sand and gravel used for the filtration process, plants use for microbial degradation (Üstün and Tırpancı 2015). This system requires 3–7 days to purify greywater (Üstün and Tırpancı 2015). Secondly, the SBR system has five cycles which are filled phase, react phase, settle phase, draw phase, and idle phase (Jefferson et al. 2010). During the phase of filling, greywater is filled into the SBR system. Carbon and suspended solid removal proceed in the react phase (Jefferson et al. 2010). Biodegradation and of pollutants occur in the settle, draw, and idle phase. RBC systems have numerous amounts of disc located on the axle that rotated by a motor (Üstün and Tırpancı 2015). Biofilms that are located on the surface of the disc capture the micro-organism and biological reactions that occur in biofilms. Lastly, the MBR system has unique filters and pumps for the purifying process. Pumps pressurise the water and push it into the filter. After this process, filtered water is purified from grains, bacteria, and viruses.

Among all these technologies, the MBR system provides the finest quality water faster, and it requires less space than other methods (Orhon and Hocaoğlu 2011). Unlike other systems, the MBR system does not require a disinfection process after the filtration stage (Üstün and Tırpancı 2015). According to those pieces of information, the MBR system is the most appropriate option for reusing and purifying greywater.

2.3.2 Waste Management

Waste management of ODIH is one of the critical elements to make this project truly sustainable. ODIH's wastes can be categorised into four main groups: domestic waste, sewer, agricultural waste, and leftover waste.

Domestic waste can be categorised into two parts, such as non-hazardous and hazardous (Busch System 2020). Domestic wastes such as plastic, food scraps, paper, clothes, etc. that produce in consequence of households' daily activities are called non-hazardous wastes and batteries, household cleaners, etc., called hazardous waste (Busch System 2020). Another type of wastes is agricultural wastes. The materials are generated at agricultural operations, such as raising animals and growing crops named agricultural waste (AQMD 2020). The main agricultural wastes are manure and fronds (AQMD 2020). Toilet wastes such as urea and faeces are called sewers. Lastly, leftover wastes are a surplus of food.

The average daily solid waste amount per person was calculated as 1.30 kg in Turkey (Türkiye İstatistik Kurumu 2017). Human waste is divided into urea and faeces. The average person produces between 500 and 1100 g of faeces each week (Encyclopædia Britannica 2015) and a healthy person urinates 1–1.5 L a day (Keith Halperin 2020). ODIH waste management mechanism aims to separate usable wastes in biomass reactors and utilise them efficiently as sustainable as possible.

Biomass energy can be used for heating or electric power generation. The most common technique for electricity generation from biomass is the direct combustion method. Biomass directly combusts using a conventional steam cycle technique that includes boiler, turbine, and condenser (PNG Biomass 2020). Working scheme of the biomass power plant illustrated in Fig. 2.8. Since biomass power plants use the conventional steam cycle, additional systems are required, such as grate process, fly and bottom ash capturing system, etc. (Siemens 2020). Due to the working principle and mechanism of a biomass power plant, ODIH does not have an appropriate area to build a biomass power plant, and the waste products of two persons are not sufficient to feed the power plant.

A biomass composting system is another option for waste treatment. The working principle of the system is pressing and fractionating organic waste to decrease its quantity. This process requires a long time to produce fertiliser (Chew et al. 2019). After this process, the remaining waste can be used as fertiliser. Human waste is rich in nitrogen, phosphorus, and potassium, the key ingredients of most fertilisers.

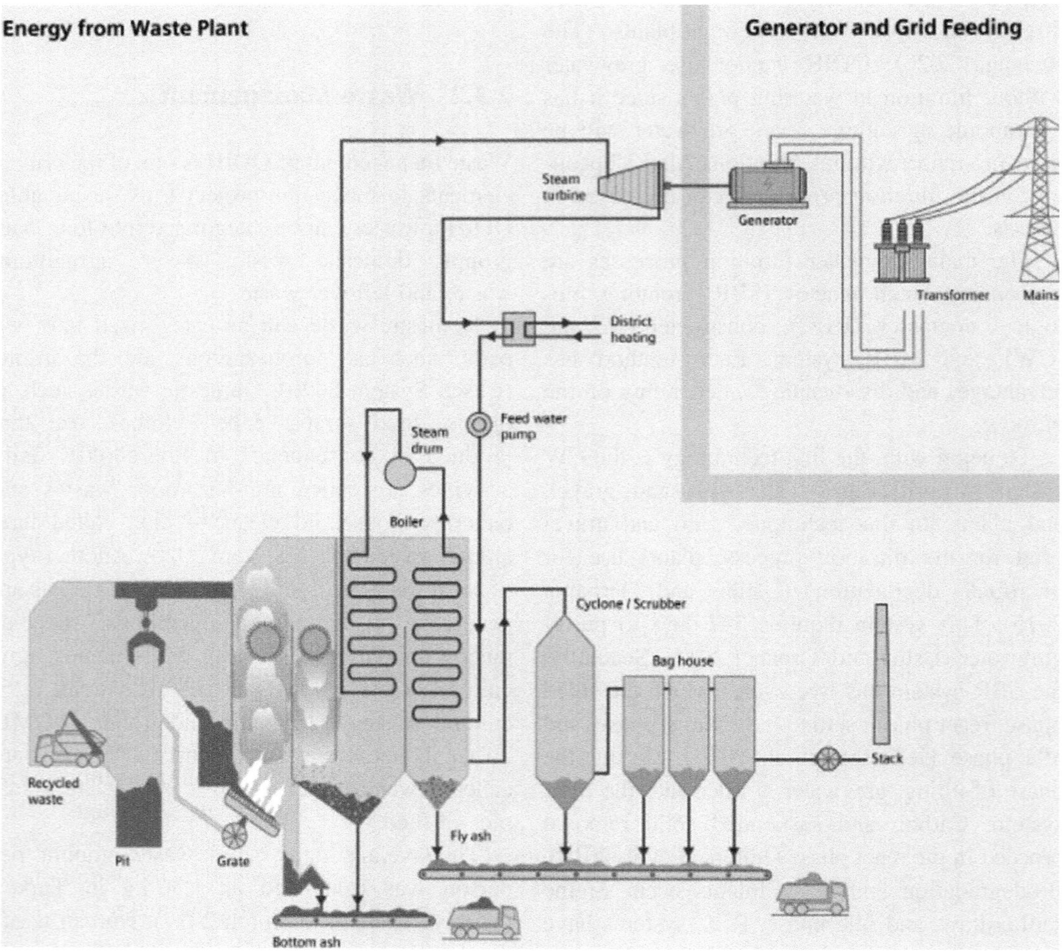

Fig. 2.8 Biomass power plant working scheme (Siemens 2020)

Table 2.5 Average temperature in Istanbul by months (MGM 2019)

Months	Jan	Feb	Mar	Apr	May	Jun	Jul	Aug	Sep	Oct	Nov	Dec
Average temperature (°C)	5.9	5.9	7.6	11.9	16.7	21.3	23.8	23.9	20.3	16.0	11.9	8.2

For hundreds of years, with varying levels of success, farmers have been using it as fertiliser (Nosowitz 2018). ODIH uses a hydroponic agricultural method for food production, so fertilisers produced by the composting process cannot be used in agriculture.

Considering the information about composting and biomass energy generation systems, they are not suitable for putting them into practice. Since ODIH has less amount of organic waste production, small-scale solutions will be more feasible. A biochemical process such as anaerobic digestion can be an example of a small-scale solution. The anaerobic digestion method can produce biogas from any biological material in warm, wet, and airless condition (IRENA 2012). In the anaerobic digestion method, methane and carbon dioxide production as biogas occurs, and these gases can be used in Stirling motors or micro-turbines. Also, some companies produce small-scale bioreactors that produce biogas with an anaerobic digestion method and use biogas on a specific burner.

A small-scale biomass energy generation system can convert organic wastes into inflammable gases, and these gases are suitable to use in the cooker. Since it is feasible to run a biomass energy generation system for years without little or no maintenance and considering the organic waste production is stable, a biomass energy generation system can run the organic waste treatment job consistently. Also, by integrating the biomass energy generation system into the sanitary installation, ODIH discharges the black water and feeds the biomass energy generation system. This makes treating the black water possible in ODIH premises and provides an amount of biogas that corresponds to using a specific burner that is connected to a biogas tank for 2 h every day (HomeBiogas 2020a, b). The tank works 24/7 without using any control mechanism. However, the only drawback of this system is that the temperature of the biogas tank must be over 15 °C to make appropriate conditions for the bacteria in the tank (HomeBiogas 2020a, b). When the climate conditions are not appropriate, such as a day of average daily temperature below 15 °C, it needs to be supported by a heat source. According to the annual average temperatures of Istanbul that can be seen in Table 2.5, six months of the year below 15 °C and this situation obliges supporting the biomass tank. The location of the biogas tank will be in a column of the house and the heating requirement of the column will be supplied by the heat pump. Putting the biogas tank in a closed area brings along a continuous ventilation need to exhaust the excess gas of the biogas tank. As a result, ODIH uses a biogas tank not only to eliminate organic wastes but also to utilise them in biogas stovetop to decrease energy consumption. Also, this reduces the carbon footprint of ODIH by burning organic gases instead of releasing the atmosphere directly. The relationship between biogas tank input and output can be seen in Fig. 2.9.

The energy consumption of biogas tank increases during winters due to heating. The ventilation of the biogas tank room must be run all year continuously, and this leads to an extra energy load. Despite using energy to operate the biogas tank safely, the tank produces more energy than its annual consumption. During the months when heating is necessary, the biogas tank consumes more energy than its production. The main aim of the biogas tank is the elimination of organic wastes. Even if the annual net energy generation was negative, using a biogas tank is still wise since it treats waste in the premises of ODIH rather than dumping it into nature.

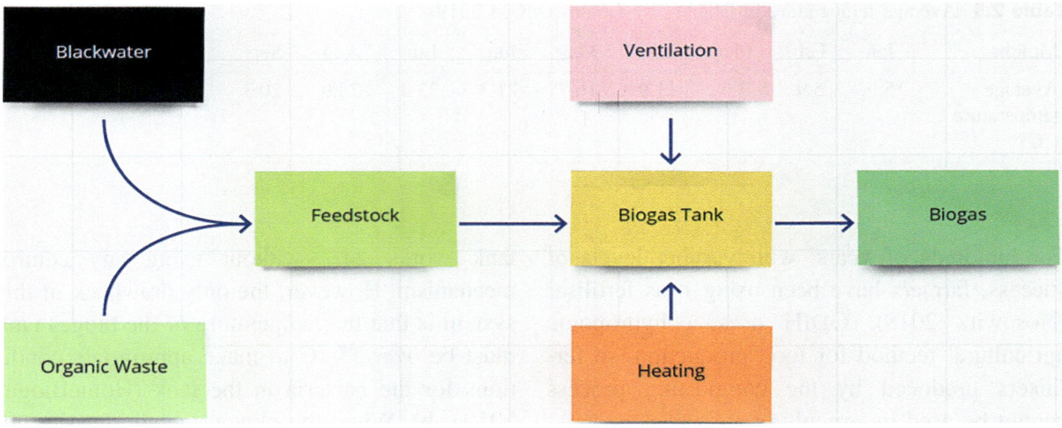

Fig. 2.9 Input and output of biogas tank

If we compare the anaerobic biogas tank with the biomass power plant, biogas tanks' working principle is more straightforward since it does not require mechanisms such as grate process, fly and bottom ash capturing system. Also, the composting system is not appropriate for ODIH since it has hydroponic agriculture. On the other hand, the amount of blackwater and organic wastes are sufficient to run an anaerobic digestion reactor to produce biogas. Upcycle of the anaerobic digestion reactor is depicted in Fig. 2.10.

The recycled materials such as paper, cardboard, glass bottles, jars, plastics, steel, and aluminium cans are planned to be stored and regularly sent out of ODIH to be recycled. Some wastes such as batteries, chemical products, electronic items, clothing, and industrial oils will be categorised as recyclable wastes. However, these wastes are extremely hazardous substances and need to be deposited in a strictly secure place, such as the closed column of ODIH. After this storage process, recyclable materials will be transmitted to appropriate recycling plants. With this process, the material footprint of ODIH will be decreased, as suggested in SDG #12 Responsible Consumption.

2.3.2.1 Toilet System

The toilet system has a critical role in ODIH because the wastewater of toilets is called blackwater, and it cannot be recycled like greywater to use again in the building. A relevant design toilet system should prevent the spread of pathogens and diseases caused by pathogens.

There are three toilet systems that can be used in ODIH. The first one is a composting toilet. The composting toilet is an environmentally friendly type of toilet that does not use water. The composting toilet processes waste materials to capture nutrients like nitrogen and phosphorus. This toilet collects urea and faeces at separate points. In this way, solid and liquid waste do not mix, and the composting process of solid waste takes a shorter time. At the location where solid waste is stored, it is mixed with organic materials or sawdust and then turned into rich hygienic fertiliser (Crennan 2007). The production process of rich sterile fertiliser will be last approximately six months. Thus, it requires a large place for a sewer storage tank (Crennan 2007). The second method is the vacuum toilet. The vacuum toilet provides the same level of comfort as traditional toilets. The main advantages of vacuum toilets are water-saving, odourless operation, and reduction in the use of detergent. The last option is Bio-toilet. To treat human waste, bio-toilet has a bio-tank that includes bacterial culture and supports. The Bio-toilet system does not need any external energy for treatment. By using a mechanical hand-pump, water fills up into the toilet. The waste goes to

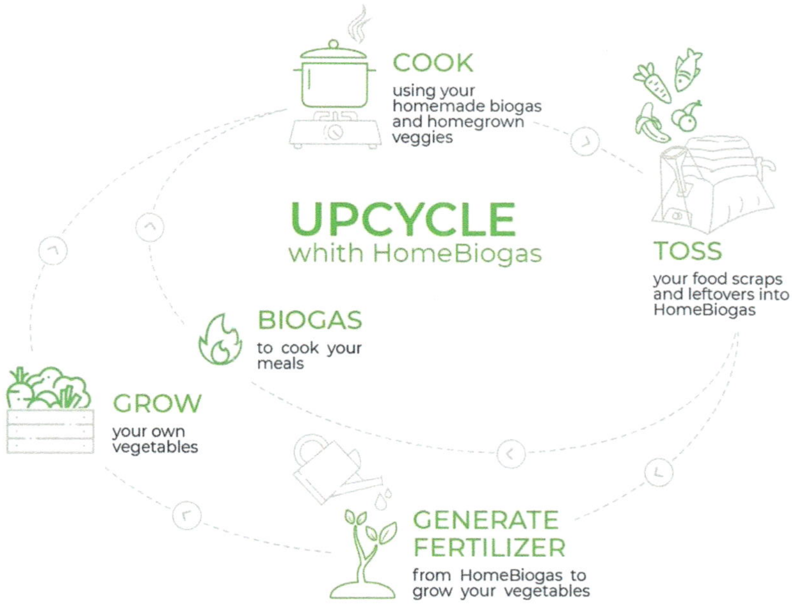

Fig. 2.10 Upcycle of Homebiogas (HomeBiogas 2020a, b)

the tank, and biodigesters break down the waste. Bio-toilet also uses human waste as an additional resource to generate the cooking gas via a specific bioreactor. If this type of bioreactor is selected to be utilised in ODIH, Bio-toilet is the most suitable option. Table 2.6 shows the advantages and disadvantages of toilet systems.

2.3.3 HVAC

The HVAC system in ODIH aims to keep indoor air quality and temperature stable. Air quality and an HVAC system with high performance are essential for the residents. Heating and ventilation in the agriculture rooms will provide optimal air quality and temperature for the plants to protect them from spore production and remove undesirable gases. Besides plants, the bacteria that live in the biogas tank and need a warm ambient condition to survive. Heating the tank will be required on cold days of the year, and it will be supported by exposing it to direct sunlight. Since biomass releases excess gases, a ventilation fan is also required.

The air conditioning of ODIH is provided by heat pumps. Heat pumps transfer heat from one area to another. This system uses a substance called refrigerant and circulates it in the system to the evaporation and condensation cycles. A compressor delivers this substance between two heat exchanger coils. In the first coil, the refrigerant evaporates with low pressure and absorbs heat. After this process, the system

Table 2.6 Advantages and disadvantages of toilet systems

Types	Advantages	Disadvantages
Composting toilet	More environmentally friendly and requires less water	Needs more maintenance than standard toilets, and it should be emptied when the compost is ready
Vacuum toilet	Odourless operation	Blockage in the piping system due to drying and crystallisation of urea
Bio-toilet	99% aerobic treatment	High logistics cost of bacteria transportation

Table 2.7 Advantages and disadvantages of heat pump sources (ICAX 2020)

Types	Advantages	Disadvantages
Air source	Budget-friendly and easy installation	Since the air has a low temperature in winter, efficiency decreases
Water source	Since water is more efficient in heat exchange, this heat pump has high efficiency	Needs a very high amount of water, Low efficiency in winter
Ground source	Low operating cost	Efficiency drops if the ground temperature decreases, high installation cost

pumps this substance to the other coil. In the second coil, this refrigerant condenses and releases the absorbed heat (Government of Canada 2017).

Heat pumps are categorised by the source they use. There are three types of heat pumps: air source, water source, and ground source. Ground source and water source heat pumps demand a suitable area to utilise heat source. On the other hand, the air source heat pump does not need that suitable area to transfer heat. The advantages and disadvantages are demonstrated in Table 2.7.

Due to the space constraints of this project, these technologies are not convenient for ODIH. Air source heat pumps directly use ambient air to transfer heat energy. This difference between the heat pump systems brings an advantage for the air source heat pump. It is the most affordable option considering the capital expenditure, the easiest for installation, and the highest for energy efficiency (Jingyong Cai 2020). Also, from the thermodynamic side, the air source heat pump has the highest Coefficient of Performance (COP) with warm air input (Bianco et al. 2017). The coefficient of performance is calculated with the ratio between energy usage and useful heat extracted from the condenser. This value depends on various factors based on the temperature difference. Briefly, COP can be explained as the efficiency of the heat pump (Industrial Heat Pumps 2020). The air source heat pump is the most suitable option for air conditioning and hot water production of ODIH. The air source heat pump working scheme is demonstrated in Fig. 2.11.

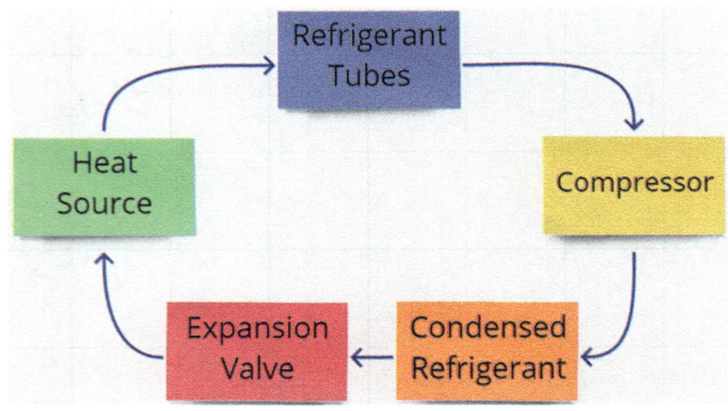

Fig. 2.11 Working scheme of air source heat pump (Donn 2020)

2.3.4 Location of HVAC, Waste Treatment and Water Circulation Systems in ODIH

On the first floor of ODIH, there are four columns for bio-tank, heat pump and water systems. The first column, which is larger than other columns, belongs to the bio-tank since humans' disposing of organic waste is continuous during the day. The second and third columns are used for water tanks, and the last one belongs to reverse osmosis and boiler of the heat pump. The location of the whole storage spaces is depicted in Fig. 2.12.

2.4 Materials

2.4.1 Reverse Osmosis System

According to data summarised in Table 2.8, the most efficient reverse osmosis system is FSM Watermaker's Nemo 2. Nemo-2 has 60 kg of dry

Fig. 2.12 Location of HVAC, waste treatment and water circulation systems in ODIH

Table 2.8 Reverse osmosis systems

Company	Product	Capacity	Power Supply	Size(mm)	Maintenance
FSM Watermaker (2020)	Nemo-2	100 L/h	220 V, 1.1 kW AC	600 × 350 × 350	Changes according to sea pollution
Samar Watermaker (2020)	Mini Compact 60	60 L/h	220 V–AC 1.1 kW	1250 × 500 × 500	Changes according to sea pollution
Delfin Denizcilik (2020)	Mini 60	60 L/h	12 V–220 V AC	500 × 250 × 250	Changes according to sea pollution
EastMarine (2020)	Rainman	60 L/h	12 V DC	690 × 150 × 220	Changes according to sea pollution

weight, and it will be 70 kg when it is working. 100 L of water can be produced by Nemo-2 with 1.1 kWh energy consumption. ODIH's predicted daily freshwater need is 355 L. If there is no recycling in ODIH, Nemo-2 must work about 3.5 h per day. However, with the greywater treatment system, 75–90% of greywater and rainwater are expected to be reused. This means Nemo-2 will be operational much less than 3.5 h per day.

Nemo-2's estimated daily working period will be 1 h after the first deployment for the summer months. Thus, Nemo-2 will consume 3.85 kWh electricity on the first day for kick-off, and then its daily consumption will be between 0.5 and 1 kWh. We estimate that ODIH will produce 1 L of freshwater from seawater using by using 0.011 kWh of energy. Table 2.8 summarises reverse osmosis systems in terms of capacity, power supply, size, and maintenance requirements.

2.4.2 Heat Pump

In Table 2.9, heat pumps are compared according to COP (Coefficient of performance), capacity, and electrical information. One of the main constraints in ODIH is energy consumption.

Table 2.9 Heat pump systems

Company	Product	COP	Heating/cooling capacity (kW)	Electrical information (V-Ph-F)
DemirDöküm (2020)	Maxiair	4.10	15.5/14.5	220&240-1-50
Alarko Carrier (2020)	61AF	3.7	20.2	400-3-50
Baymak (020)	Iotherm	4.06	15.5/13.8	220&240-1-50
Daikin (2020)	Altherma	4.80	16/17.8	230-3-50

Therefore, COP and the capacity of the device are the most important factors to select the most suitable product. When we consider the COP value and the capacity, Daikin-Altherma steps forth as the most suitable option. It has a higher COP and less energy consumption than other options.

2.4.3 Water Capturing System

As we mentioned earlier, we would like to install a supplementary water supply system that utilises humidity in the air. In Table 2.10 two water capturing systems are compared in terms of capacity and energy consumption. As the air to the freshwater system, Watergen-GENNY is the best option for the ODIH. This product can produce 30 L of water per day. GENNY consumes 0.35 kWh to produce 1 L water from air which is less than most of any other products. Since ODIH is a floating platform, weight is a constraint in the design. GENNY is 80 kg, and it is lighter than its competitor. The dimensions of this product are also smaller than the other one.

2.4.4 Biogas Reactor

A small-scale biogas reactor is the most appropriate option for organic waste and black water treatment. Homebiogas reactor can convert organic wastes and black water into gas that can burn in a cooker. The system can provide 600 L of gas with 6 L of organic waste and 30 L of black water. 600 L of gas can operate the stove for two hours. Gas production occurs by an anaerobic digestion method with the aid of animal fertilisers. Additional probiotic tablets increase the efficiency of the reaction. Also, a stable temperature of about 20 °C is needed to maintain the process since bacteria are more efficient in warm conditions to break down organic wastes. Related information about Homebiogas 2.0 can be seen in Table 2.11. To initiate a biomass reactor, 100 L animal fertiliser is required (Homebiogas 2020a, b). Since the reactor is in an enclosed space, a fan is needed to prevent excess gas accumulation in the room. The electrical information of the fan is 220–240 V, 50 H, 1 Ph (Blauberg 2020). The estimated energy consumption of the fan is 0.14 kWh per day.

Table 2.10 Water capturing systems

Company	Weight (kg)	Dimension (cm)	Capacity	Power consumption (nominal-peak)	Electrical information (Ph-F-V)
Watergen-Genny (2020)	80	54 × 42 × 130	30 L per day	0.7–3 kW	1 ~ /50 Hz/230 V
AKVO 36 k (Watergen 2020)	105	90 × 69 × 140	100 L per day	1.1–1.5 kW	1/50 Hz/220 V

Table 2.11 Biogas reactor system

Company	Product name	Weight	Dimension (L × W × H)	Water input per day	Energy output daily explicit-Max.)
Homebiogas (2020a, b)	Homebiogas 2.0	1200 kg	210 × 115 × 125 cm	36 L	4.004–4.4 kWh

2.4.5 Water Tanks

For water storage, four tanks are required. Two tanks will store the clean water produced by the reverse osmosis system and wastewater filtration system. The third tank will be used for greywater storage, and the last one will store the rainwater. For balanced weight distribution, clean water will be stored in two tanks with 1.25 m^3 capacity each. Galvanic and stainless types of the water tank will be used for storage. Galvanic water tanks are optimal types of tanks for the fire-protection water system and domestic water storage. Also, they have a modular structure; thereby, they can montage in a confined space, and they can be used in a humid environment (Mekser 2020). Thus, two pieces of galvanic water tanks, each with 1.25 m^3 capacity, will be used for drinkable water storage. Other types of water tanks are stainless water tanks. Stainless water tanks will be used for greywater and rainwater storage. Contrarily the plastic water tanks prevent algal and bacterial growth (Otank 2020). Two stainless tanks will be used in ODIH, one of them is for rainwater which has 1.5 m^3 storage capacity, and another one is for grey water which has 630 L capacity. Water tanks cannot be located on the roof. Consequently, a hydrophore will be used to pump the water. The electrical information of the hydrophore is 0.75 kWh nominal power and 220 V. The hydrophore controls the pressure of the tank, and if it is decreased, the hydrophore cuts-in (Etna 2020). Detailed information about the tanks is shown in Table 2.12.

Also, ODIH uses special LEED, and BREAM certificated vanes that use less water than conventional vanes. In the bathroom, Artema's Sense shower set will be used. This product saves 14 L of water per minute compared to an ordinary vane, and it has a green building certification (Artema 2020). Secondly, Style X will be used as a lavatory faucet. This armature avoids wasting 7 L of water per minute, and it has LEED and BREAM certification (Artema 2020). Besides, this product has anti-limescale aspects (Artema 2020). Lastly, Nest Trendy kitchen faucet will be used in ODIH's kitchen. This product also has LEED and BREAM certification and an anti-limescale aspect (Artema 2020).

2.4.6 Toilet System

Since Homebiogas 2.0 will be used in ODIH, Bio-toilet is the most appropriate option for the toilet system. Other options like the vacuum and composting toilets require less water, but they cannot integrate into the reactor directly. Because the Bio-kit toilet has specific water usage for density regulation, the density of wastes that digested in the reactor affects the efficiency of the reaction. Detailed information about the Bio-toilet kit is illustrated in Table 2.13.

Table 2.12 Water tank companies

Company	Products name	Dimensions	Weight	Volume (m^3)
Mektank (2020a, b)	Modular galvanic tank	108 × 108 × 108 cm	130 kg	1.25
Mektank (2020a, b)	304 Cr–Ni stainless cylindrical tank	92.8 × 100 cm	75 kg	0.630
Otank (2020)	97/1.5 A stainless cylindrical tank	97 × 235 cm	120	

Table 2.13 Toilet system

Company	Product	Dimension (W × L × H in cm)	Water per flush	Nominal daily effluent output
Homebiogas (2020a, b)	Homebiogas 2.0 Bio-toilet kit	51 × 53 × 43	1.2 L	Up to 30 L

Table 2.14 Reuse of grey water

Company	Product	Dimension (W × L × H in m)	Nominal and Peak Power	Capacity
Norm Çevre	Special R&D project for ODIH	72 × 60 × 190	1.2–3 kW	500 L/day

2.4.7 Reuse of Greywater

There are numerous solutions and systems for greywater reuse. But most of the companies provide this system for larger greywater production. The amount of greywater limits effectiveness. There are two companies that we reached and agreed to provide a greywater recycling system for ODIH. One of them is Eurosan, and the other one is Norm Çevre. Even Eurosan can recycle the only greywater that has come from the washbasin and bathroom shower; nevertheless, Norm Çevre can recycle all wastewater except black water. Table 2.14 provides detailed information on Norm Çevre's product.

2.5 Results

All necessary equipment for water systems is selected in the previous section. This section presents the performance of each water mechanism in ODIH. Initial constraints are chosen as follows: for bio-tank, 100 L of animal manure is needed, and three days' worth of freshwater required. As we mentioned before 355 L of freshwater water used in one day. Since there is no greywater and rainwater in the system initially, 1065 L of water supplied by reverse osmosis for the water demand of ODIH.

We should begin with the cheapest water source: harvested rainwater. Harvestable rainwater capacity is illustrated in Fig. 2.13. The blue column shows the utilised rainwater quantity, and the orange column shows surplus rainwater.

The rainwater capacity is calculated for each month in Istanbul and analysed for ODIH's rainwater tank capacity. Due to excessive rainfall in the winter, spring, and autumn seasons, tank capacity was exceeded in these seasons. Therefore, this surplus will be stored in the rainwater tank. Also, in summer, rainwater cannot provide the necessary supply due to low rainfall. Total harvestable rainwater can satisfy 41% of total demand; however, utilised rainwater meets 35% of total demand. Since ODIH has limited space, we cannot store the whole surplus rainwater.

Figure 2.14 indicates the water generation share of reused greywater, rainwater, reverse osmosis water, and water from humidity.

We should remind that at the opening phase of ODIH Reverse Osmosis System will be operated to supply the building. After a sufficient amount of greywater production, the Reverse Osmosis System will be used as a backup solution mainly from April to September each year. More precise values for annual generation can be seen in Table 2.15.

The monthly water requirement is calculated according to 355 L of daily consumption. Greywater and rainwater supply 87.76% of the total annual requirement. The amount of reused greywater is stable for the whole year. The reverse osmosis system will support the water

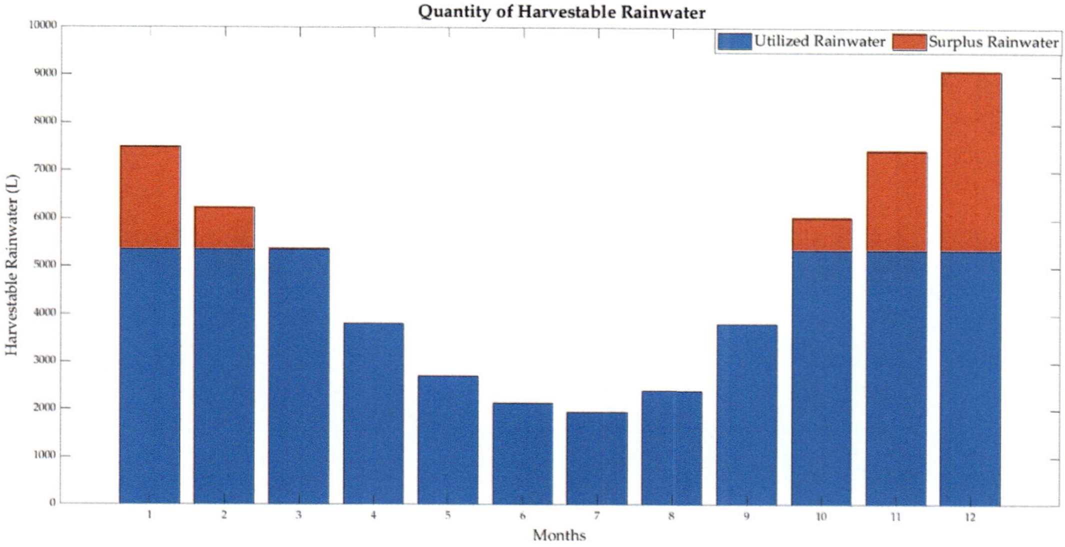

Fig. 2.13 Quantity of harvestable rainwater

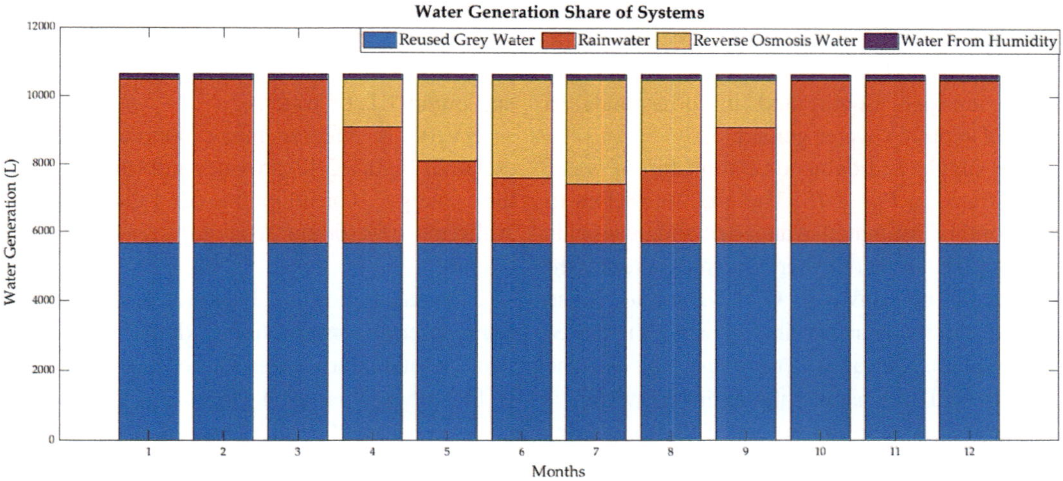

Fig. 2.14 Water generation share of systems

Table 2.15 Annual water generation by sources

Source	Water amount (L)	Share (%)
Greywater	68,040	53.24
Rainwater	44,112	34.52
Reverse Osmosis	13,848	10.84
Humidity to water	1800	1.4
Total	127,800	100

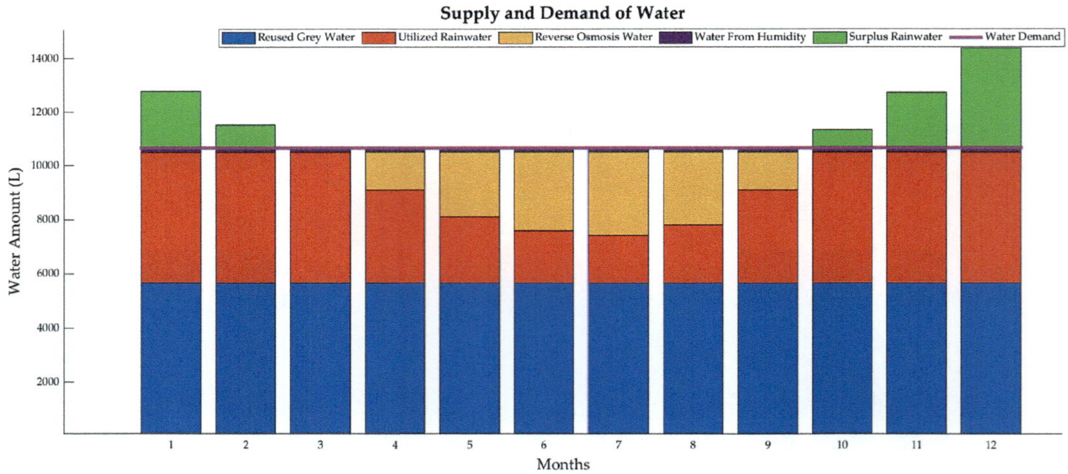

Fig. 2.15 Supply and demand for water

generation system when there is low rainfall. In case of any emergency, the reverse osmosis system will be activated. On the other hand, the contribution of water from humidity cannot be seen clearly in Fig. 1 because it has 150 L of production per month, and it corresponds to 1.4% of monthly water generation.

The supply and demand relationship of water is illustrated in Fig. 2.15. The pink line refers to the monthly estimated water demand, which is constant for the whole year. Green bars represent the surplus of rainwater. ODIH aims to utilise rainwater as much as possible by using the HMS. Nevertheless, the rainwater storage tank volume restricts ODIH to harvest the entire rainwater.

Energy consumption is one of the main concerns of this project. Therefore, we need to calculate and observe this consumption carefully. The energy consumption of water generation systems and Hydrophore is shown in Fig. 2.16.

Annual energy consumption values are given explicitly in Table 2.16.

The energy consumption of water generation is stable except for the summer season. In April, May, June, July, August, September, the reverse osmosis system works, and this increases the energy consumption by 13%, 20.9%, 24.6%, 25.8%, 23%, 13%, respectively. Purifying greywater and rainwater correspond only to 25.13% of total energy demand. The monthly average energy consumption is 116.46 kWh for water generation and pumping. According to Tables 5. 1 and 5.2, 127.800 L of water will be produced and pumped by consuming 1397.5 kWh. Therefore 0.0109 kWh energy is required to generate and pump 1 L of water.

HVAC is another major source of energy demand. Figure 2.17 provides information about the energy consumption of the Daikin's Altherma heat pump separately as heating and cooling.

The calculation of Daikin's Altherma heat pump energy consumption is done by assuming that the total internal area for cooling and heating is 144 m^2 and the Heat Pump is active continuously, set at 20 °C heating and 24 °C for cooling. During spring and autumn, since the monthly average temperature of 12 and 16.1 °C is close to the heat pump's set temperature, the energy consumption drops when it is compared to the other seasons.

In ODIH, the Biogas tank generates 0.11 kWh of energy while it treats 1 L of organic waste. Nevertheless, ventilation of biogas room consumes 0.14 kWh energy per day, and we included this figure in the heat pump energy consumption estimation. The Biogas generation system is illustrated in Fig. 2.18.

It can be observed that reusing greywater and harvesting rainwater reduces the dependence on

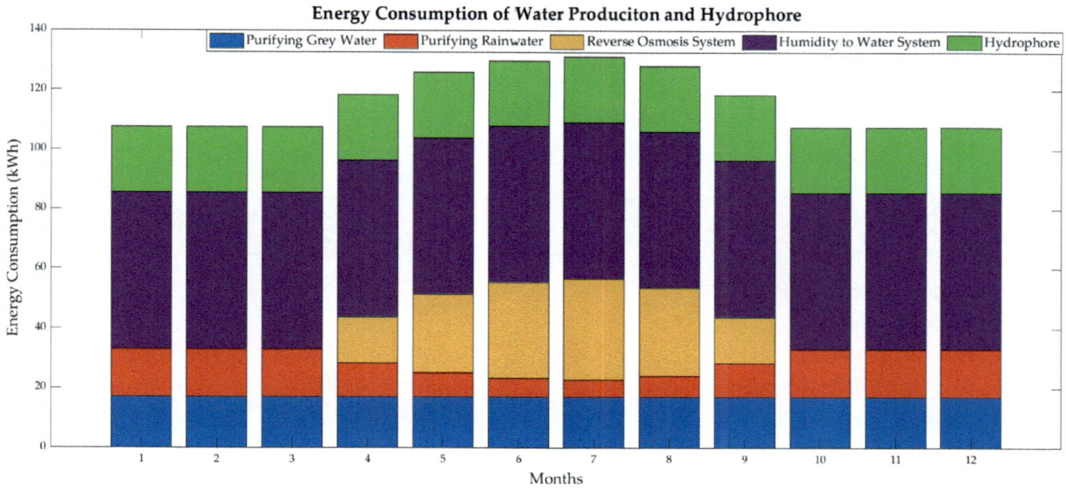

Fig. 2.16 Energy consumption of water generation and hydrophore

Table 2.16 Annual energy consumption share

Source	Energy (kWh)	Share (%)
Purifying greywater	226.8	15.97
Purifying rainwater	147.041	10.35
Reverse Osmosis System	152.328	10.73
Humidity to water system	630	44.36
Hydrophore	264	18.59
Total	1420.169	100

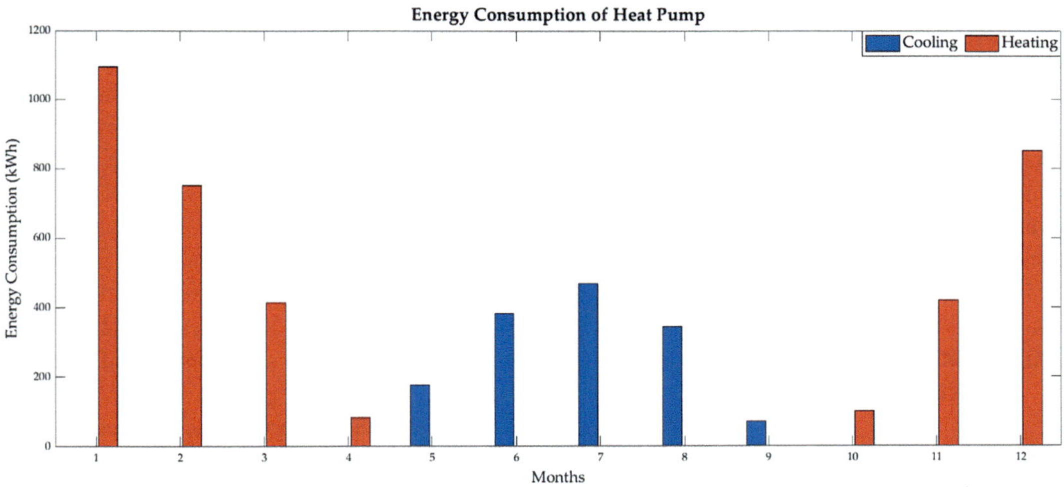

Fig. 2.17 Energy consumption of heat pump

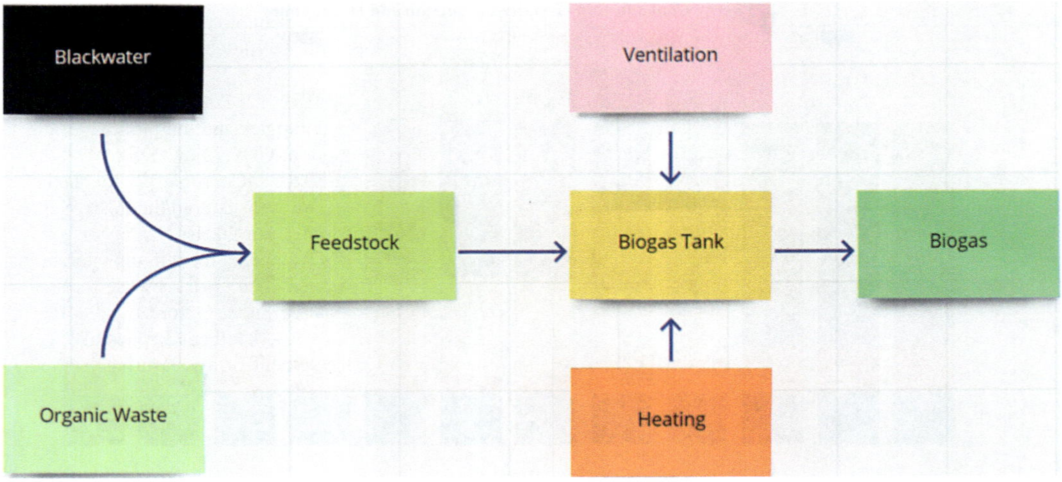

Fig. 2.18 Biogas generation diagram

reverse osmosis and decrease the energy consumption of freshwater generation. Filtration of greywater is more environmentally friendly since the wastewater generation is reduced. For design concerns of ODIH, tanks are located on the first floor. Thus, the energy consumption of Hydrophore will increase by pumping the water upstairs. Since the rainfall is not consistent for the whole year in Istanbul, ODIH's supply and demand gap of water will be covered by reverse osmosis in April, May, June, July, August, and September. The heat pump system has the biggest share in energy consumption when it is compared to other systems.

To sum up, these results are encouraging using renewable energy sources for air conditioning and reusing of solid and liquid wastes. Eventually, this will have a significant impact on achieving self-sustaining operation.

2.6 Discussion and Policy Recommendations

United Nations SDG #6 Clean Water and Sanitation aim to provide clean water to all people regardless of geography. In ODIH, clean water will be supplied through reverse osmosis and humidity to water systems. We will also harvest rainwater and purify greywater to be used in domestic usage. By being the building block of future smart cities, buildings like ODIH will be able to meet their own clean water need in a sustainable manner. SDG #14 is about Life Below Water. ODIH is designed to be a floating platform. Similar floating objects such as ships and boats dump their liquid waste into the sea, thus contaminating the ecosystem and harming the life below water. ODIH recycles its liquid waste and therefore protects life below the water by not polluting the seas or oceans.

If ODIH is built on land, we might have the opportunity to use more land and therefore a larger rooftop. This means we could harvest and store more rainwater; therefore, natural sources such as the humidity of the air can be used less. Such a situation could occur when ODIH is not floating on water but is on land with a side connected to the sea. In this case, the probability of the need to use seawater would be less; however, it would still be available while enabling more area to store treated and untreated water.

Although the history of the heat pump dates back 150 years, it is still unknown to many. One of the main sources that can rival natural gas in the air conditioning process is electric energy. Heat pumps consume less energy than technologies such as electric heater and air conditioner, which can be used to air conditioning the

144 m^2 of interior space of ODIH. However, the initial investment cost is excessive. Also, the demand for the product is affected by a deficiency in the number of technical staff. Those are the main problems of the heat pump system. To address these problems, policymakers should incentivise heat pump technology to bring down capital expenditures. They can also help to increase public awareness by promoting the product.

Although waste to biogas is a method that can be used to dispose of toilet waste and organic waste, it also has some disadvantages. The first disadvantage is that the exposed excess gas needs ventilation to prevent gas from accumulating indoors, which increases energy consumption while the waste is digested. Secondly, the heating required to maintain the anaerobic digestion process of bacteria also decreases the energy output that is already minimal. Lastly, the system of which main activity is organic waste disposal, has a small production of gas which can be used in a unique stove. However, it does not create a significant change in electricity consumption. For a single household, it allows us to use biogas as heat conversion while collecting and recycling more waste. It also provides electricity generation as a by-product. With government incentives, it will be possible to generate electricity in small-scale biomass plants by using people's organic wastes, which can be collected in public areas such as parks and it will help to expand the use of biogas.

As a result of domestic water consumption, a considerable amount of greywater is produced. In this case, ensuring the effective use of resources is of great importance in countries experiencing water scarcity. 53.24% of total water demand in ODIH is met by recovered greywater. A detailed and high-cost filtration process is required because the greywater released here will be used again in the house as a source of clean water. However, in certain places, greywater can also be used without the need for extensive filtration. For example, since water collected from toilet and bathroom sinks don does not have contaminants such as oil, unlike a kitchen sink, it can be used in siphons with a simple filtration, or in addition to these waters, wastewater from showers can also be collected in a common tank and used in garden irrigation operations. Since ODIH does not have a garden, it instead operates hydroponic agriculture that uses a smaller amount of high-quality water, so it cannot directly be used in ODIH. Besides, since ODIH will not use greywater in the car wash process, it can only be used in the toilet reservoir after a simple and cost-free pre-treatment. In this case, only 16.1% of the total greywater can be used without the need for advanced filtering. Complete recycling of greywater in homes is not feasible due to high costs; the main reason for the increase in cost is the use of membrane technologies due to insufficient available space. The main reason for using this system in ODIH is that greywater cannot be stored for long periods like rainwater due to the problem of smell and pathogens. Another reason is that there is a legal obstacle to releasing water into the sea without sailing three miles and initialising the impact on underwater life. As sustainability becomes more important in the future, the expansion of these systems can reduce the cost. Less costly systems cannot be used in ODIH since they need more space. However, these systems can be easily used in public institutions such as schools and student dormitories. Besides, making the systems that can assist in regaining greywater by obligating basic filtering has high importance to avoid water scarcity in our world.

Another alternative to a limited number of resources is rainwater. Although rainwater is not suitable for direct use, it can be treated by undergoing a simpler process compared to greywater treatment. ODIH is planned to harvest 44,112 L amount of water per year from its 120 m^2 rooftop, and this is equal to 36% of the total need. Due to space constraints, an adequately sized rainwater tank cannot be used, causing only 87.8% of total rainwater capacity to be used.

A rise in rooftop size or establishing simple systems on unused areas would increase the benefit. Rainwater can be used directly in operations such as balcony washing or garden irrigation. In newly constructed buildings, a simple Eaves system and rainwater systems with simple sand filters should be mandatory. Collecting this water in a typical tank will also be a solution for water consumption caused by other activities such as car washing and cleaning of the grounds.

Though seawater treatment is a method with a high treatment cost and produces a byproduct with a high concentration of salt, it is still used as a back-up water source. Even when seasonal changes are taken into account, seawater can meet water requirements easily; thus, it is seen as a reliable water source. When big treatment plants treat seawater, they might cause an impact on the sea ecosystem. However, ODIH, as a small-scale residential consumer, supplies freshwater mostly with the treatment of greywater and harvested rainwater. Seawater treatment is needed only when the amount of freshwater stored is lower than required. This algorithm is shown in Fig. 2.5. With every 4000 L of water treated, filters must be changed. In our case, this means changing filters three times a year.

Another technology that can be an alternative to freshwater sources is the humidity-to-water system. For every produced litre of water, it consumes more energy compared to other methods. For this reason, instead of using humidity to water as the primary water supply, it will be used for activities that need a higher quality of water, such as drinking and cooking water. Even though the mechanism depends on the humidity in the air, it can work until low humidity levels as low as 30%. The mechanism is easy to set up, and water produced by the system is suitable to use directly since the process does not need disinfection or the addition of minerals. Another advantage of the system is that it has 30 L daily capacity, and it requires relatively less space.

Every drop of water we consume leaves a footprint behind us. When it comes to water consumption, it will not be enough to think only about the amount we see in bills and take action to reduce it. The majority of the water footprint indirectly depends on many parameters, such as eating habits, textile product consumption, and lifestyle. The lifestyle of two people who will live at ODIH will also be aimed at reducing their water footprint. Policymakers can implement the agricultural and water production methods used in ODIH as measures to reduce the water footprint of the residents of future smart cities. ODIH has adopted a lifestyle aimed at responsible consumption and sustainable production. Instead of transporting freshwater through long distances, we promote and test on-site water production. ODIH recommendations for water management in accordance with the smart city concept, can be listed as follows:

- Acting upon water footprint billing.
- Subsidies for water supply transformation.
- Strict regulations for dam and water pipeline projects and gradual reduction of dam numbers.
- On-site water supply transformation has a huge potential to support innovative technologies and create new jobs.
- Using digital water flow meters to increase awareness in specific consumption and detect leakage in pipelines by gathering instantaneous consumption data of districts or households.
- The digitalisation of sewage systems can provide insight into the detection of epidemics or drug use.

2.7 Conclusion

ODIH's mission is to raise awareness and bring solutions to the water supply–demand imbalance and waste problem that awaits the world. Another goal of ODIH is to reduce the share of end consumers in energy use, especially find solutions that will reduce excess electricity spent on HVAC. Smart cities, which integrate with the concept of self-sustaining, cannot be considered independently of activities such as water, waste, and air conditioning management. In addition, waste and water terms will be included in the prosumer concept. ODIH has selected specific technologies to make air conditioning, water consumption, and waste disposal needs with optimum resource consumption. ODIH has designed a compound using all these methods in a way that targets alternative life above water. This will successively bring a solution to communities that will be affected by rising water levels in the future.

Preferred technologies are the most feasible ones when the conditions of ODIH are taken into account. They may lose their advantage under different circumstances. Due to the fact that ODIH will be a floating compound, in the implementation of these products, the lack of network connectivity and limited available space are the main obstacles. While selecting these technologies, the concept of self-sufficiency was taken into consideration. Therefore, mechanisms that consume a low amount of energy and use resources efficiently were preferred. Given all this, the most common problem has been the initial investment costs, or the capital expenditure, of products. However, since renewable resources are used during the operation, the operational expenses are limited to electricity consumption. In addition, innovative and highly independent revolutionary technologies such as capturing moisture from the air have also shown that conventional houses may also become increasingly decentralised and become self-sustaining in the future.

This hub is designed for two people to live in. 355 L of water would be enough for two adult humans to maintain daily their life. Annually 34.52% of this water is supplied from rainwater, 53.24% from greywater, 10.84% from seawater, and lastly, 1.4% from humidity. By merging these sources, the hybrid system is created. The energy consumption of this hybrid supply system is 3.89 kWh per day. Humidity to water system is the one that produces the least amount of water while consuming the highest amount of energy by 44.36%. Rainwater and greywater are put into the same treatment system, which makes 26.32% of overall energy consumption. A reverse osmosis system that uses seawater to produce potable freshwater makes 10.73% of the overall energy consumption. The system that is used to transport produced water consumes 18.59% of overall energy consumption.

The combination of these innovative technologies with digital tracking systems such as IoT and HMS has produced promising results. For example, it has become possible to get an output of 4 kWh of energy per day with organic waste and meet 87.76% of water production from sources such as rainwater and greywater. In addition, it will be possible to collect the data of all digitalised systems, which will allow residents to shave or shift peak load, study the consumption of households or regions, and prevent accidents/failures that are difficult to detect. These results are proof that the idea of self-sustaining compounds is not too far from reality.

A significant cost reduction is expected in the future in the aforementioned technologies. In particular, the expansion of systems that provide high-quality purification of grey and rainwater and the use of heat pump will significantly reduce the burden of small consumers at the network. Additionally, there are many drawbacks of long-distance water and electricity transmission lines installed to provide services to these consumers. The elimination of costs, such as losses during transmission, maintenance, and installation is an important advantage of the decentralised future.

Acknowledgements The author would like to acknowledge the help and contributions of Ali Barışcan Kaya, Ali Kemal Dikmecli, Doğu Mert Özkan, Efe Durmazkul, Güney Yurtsever, Yasemin Beril Kılıç in completing of this chapter.

Appendix

Appendix 2.1 Harvestable Rainwater (Area*rainfall*0.72)

Months	Jan	Feb	Mar	Apr	May	Jun	Jul	Aug	Sep	Oct	Nov	Dec	Total
Litres	7499.52	6229.44	5382.72	3810.24	2704.32	2142.72	1952.64	2393.28	3810.24	6048	7439.04	9097.92	58,510.08

Appendix 2.2 Harvestable Rainwater After Purification

Months	Jan	Feb	Mar	Apr	May	Jun	Jul	Aug	Sep	Oct	Nov	Dec	Total
Litres	6749.568	5606.496	4844.448	3429.216	2433.888	1928.448	1757.376	2153.952	3429.216	5443.2	6695.136	8188.128	52,659.072

Appendix 2.3 Used Rainwater

Months	Jan	Feb	Mar	Apr	May	Jun	Jul	Aug	Sep	Oct	Nov	Dec	Total
Litres	5366.667	5366.667	5366.667	3810.24	2704.32	2142.72	1952.64	2393.28	3810.24	5366.667	5366.667	5366.667	49,013.442

Appendix 2.4 Surplus Rainwater

Months	Jan	Feb	Mar	Apr	May	Jun	Jul	Aug	Sep	Oct	Nov	Dec	Total
Litres	2132.853	862.773	16.053	0	0	0	0	0	0	681.333	2072.373	3731.253	9496.638

Appendix 2.5 Used Rainwater After Purification (Used*0.9)

Months	Jan	Feb	Mar	Apr	May	Jun	Jul	Aug	Sep	Oct	Nov	Dec	Total:
Litres	4830	4830	4830	3429.216	2433.888	1928.448	1757.376	2153.952	3429.216	4830	4830	4830	44,112.096

Appendix 2.6 Greywater Production (Per day: 355 * 0.75 * 0.8 ≌ 210 L)

Months	Jan	Feb	Mar	Apr	May	Jun	Jul	Aug	Sep	Oct	Nov	Dec	Total
Litres	6300	6300	6300	6300	6300	6300	6300	6300	6300	6300	6300	6300	75,600

Appendix 2.7 Greywater Amount After Purification (Greywater Production*0.9)

Months	Jan	Feb	Mar	Apr	May	Jun	Jul	Aug	Sep	Oct	Nov	Dec	Total
Litres	5670	5670	5670	5670	5670	5670	5670	5670	5670	5670	5670	5670	68,040

Appendix 2.8 Total Water by Sources

Months	Jan	Feb	Mar	Apr	May	Jun	Jul	Aug	Sep	Oct	Nov	Dec	Total
Greywater	5670	5670	5670	5670	5670	5670	5670	5670	5670	5670	5670	5670	68,040
Rainwater	4830	4830	4830	3429.216	2433.888	1928.448	1757.376	2153.952	3429.216	4830	4830	4830	44,112.096
Reverrse osmosis	0	0	0	1400.784	2396.112	2901.552	3072.624	2676.048	1400.784	0	0	0	13,847.904
Water from humidity	150	150	150	150	150	150	150	150	150	150	150	150	1800
Total	10,650	10,650	10,650	10,650	10,650	10,650	10,650	10,650	10,650	10,650	10,650	10,650	127,800

Appendix 2.9 Water from Humidity (5 L * 30)

Months	Jan	Feb	Mar	Apr	May	Jun	Jul	Aug	Sep	Oct	Nov	Dec	Total
Litres	150	150	150	150	150	150	150	150	150	150	150	150	1800

Appendix 2.10 Reverse Osmosis

Months	Jan	Feb	Mar	Apr	May	Jun	Jul	Aug	Sep	Oct	Nov	Dec	Total
Litres	0	0	0	1400.784	2396.112	2901.552	3072.624	2676.048	1400.784	0	0	0	13,847.904

Appendix 2.11 Energy Consumptions (Purification: 3 kWh/m³, Reverse Osmosis: 11 kWh/m³, Water from Humidity: 350 kWh/m³, Hydrophore: 2.11 kWh/m³)

Months	Jan	Feb	Mar	Apr	May	Jun	Jul	Aug	Sep	Oct	Nov	Dec	Total
Greywater	18.9	18.9	18.9	18.9	18.9	18.9	18.9	18.9	18.9	18.9	18.9	18.9	226.8
Rainwater	16.1	16.1	16.1	11.431	8.113	6.428	5.858	7.18	11.431	16.1	16.1	16.1	147.041
Reverse Osmosis	0	0	0	15.409	26.357	31.917	33.799	29.437	15.409	0	0	0	152.328
Water from humidity	52.5	52.5	52.5	52.5	52.5	52.5	52.5	52.5	52.5	52.5	52.5	52.5	630
Hydrophore	22	22	22	22	22	22	22	22	22	22	22	22	264
Total	109.5	109.5	109.5	142.24	127.87	131.745	133.057	130.017	120.24	109.5	109.5	109.5	1420.169

References

Abundant Water (2020) The importance of water [Online]. Available at: https://www.abundantwater.org/the-importance-of-water. Accessed 12 Aug 2020

Advancing Global Change Science and Solutions (2020) What is the hydrosphere? [Online]. Available at: https://www.agci.org/earth-systems/hydrosphere#. Accessed 11 Aug 2020

Aksoy A, Öktem A (2014) Türkiye'nin Su Riskleri Raporu. Ofset Yapımevi, WWF-Türkiye

AKVO (2020) The science of making water from air [Online]. Available at: https://akvosphere.com/air-to-water-technology/. Accessed 17 Aug 2020

Alarko Carrier (2020) Carrier AquaSnap 61AF Isı Pompaları [Online]. Available at: https://www.alarko-carrier.com.tr/tr/urun/Carrier-AquaSnap-61AF-Isi-Pompalari. Accessed 04 Oct 2020

Alsulaili AD, Hamoda MF, Al-Jarallah R, Alrukaibi D (2017) Treatment and potential reuse of greywater from schools: a pilot stud. Water Sci Technol 75 (9):2119–2129

Antonyan M (2019) Energy footprint of desalination. University of Twente, Enschede

AQMD (2020) Agricultural waste [Online]. Available at: https://www.aqmd.gov/home/rules-compliance/compliance/open-burn/agricultural-waste#: ~ :text=Agricultural%20Waste%20is%20unwanted%20or, Grape%20Vines. Accessed 1 Nov 2020

Artema (2020) Artema [Online]. Available at: https://www.artema.com.tr/armaturler-dus-sistemleri-aksesuarlar/armaturler/mutfak-bataryalari/nest-trendy-eviye-bataryasi-sku-a42114. Accessed 1 Nov 2020

Baymak (2020) Yenilenebilir Enerji Sistemleri [Online]. Available at: https://www.baymak.com.tr/urunler/yenilenebilir-enerji-sistemleri/isi-pompalari. Accessed 04 Oct 2020

Busch System (2020) What is household wastes [Online]. Available at: https://www.buschsystems.com/resource-center/knowledgeBase/glossary/what-is-household-waste. Accessed 1 Nov 2020

Cambridge Dictionary (2020) Cambridge dictionary [Online]. Available at: https://dictionary.cambridge.org/tr/. Accessed 11 Aug 2020

Central C (2020) Land projected to be below annual flood level in 2050 [Art]

Chew KW et al (2019) Transformation of biomass waste into sustainable. Sustainability 11(2266):1–20

Crennan L (2007) Sustainable sanitation manual and construction guidelines for a waterless composting toilet. s.l.: SPREP

Daikin (2020) Daikin Altherma [Online]. Available at: https://altherma.daikin.com.tr/. Accessed 04 Oct 2020

Demir Ö, Yıldız M, Sercan Ü, Arzum CŞ (2017) Atıksuların Geri Kazanılması ve Yeniden Kullanılması. Harran Univ J Eng 2:1–14

DemirDöküm (2020) DemirDöküm MaxiAir Isı Pompası [Online]. Available at: https://www.demirdokum.com.

tr/urunler/demirdokum-maxiair-s-pompas-29313.html. Accessed 04 Oct 2020
Dekfin Denizcilik (2020) Dekfin Denizcilik [Online]. Available at: https://www.delfindenizcilik.com/urun/31/mini-45-60-dc. Accessed 5 Oct 2020
Donn D (2020) Deely house [Online]. Available at: https://www.deelyhouse.com/what-is-a-heat-pump/. Accessed 24 Oct 2020
EastMarine (2020) EastMarine [Online]. Available at: https://www.eastmarine.com.tr/tesisat-ve-havalandirma#/manFilters=12621&pageSize=20&viewMode=grid&orderBy=0&pageNumber=1. Accessed 5 Oct 2020
Encyclopædia Britannica (2015) Feces [Online]. Available at: https://www.britannica.com/science/feces. Accessed 08 Sept 2020
Environment Agency (2020) Meeting our future water needs: a national framework for water resources [Online]. Available at: https://www.gov.uk/government/publications/meeting-our-future-water-needs-a-national-framework-for-water-resources/meeting-our-future-water-needs-a-national-framework-for-water-resources-accessible-summary. Accessed 11 Aug 2020
Ertem O, Doğan A (2016) İstanbul İçin Nüfus ve Su Tüketimi Artışlarının İncelenmesi ve Talebin Değerlendirilmesi. Euro J Sci Technol 5(9):7–27
Etna (2020) Küçük Konutsal Hidrofor Tipleri [Online]. Available at: https://www.etna.com.tr/urunlerimiz/kucuk-konutsal-hidrofor-sistemleri. Accessed 06 Oct 2020
European Federation of Bottled Waters (2020) Importance of water [Online]. Available at: https://www.efbw.org/index.php?id=46. Accessed 11 Aug 2020
FSM Watermakers (2020) FSM watermakers [Online]. Available at: https://fsmwatermakers.com/fsm-nemo/. Accessed 5 Oct 2020
Government of Canada (2017) Government of Canada [Online]. Available at: https://www.nrcan.gc.ca/energy-efficiency/energy-star-canada/about-energy-star-canada/energy-star-announcements/publications/heating-cooling-heat-pump/what-heat-pump-and-how-does-it-work/6827. Accessed 24 Oct 2020
Halperin K (2020) Water: how much should I drink? [Online]. Available at: https://www.keithhalperin.com/. Accessed 08 Sept 2020
Hay CC, Morrow E, Kopp RE, Mitrovica JX (2015) Probabilistic reanalysis of twentieth century sea-level rise. Nature 517:481–484
HomeBiogas (2020a) HomeBiogas 2 [Online]. Available at: https://www.homebiogas.com/Products/HomeBiogas2. Accessed 05 Oct 2020
HomeBiogas (2020b) Main parts of the system [Art]. (Homebiogas Inc)
Horspool S (2019) Why water is important to life [Online]. Available at: https://owlcation.com/stem/The-Importance-of-Water-to-Life#. Accessed 12 Aug 2020
ICAX (2020) ICAX [Online]. Available at: https://www.icax.co.uk/. Accessed 24 Oct 2020
Industrial Heat Pumps (2020) Industrial heat pumps [Online]. Available at: https://industrialheatpumps.nl/en/how_it_works/cop_heat_pump/. Accessed 24 Oct 2020
IRENA (2012) Renewable energy technologies: cost analysis series. IRENA Innovation and Technology Center, Germany
Izzati T et al (2016) An initial study of industrial area's effects for the air pollution through rainwater in East Jakarta. IOSR J Mech Civ Eng 13(4):159–162
Jefferson B et al (2000) Technologies for domestic wastewater recycling. Urban Wtaer 1(4):285–292
Jefferson B, Kraume M, Geißen S (2010) Greywater treatment with a submerged membrane sequencing batch reactor. Von der Fakultät III – Prozesswissenschaften der Technischen Universität Berlin, Berlin
Joanna SS et al (2020) The quality of stored rainwater for washing purposes. Water 12(252):1–17
Khan SU-D et al (2018) Nuclear energy powered seawater desalination. In: In Renewable energy powered desalination handbook. COMSATS Institute of Information Technology, Pakistan, pp 225–264
Kulp SA, Strauss BH (2019) New elevation data triple estimates of global vulnerability to sea-level rise and coastal flooding. Nat Commun 29(10):10(4844)
Li Z, Boyle F, Reynolds A (2010) Rainwater harvesting and greywater treatment systems for domestic application in Ireland. Desalination, 1–8
Loh M, Coghlan P (2001) Domestic water use study. Water Corporation, West Leederville
Lu D, Flavelle C (2019) Rising seas will erease more cities by 2050, new research shows [Online]. Available at: https://www.nytimes.com/interactive/2019/10/29/climate/coastal-cities-underwater.html. Accessed 11 Aug 2020
Mekser (2020) Galvaniz prizmatik depo [Online]. Available at: https://www.mek-ser.com/urunler/galvaniz-prizmatik-depo. Accessed 1 Nov 2020
Mektank (2020a) Mektank [Online]. Available at: https://www.mektank.com/2-galvaniz-su-deposu-urun-detay. Accessed 5 Oct 2020
Mektank (2020b) Mektank [Online]. Available at: https://www.mektank.com/8-domestik-silindirik-su-depolari-urun-detay. Accessed 5 Oct 2020
MGM (2019) İllere Ait Mevsim Normalleri [Online]. Available at: https://www.mgm.gov.tr/veridegerlendirme/il-ve-ilceler-istatistik.aspx?k=undefined&m=ISTANBUL. Accessed 3 Oct 2020
Monika Z, Justyna Z, Dorota P, Daniel S (2020) The quality of rainwater collected from roofs and the possibility of its economic use. Rzeszow University of Technology, Rzeszow
Mughalles A et al (2012) Household greywater quantity and quality in Sana'a, Yemen. EJGĖ, 1026–1034
Mullen K (2020) Information on earth's water [Online]. Available at: https://www.ngwa.org/what-is-groundwater/About-groundwater/information-on-earths-water. Accessed 11 Aug 2020
Nexus-The Water, Energy & Food Security Platform (2011) Introduction (Bonn2011 conference) [Online].

Available at: https://www.water-energy-food.org/about/introduction/. Accessed 11 Aug 2020
Nexus-The Water, Energy & Food Security Platform (2020) Bonn2011 conference [Online]. Available at: https://www.water-energy-food.org/about/bonn2011-conference/. Accessed 20 Aug 2020
Nosowitz D (2018) Where to use human waste as fertilizer and irrigation [Online]. Available at: https://modernfarmer.com/2018/08/where-to-use-human-waste-as-fertilizer-and-irrigation/. Accessed 05 Sept 2020
Nowak T (2018) Heat pumps integrating technologies to decarbonise heating and cooling [Online]. Available at: https://www.ehpa.org/fileadmin/user_upload/White_Paper_Heat_pumps.pdf. Accessed 11 Aug 2020
OECD (2017) Water risk hotspots for agriculture, OECD studies on water. Paris: s.n.
Onubi HO (2019) International journal of sustainable building technology and urban development [Online]. Available at: https://www.researchgate.net/profile/Hilary_Onubi/publication/341070047_durabi-2019-010-04-4/links/5eaba6a892851cb267692a7d/durabi-2019-010-04-4.pdf. Accessed 11 Aug 2020
Orhon D, Hocaoğlu SM (2011) Ayrık evsel atıksuyun membran biyoreaktörde ayrışma mekanizmaları. itüdergisi/e 21(1):35–44
Oron G et al (2014) Greywater use in Israel and worldwide: standards and prospects. In: Water research. s.l.:Elsevier, pp 92–101
Otank (2020) Paslanmaz Çelik Depo Avantajları [Online]. Available at: https://otank.com.tr/paslanmaz-celik-depo-avantajlari/. Accessed 1 Nov 2020
PNG Biomass (2020) How a biomass power plant works: technology & design [Online]. Available at: https://pngbiomass.com/how-a-biomass-power-plant-works/. Accessed 25 Oct 2020
Pradeep P, Lijin Z (2017) Water-energy nexus in the People's Republic of China and emerging issues [Online]. Available at: https://www.adb.org/sites/default/files/publication/384291/water-energy-nexus-prc.pdf. Accessed 11 Aug 2020
RainMaker (2020) How Rainmaker's air-to-water technology works [Online]. Available at: https://rainmakerww.com/technology-air-to-water/. Accessed 27 Oct 2020
Rathnayaka K et al (2015) Seasonal demand dynamics of residential water end-uses. Water 7:202–216
Rosegrant MW, Cal X, Cline SA (2002) World water and food to 2025: dealing with scarcity. s.l.:Intl Food Policy Res Inst.
Seamar Watermakers (2020) Seamar watermakers [Online]. Available at: https://seamarwatermakers.com/boat-group/seamar-mini-compact-60. Accessed 5 Oct 2020
Shiklomanov I, Rodda JC (2003) World water resources at the beginning of the twenty-first century. The Press Syndicate of The University of Cambridge, Cambridge
Siemens (2020) Process in biomass power plant [Online]. Available at: https://new.siemens.com/global/en/markets/chemical-industry/continuous-processes/green-refinery/biomass-power-plant.html. Accessed 25 Oct 2020
Solutions A. G. C. S. a. (2020) Water on earth [Art]
Somalı B, Ilıcalı E (2020) Leed ve Breeam Uluslararası Yeşil Bina Değerlendirme Sistemlerinin Değerlendirilmesi [Online]. Available at: https://www1.mmo.org.tr/resimler/dosya_ekler/5464e0031fd7f46_ek.pdf. Accessed 11 Aug 2020
Statista (2017) [Art]
Taler D, Pitry R, Taler J (2019) Operation assessment of hybrid heat source for heating the building and preparation of hot water in the fire brigade building. In: Cleaner production. s.l.:Elsevier, pp 962–974
TEMA (2020a) Çatı Suyu Hasadı [Online]. Available at: https://sutema.org/gelecegin-suyu/cati-suyu-hasadi.19.aspx. Accessed 18 Oct 2020
TEMA (2020b) Evsel Su Tüketimi [Online]. Available at: https://sutema.org/gelecegin-suyu/evsel-su-tuketimi.18.aspx. Accessed 11 Aug 2020
The Reenage (2020) Direct use of grey water [Online]. Available at: https://www.thegreenage.co.uk/tech/greywater-recycling/. Accessed 24 Oct 2020
The World Bank (2020a) Employment in agriculture (% of total employment) (modeled ILO estimate) [Online]. Available at: https://data.worldbank.org/indicator/SL.AGR.EMPL.ZS. Accessed 11 Aug 2020
The World Bank (2020b) Water [Online]. Available at: https://www.worldbank.org/en/topic/water/overview. Accessed 11 Aug 2020
The World Bank (2020c) Water in agriculture [Online]. Available at: https://www.worldbank.org/en/topic/water/overview. Accessed 11 Aug 2020
Tıryak İ, Çiçekalan B, Öztürk İ (2020) Cost analysis in drinking water distribution and sewer networks: the case of Istanbul Asian Side. Dokuz Eylül Üniversitesi Mühendislik Fakültesi Fen Ve Mühendislik Dergisi 22 (64):233–246
TMMOB (2019) Dünya Çevre Günü Türkiye Raporu, s.l.: TMMOB Çevre Mühendisleri Odası
Tolson M (2011) Green homes vs traditional homes [Online]. Available at: https://buildipedia.com/at-home/design-remodeling/green-homes-vs-traditional-homes#: ~ :text=As%20you%20can%20see%2C%20green,local%20real%20estate%20tax%20savings. Accessed 11 Aug 2020
Turkey Ministry of Environment and Urbanisation (2016) Turkey ministry of environment and urbanisation [Online]. Available at: https://webdosya.csb.gov.tr/db/ced/icerikler/gostergeler-2016-20180618144826.pdf. Accessed 11 Aug 2020
Türkiye İstatistik Kurumu (2017) Belediye Katı Atık Temel Göstergeleri, 2004. Haber bülteni 29(12):1
UNICEF (2017a) Thirsting for a future: water and children in a changing climate. [Online]. Available at: https://www.unicef.org/publications/index_95074.html. Accessed 11 Aug 2020
UNICEF (2017b) Thirsting for a future:water and children in a changing climate [Online]. Available at: https://www.worldbank.org/en/topic/water/overview. Accessed 11 Aug 2020
Unicef Data (2019) Progress on household drinking water, sanitation and hygiene, 2000–2017 [Online].

Available at: https://data.unicef.org/resources/progress-drinking-water-sanitation-hygiene-2019. Accessed 20 Aug 2020

Üstün GE, Tırpancı A (2015) GRİ SUYUN ARITIMI VE YENİDEN KULLANIMI. Uludağ Üniversitesi Mühendislik Fakültesi Dergisi 20(2):119–139

Uyar V (2018) Dünya Su Günü [Online]. Available at: https://www.dogrulukpayi.com/bulten/turkiye-su-fakiri-bir-ulke-olma-yolunda-ilerliyor. Accessed 11 Aug 2020

Vahaa (2020) Vahaa Geleceğiniz için Sürdürülebilir Tarım [Online]. Available at: https://www.vahaa.co/product-page/vahaa-dikey-bah%C3%A7e. Accessed 1 Nov 2020

Watergen (2020) Water-from-air, at home or in the office [Online]. Available at: https://www.watergen.com/product/genny/. Accessed 04 Oct 2020

WBCSD (2009) Binalarda Enerji Verimliliği. G.M. Matbaacılık, Istanbul

WHO (2019) Drinking-water [Online]. Available at: https://www.who.int/news-room/fact-sheets/detail/drinking-water. Accessed 17 Aug 2020

Wong TE, Keller K, Bakker A (2017) Impacts of Antartic fast dynamics on sea-level projections and coastal flood defense. Clim Change 144:347–364

World Economic Forum (2017) Closing the water gap [Online]. Available at: https://www.weforum.org/our-impact/closing-the-water-gap. Accessed 11 Aug 2020

World Health Organization (2013) How much water is needed in emergencies [Online]. Available at: https://www.worldbank.org/en/topic/water/overview. Accessed 11 Aug 2020

Xhensila T, Vedat T (2020) Credit success rates of certified green buildings in Turkey. Teknik Dergi 580:10063–10084

3 Energy

Abstract

One of this book's main ambitions is to create a sustainable ecosystem including water generation and sanitation, food production, and renewable energy generation by producing minimum pollution to the environment. This chapter completes the big picture by providing the design of crucial components such as solar and wind energy generation systems, energy storage system and developing a strategy for power supply system for the Open Digital Innovation Hub (ODIH). Long-term and short-term energy production and consumption estimations were carried out, considering the equipment's technical characteristics and the house model's geographical location. To achieve a continuous power supply, system needs are gathered, and a requirement list is created. As a result of detailed market research, the necessary solar panels, auxiliary wind turbines and domestic batteries were selected by paying attention to ODIH's monthly energy consumption Peak and continuous power characteristics of every device used in ODIH are considered as well as their energy requirements. The energy generation system is then structured by mechanically integrating renewable energy generation and storage systems in ODIH. Additionally, we show the carbon emissions comparison between a typical house fed by a traditional energy system and renewable sources.

The author would like to acknowledge the help and contributions of Mehmet Akif Ekrekli, Alaattin Canpolat, Berk Ürkmez, Enes Ürkmez, Eymen Ipek, İlayda Zeynep Mert, Melike Esin Sert, Mert Yavaşca, Tayfun Ozan Koç, and Yade Kilimci in completing of this chapter.

3.1 Introduction

3.1.1 Water-Energy-Food (WEF) Nexus

Demands for energy, freshwater, and food will increase significantly by the forthcoming years because of the growth of the economy and global population, mobility, cultural and technological change, dietary differentiation, and global climate change (Hoff 2011). Water is significantly important for agriculture and food. Water consumption for agriculture irrigation is responsible for 70% of the total global freshwater withdrawal (Food and Agriculture Organization of the United Nations 2017). Water is used for different purposes. For example, daily household use, industrial purposes, and agricultural production. Also, the global agricultural food production and supply chain composed of 30% of the total energy demand (Food and Agriculture Organization of the United Nations 2017). WEF Nexus, plays an essential role in the sustainable development strategies of countries. The ability of existing

S. Küfeoğlu, *The Home of the Future*, Sustainable Development Goals Series,
https://doi.org/10.1007/978-3-030-75093-0_3

water, energy and food systems to meet this growing demand is constrained given the competing needs for limited resources. Thanks to the Nexus concept, positive progress can be made in meeting the increasing demand with the effects of global climate change and different challenges.

The WEF Nexus is the study of the links between water, energy, and food together with the conflicts and trade-offs (Simpson and Jewitt 2019). In this concept, every part is related to each other. This relation creates a sustainable environment. Sustainability focuses on using existing resources without undermining the right of future generations to access the same resources (Mitchell Grant 2020). The United Nations General Assembly (UNGA) adopted the Sustainable Development Goals (SDGs) in 2015 to ensure global cooperation in order to achieve a sustainable future for our planet. "Agenda 2030" contains 17 SDGs and 169 sub-targets, determine a path to end inequality, injustice and poverty, and protect the environment. Clean and affordable energy, SDG#7, and sustainable cities and communities, SDG#11, are taking a vital role in the success of Agenda 2030.

The WEF Nexus model is a complex structure, where there is almost no waste, and everything works interconnected. The WEF Nexus concept is designed to continue life without any external support. Thus, sustainable and self-sufficient life can be created. This concept is an idea that identifies the connection between water, energy and food systems (Albrecht et al. 2018).

Figure 3.1 represents modelling the interactions between water, energy and food end-uses at a household scale (Hoff 2011).

Energy is well known to be essential for human development. It is evident that many critical activities like water supply for human usage. Irrigation, as well as food conservation and

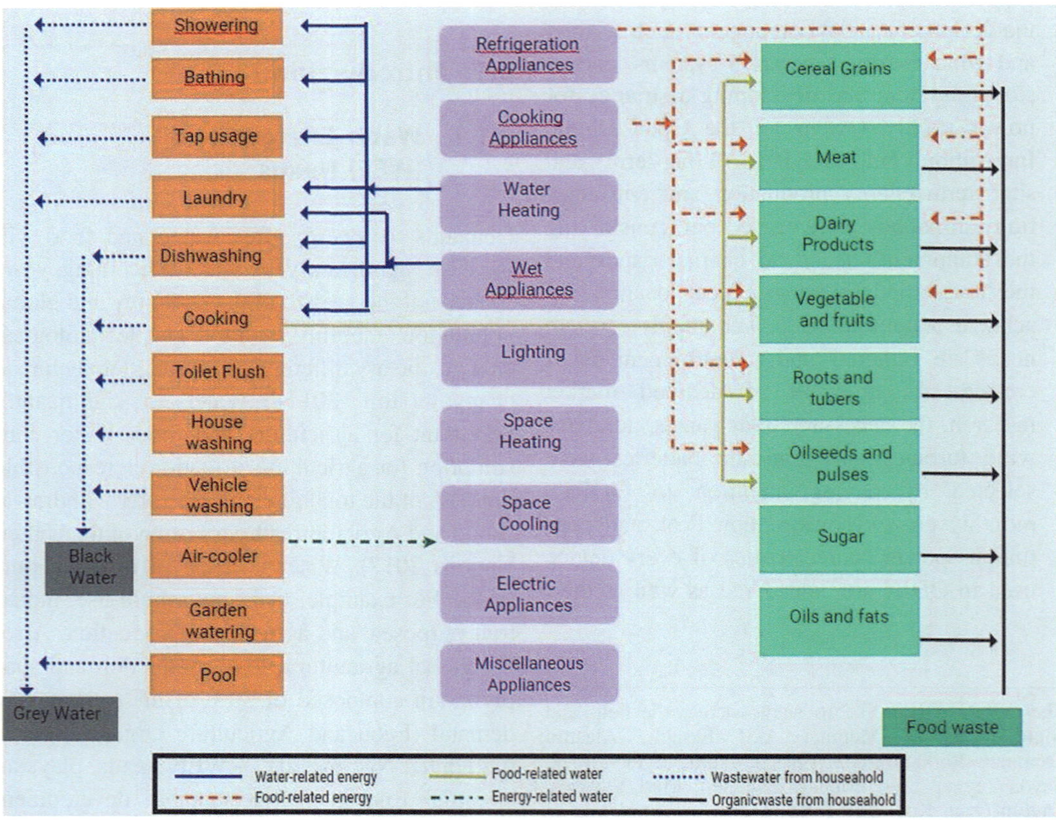

Fig. 3.1 Modelling the interactions between water, energy and food end-uses at a household scale (Hoff 2011)

production, cannot be sustained without a continuous supply of energy. Energy's position in the WEF model is critical as it is the essential element to enable other activities (Foundation 2015).

The world population is growing at a remarkable rate. According to the World Bank, in 1960, the world's population was 3.1 billion, and in 2019 it was 7.67 billion (The World Bank 2019). As the population grows, energy demand is increasing, which raises the consumption of energy resources. We can categorise energy sources into two types: renewable and non-renewable energy sources. Non-renewable energy sources are related to limited supplies as it takes a long period of time to be replenished. Renewable sources are replenished in relatively short periods of time without human contribution (U.S. Energy Information Administration 2020). According to the way of production, energy sources can also be divided into two categories: primary and secondary sources. Primary sources can be used directly because they are readily available in the natural environment. Secondary sources derive from the processing of primary energy sources (Eni 2020). Table 3.1 summarises energy sources and categorisation.

Providing a sufficient energy source is one of the critical issues in a growing society. The urban population increased by three times worldwide from 1960 to 2018. In 1960 1.01 billion people were living in urban areas, whilst in 2019, this number increased to 4.27 billion (The World Bank 2019). The energy consumption continues to increase dramatically with the population growth in urban areas. Increasing decentralisation and digitalisation enable small-scale solutions for energy production. These small-scale producers, especially in the household level or residential ones, are called prosumers. The prosumers might consume energy from the grid or inject their excess power back to the grid whenever conditions call. In urban areas, solutions such as rooftop solar energy, small-scale wind energy and biomass are options to respond to the energy demand.

3.1.2 Solar Energy

The sun is the primary energy source on earth. People used solar radiation for long times for heating and drying meat, fruit, and grains. As time passed by, people developed technologies to harness solar energy for warmth and then for converting it into electricity. Sunlight consists of tiny energy packets called photons. Every hour, enough photons fall on our planet to theoretically satisfy almost all global energy need for an entire year (Eagles, Diamond 2020). Turkey has enough potential for solar energy. Figure 3.2 shows the total solar radiation levels for all cities in Turkey.

People currently use several ways to collect and convert solar radiation into usable energy for different purposes. Solar energy can be used either to provide heat or to generate electricity (Ekins-Daukes 2006). Solar thermal energy utilises sunlight to generate thermal energy, and solar photovoltaic turns sunlight into electricity to energise electrical devices (The Solar Nerd 2019).

According to a report from the International Energy Agency (IEA), solar energy could become the largest global source of electricity by

Table 3.1 Energy sources

Energy sources	Primary energy sources	Secondary energy sources
Renewable energy sources	Wind Solar Biomass Waves Geothermal	Biofuel
Non-renewable energy sources	Crude oil Hard coal Natural gas	Petroleum products Manufactured solid fuels Gases

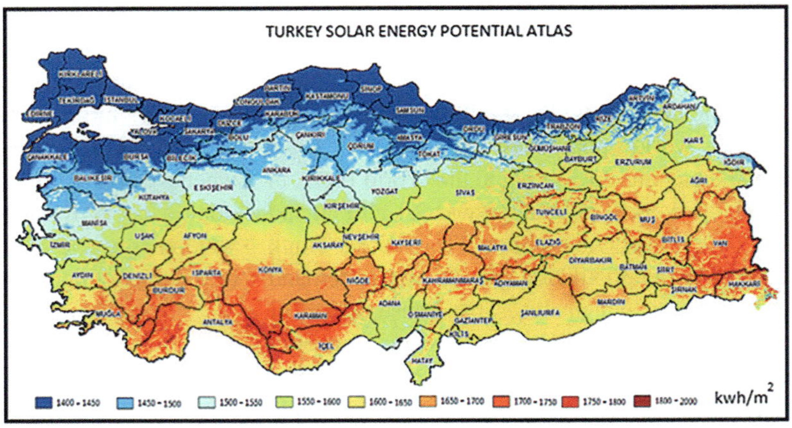

Fig. 3.2 Turkey solar energy potential Atlas (Solars 2016)

2050 (International Energy Agency 2014). Urban use of solar energy is a growing trend. London, for example, hosts more than 16,000 solar photovoltaic (PV) panels on its roofs (Stoker 2015).

Using solar energy has many benefits. One of the key advantages is no air contaminant or carbon dioxide are generated by solar energy structures, and these systems on buildings have minor environmental impacts (EIA 2019). Solar panels (hereafter PV panels) use photons produced by sunlight to generate direct current (DC). Figures 3.3 and 3.4 show solar PV panels in a solar power plant and on a rooftop.

Solar cells are made from elements called silicon, and each cell, like the batteries, consists of a positive and a negative layer to create an electric current (Ogden and Williams 1989). Figure 3.5 shows the construction of a solar PV panel.

PV panels are used for the production of electricity both on large and small scales. PV panels are assembled for small scale solar energy generation into flat plate systems that can be mounted on house rooftops or sunny areas (The Clean Energy Regulator 2019). PV panels operate quietly with no emissions, no fuel, and require little maintenance (Fonash 2020).

Solar PV can be used for on-gird or off-grid systems (Newkirk 2016, b). By far, the most popular and commonly used by homes and fields are on-grid or grid-bound solar systems. These systems do not need batteries and, using either solar inverters or microinverters, are connected to

Fig. 3.3 PV panels in solar power plant

Fig. 3.4 Rooftop PV panel model

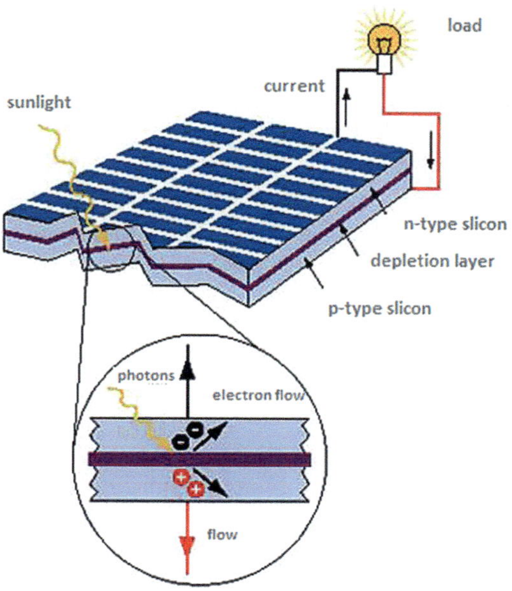

Fig. 3.5 Construction of a PV panel

the public electricity grid. Any surplus solar power one produces traded to the grid, and the owner is frequently charged a feed-in tariff (FIT) or the electricity you trade credit. If needed, batteries could be connected to on-grid systems. (Newkirk 2016, b).

The off-grid power system has no grid connection. The power-providing utility company is entirely separate from any national networks and is therefore solely responsible for servicing consumers in the region. The surplus of the supplied energy is stored in batteries, which can be reused at night (Newkirk 2016a, b). Figure 3.6 shows a rooftop system connection in two way, on-grid and off-grid with storage.

3.1.2.1 Working Principle and Components of a Photovoltaic System

1. Photovoltaic module: Generate direct current (DC) with solar radiation.
2. Maximum Power Point Tracker (MPPT) and charge regulators: Charges batteries by following the maximum power point.
3. Accumulator: Stored DC energy from the photovoltaic module.
4. Inverter and control unit: By transforming DC to AC, it supplies energy to the loads in the household.

3.1.3 Wind Energy

Wind energy is one of the natural, clean, and renewable energy sources. Pressure and temperature differences occur because the sun is not heating the earth evenly. This difference creates airflow. The warmer air mass rises, and with rise, the cooler air mass settles in the space. The result of this process is wind (YEGM 2020).

Wind energy is an important energy source for the world because it is relatively environmentally friendly and renewable. One of the most critical indicators showing the development levels of countries is that they have trouble-free, continuous and clean energy resources. The use of wind energy will increase in many folds in the coming years as the pressure for tackling climate change grows.

Turkey is a rich country in terms of renewable energy sources. Wind speeds and wind power densities are high in Marmara, Aegean and South-

Fig. 3.6 On-grid and off-grid with storage in rooftop PV systems

Table 3.2 Wind energy potential of Turkey by regions (Hepbasli et al. 2010)

Region	Average wind density (W/m^2) per year	Average wind speed (m/s) per year
Eastern Anatolia region	13.19	2.12
Middle Anatolia region	20.14	2.46
Black sea region	21.31	2.38
Mediterranean region	21.36	2.45
Aegean region	23.47	2.65
South-Eastern Anatolia region	29.33	2.69
Marmara region	51.91	3.29
Average	25.82	2.58

Fig. 3.7 Vertical-axis wind turbine (Markham 2018)

Fig. 3.8 Horizontal-axis wind turbine (Top Alternative Energy Sources 2020)

Eastern Anatolia regions, and thus, wind power plants are mainly located in these regions (Doymaci and Yilmaz Ulu 2018). Turkey's wind energy potential by region is summarised in Table 3.2.

Constructing wind turbines in urban settlements is a challenge. To overcome this problem, small wind turbines are designed. The Wind-electric system consists of a wind turbine which is placed on the top of the tower to reach higher wind speeds. (Energy 2020). There are basically two types of wind turbines. These are divided into two groups as the vertical axis and horizontal axis wind turbines. Figures 3.7 and 3.8 demonstrates the vertical and horizontal axis wind turbine models.

3.1.3.1 Horizontal-Axis Turbines

Horizontal-axis turbines usually have three blades, and the blades are similar to aeroplane propellers. The longer the blades and the turbine are, the more electricity they produce. It is the most used type of turbine today.

3.1.3.2 Vertical-Axis Turbines

Vertical-axis turbines have blades that are connected to a vertical rotor above and below. Since they are not efficient as well as horizontal-axis turbines, they are not preferred today (U.S. Energy Information Administration 2019).

Most of today's small wind turbines are horizontal-axis. Typically such blades are made from composite material, such as fibreglass. Since wind velocity is directly proportional to the altitude, small wind turbines are placed onto towers. A small wind turbine is mounted on a commonly two type of tower which is free-standing and guyed (Energy 2020).

3.1.4 Biogas

Decomposition of animal waste or manure can be used for the production of biogas. Greenhouse emissions can be mitigated by the production of biogas from poultry, livestock and pig farms, and the exploitation of methane from biogas could provide a renewable source of energy. Retrieved biogas may be utilised for transporting fuel, energy or warming as a source of energy. Recovering biogas is a validated method that is commonly preferred in food processing and wastewater management (United States Environmental Protection Agency 2019). There are numerous benefits that biogas provides to the World we live in. Preventing global warming, reducing dependence on fossil fuels, and leading to hygiene improvements are some of these advantages. Biogas can be used for the same purposes as natural gas, including heating, production of electricity and, after updated, as a transportation fuel. Therefore, it is a too prominent resource. Additionally, biogas is obtained as a result of anaerobic treatment and can be used as an energy source (Ozturk 1999).

3.1.4.1 Anaerobic Digestion

The name of the system in which microbes break down biological material without oxygen is called anaerobic digestion. Closed structures that use this natural technique to produce biogas and other vital co-products are called anaerobic digesters. Microbes, pollutants, and odorous are also minimized by these structures (United States Environmental Protection Agency 2020). Biogas production fundamentally depends on methane, which is the main component of natural gas. In order to use methane as a source of energy, biogas non-methane ingredients are separated. A higher portion of digestible organic compounds leads to more biogas production. Co-digestion is resulted from the utilisation of anaerobic digestion in an individual anaerobic digester to dissolve several forms of organic waste. The production of biogas is possible by using organic waste, which is low-level or hard to digest if co-digestion is done (United States Environmental Protection Agency 2020). Figure 3.9 summarises the Biogas recovery system and its components.

Waste treatment is a growing concern in urban regions, thus making anaerobic digestion and biogas production as a viable option for city planners. In addition to electricity generation, heating and transport fuel production, recycling

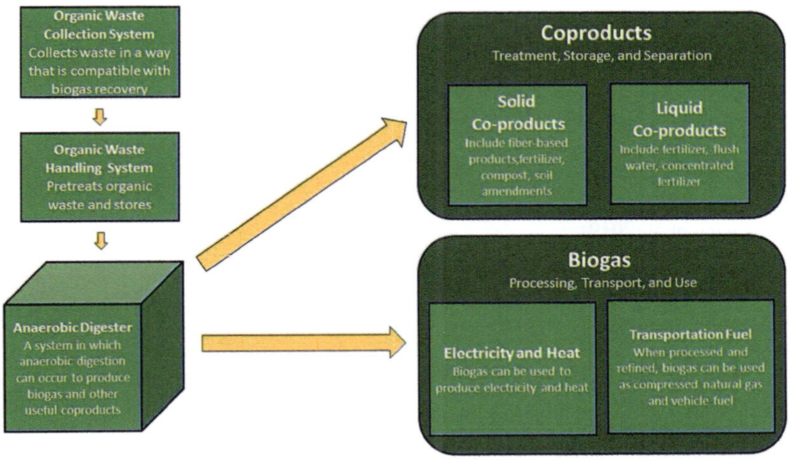

Fig. 3.9 Biogas recovery system

waste and reusing by-products such as waste protein are some of the factors that favour this technology.

3.1.5 Energy Storage Systems

Despite certain advantages of renewable energy sources, there is a critical drawback: the intermittency of supply. Intermittent electricity is electrical energy that is not available on a continuous basis due to external factors that cannot be controlled. Although the energy supply provided by renewable energy sources along the day is not stable, the demand is continuous. Intermittency might cause unexpected fluctuations of power putting more stress on the frequency response. This problem is even more eminent within microgrids or isolated grid systems. Energy storage systems are needed in order to overcome this problem. The stored electricity can be given to any system at a stable, desired rate at any time. Figure 3.10 shows a typical storage role between source and end-user.

Over the past few decades, power generation from renewable energy has risen significantly. Energy-storage technology will ultimately be crucial to achieving a clean-energy environment, providing safe and consistent access to the client from a more fragmented and unstable resource base (Hall and Bain 2008).

Energy storage systems (ESS) are frequently used for electricity generation systems based on renewable energy due to the intermittent nature. It can be claimed that ESS is an inevitable solution to provide reliable and continuous power to the system. Therefore, different solutions are developed throughout the years for optimising energy storage performance as well as reducing the cost of the overall design.

The storage mechanism may be mechanical, hydraulic, electrochemical, electrical or a combination of these systems. Table 3.3 represents an overview of frequently used energy storage systems based on their power and energy levels with respect to storage time.

The most critical metrics of these systems may be mentioned as energy efficiency, energy density, power density, cycle life, lifespan and cost. Table 3.4 represents a comparison of different technologies in terms of these metrics.

Firstly, hydraulic energy storage systems are preferred for the systems that require colossal power and energy demands, such as more significant than 1 MW, as shown in Table 3.4. Besides, the investment cost of these systems will be higher for small applications. Second, despite the advantages of mechanical ESS, such as high-

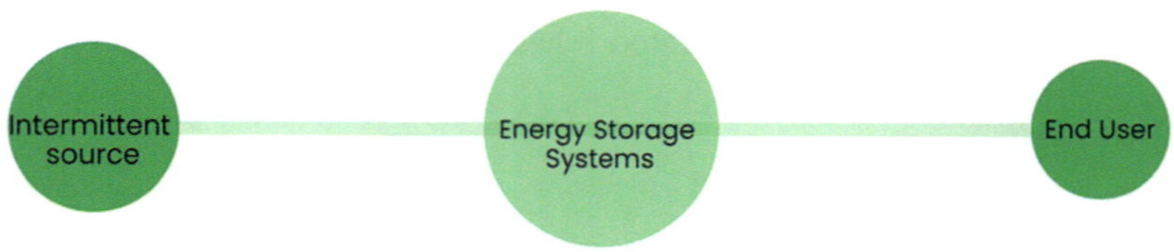

Fig. 3.10 Electricity transmission from an intermittent source

Table 3.3 Energy storage systems overview (Baumann et al. 2019)

Technology	Power rating	Storage time	Capacity
Capacitors	10 kW–10 MW	Milliseconds–seconds	Low
Supercapacitors	1 kW–10 MW	Milliseconds–hours	Low
Battery systems	1 kW–100 MW	Seconds–days	Medium
Pumped storage	1 MW–1 GW	Hours–days	Medium
Hydro reservoir storage	10 MW–1 GW	Days–weeks	High
Hydrogen	10 kW–1 GW	Weeks–months	High

Table 3.4 Technical specification of grid energy storage systems (Baumann et al. 2019)

Technology	Efficiency (%)	Energy density (W/kg)	Power density (W/kg)	Cycles (1000)	Lifetime (years)	Investment cost (cell) (Euro/kWh)
Pumped hydro storage	65–85	N/A	0.5–1.5	10–50	30–60	46–500
Compressed air	54–88	3.8–6	–	Jun-20	20–40	3–300
Flywheels	85–87	12.5–82	950–5700	28–93	15–20	537–1543
Super capacitors	90–97.5	5.2–21.7	1450–10,000	21–100	15–20	570–6800
Fuel cell	20–35	500+	500	1000	5–15	*per kW 10,000+
Li-ion batteries	81–98	84–145	253–1300	0.73–8	7.5–20	376–696

power capability and relatively lower cost with complexity, mechanical ESSs have poor energy densities and high self-discharge rate that is why they are utilised for very short-term energy storage. Fuel cell technology is also becoming widespread by its usage in electric vehicles. However, it has drawbacks considering its poor power density that requires additional power source such as capacitors or batteries. Finally, comparing capacitors with the electrochemical ESS, electrochemical ESS is much more capable of storing energy. Also, the energy densities are remarkably higher than the capacitors.

3.1.5.1 Batteries

As electrochemical ESS, batteries are usually used as home energy storage systems. Different types of batteries are used by application needs, such as lead-acid, nickel or lithium-based batteries (Dehghani-Sanij et al. 2019). Installed battery systems may require power, energy, or both.

Batteries can be configured to supply power, energy, or both at the same time. However, the battery system must be optimised in terms of system requirements. Although lead-acid and nickel-based batteries have a long history in the storage systems, Lithium-ion batteries (LIB) are being the most preferred choices in most cases, considering their superiorities on lifetime, efficiency, power and energy density. Besides, despite the challenges of LIB regarding safety, cost and complexity, it is becoming a mature solution with electric vehicles.

There are different types of LIB developed to optimise key metrics. Table 3.5 shows a comparison of different kinds of LIB in the market commercially available (Moseley and Garche 2015).

Table 3.5 Comparison of different types of LIB (Miao et al. 2019)

Cathode Chemistry	Energy	Power	Safety	Durability	Maturity	Cost
NMC	Good	Fair	Fair	Fair	Good	Good
LFP	Bad	Good	Good	Good	Good	Fair
NCA	Good	Fair	Bad	Bad	Good	Good
LMO	Fair	Good	Fair	Bad	Fair	Fair
LCO	Fair	Fair	Bad	Fair	Good	Fair

3.2 Aim of the Study

Open Digital Innovation House (ODIH) is a research project that aims to establish a fully smart and digitised self-sustaining compound by collecting the required energy from renewable sources and trading residual energy. This study aims to design energy generation and storage systems of ODIH by paying attention to WEF nexus. We shall focus on the following subjects:

- Rooftop PV systems
- Small-scale wind turbines
- Biomass-anaerobic digestion (zero-waste)
- Storage systems and battery options
- Impact of ODIH on the environment.

The project is at the intersection of WEF nexus and self-sustaining systems concepts. Digitalisation and smart systems will be supporting elements to achieve this project.

3.3 Methodology and Materials

In ODIH, energy is produced using renewable energy sources such as rooftop PV panels, wind turbines and biogas to meet the energy demand. First, monthly energy consumption calculations of ODIH were made to calculate energy production. Market research was conducted for the acquisition of PV panels and wind turbines to meet this need according to the calculations. To calculate PV generation, the daily average radiation data of Istanbul were examined, and a suitable PV panel model is selected accordingly. For the wind turbines installation, monthly average and maximum wind speeds in the Bosphorus area were obtained from the Turkish State Meteorological Service, and the appropriate wind turbine is chosen. The biogas production process and biogas potential in Turkey were investigated to transform organic materials into energy that we can use in ODIH. Nevertheless, biogas technology will only be used in the water sanitation process in a limited amount due to the space constraint of the project.

The battery was used to support the electrical system, energy savings and energy storage of ODIH. By comparing the advantages and disadvantages of different battery technologies, the appropriate battery type to be used in ODIH is determined. At the end of the process, energy planning of ODIH is made with the aid of turbines, panels and batteries using the necessary inverters. In Appendix A, it is seen that the floating compound consumes approximately 18.000 kWh of energy per year.

3.3.1 PV Panel

Solar energy is one of the most relevant sources of renewable energy options in Turkey. Turkey's geographical location is quite advantageous in terms of solar energy potential (Balat 2005). 307 terawatt-hours (TWh) of electricity was produced in 2019 in Turkey. 15% of the electricity generation was provided by renewable energy sources (Statista 2020). In addition, Turkey has an average yearly solar radiation potential of approximately 4.52 kWh/m^2—day and the average daily sunshine duration is approximately 7.58 h (Ekmekci 2016). The main objective of this chapter is to evaluate the solar PV systems in Istanbul. Istanbul (latitude = 40.58 N, longitude = 29.05 E and elevation = 39 m) is Turkey's most crowded city and has a rapidly growing economy for decades. It is located north-western part of Turkey (Batman et al. 2012). Figure 3.11 and Table 3.6 show the solar radiation levels in Istanbul.

Photovoltaic panels can produce electricity in a clean and environmentally way. PV systems consist of photovoltaic cells that transform sunlight directly into electricity, and since the source of energy used to generate electricity is the sun, it is generally referred to as a "solar cell". There are three major types of solar panels: monocrystalline, polycrystalline, and thin film. Table 3.7 compares these types in various aspects.

The solar panels of the company Schmid Pekintas Energy were chosen for this project by considering the location and physical properties

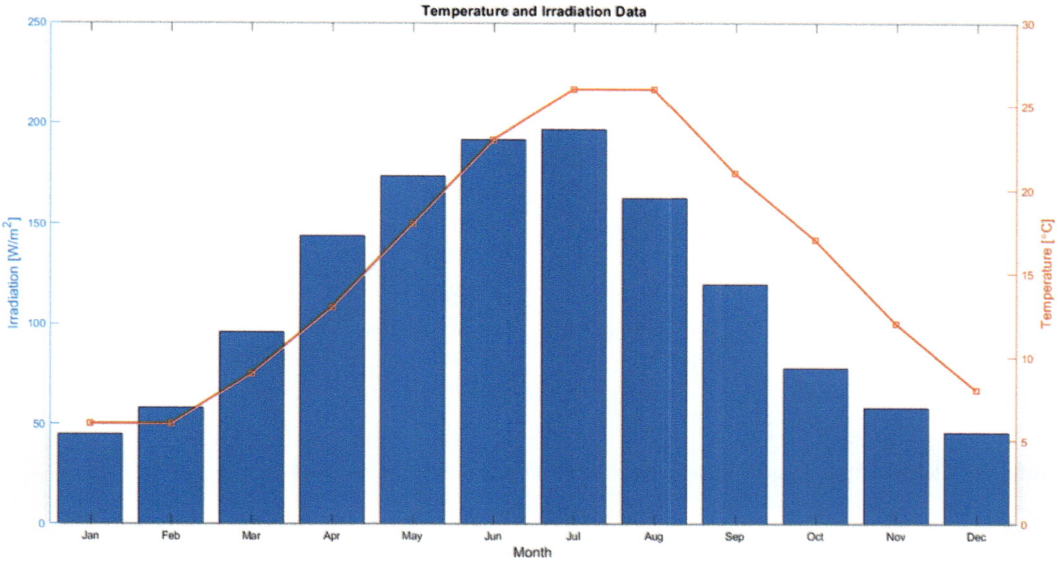

Fig. 3.11 Istanbul global radiation level (kWh/m^2-day) Turkey global radiation level (kWh/m^2-day) (YEGM 2020)

Table 3.6 Istanbul monthly average daily radiation and sunshine data (YEGM 2020)

Months	kWh/m^2-day	Sunshine duration (h)
January	2	3.46
February	2.57	4.43
March	4.2	5.32
April	5.28	6.85
May	6.3	8.61
June	6.79	10.51
July	6.79	11.17
August	6.07	10.14
September	5.09	7.83
October	3.74	5.22
November	2.37	3.85
December	1.8	2.96
Mean	4.42	6.7

of ODIH. The electrical features of the possible three models of selected panels are demonstrated in Table 3.8.

Considering the dimensions of the ODIH's roof, forty-five solar panels are planned to be placed with two small wind turbines. A detailed plan could be observed in Fig. 3.12.

3.3.1.1 Solar Inverter

In a solar system, an inverter is typically the part that transforms the DC power from the panels into AC power for on-grid or off-grid home appliances and the main grid. This function is fulfilled by a hybrid inverter while also transferring DC power to a battery to maintain it for

Table 3.7 Comparison of PV types (El-din et al. 2014)

	Monocrystalline	Polycrystalline	Amorphous
Composition	Single-crystal silicon	Multi-crystal silicon	Thin silicon layer
Size	Small	Larger	Largest
Efficiency (%)	15–20	13–16	6–9
Durability	High	High	Shorter
Price	Highest	High	Lowest
Appearance	Black or dark blue cells with rounded corners	Blue rectangular cells	Black or blue uniform surface

Table 3.8 Electrical specifications of selected panel (Schmid Pekintas Energy 2018)

Manufacturer	Schmid Pekintas energy		
Cell type	Monocrystal PERC		
Model	SPE380	SPE385	SPE390
Voltage at Pmax (V)	39.65	40.11	40.56
Current at Pmax (A)	9.59	9.60	9.62
Open circuit voltage (V)	48.33	48.58	48.83
Short circuit current (A)	10.14	10.16	10.18
Module efficiency (%)	19.11	19.35	19.61
Tolerance of maximum power rating (W)	±5	±5	±5
Max. power (Wp)	287	291	295
Voltage at Pmax (V)	38.61	38.81	39.12
Current at Pmax (A)	7.44	7.50	7.55
Open circuit voltage (V)	47.51	47.72	48.05
Short circuit current (A)	7.89	7.96	8.03
Dimensions (mm)	1990 × 1000 × 40	1990 × 1000 × 40	1990 × 1000 × 40
Weight (kg)	22 ± 1	22 ± 1	22 ± 1

later use and, if necessary, from the battery. Many hybrid inverters are designed to be high-voltage battery compatible but also incorporate hybrid inverters directly into a battery (Spirit Energy 2020). The hybrid inverter converts DC which comes from solar PV panels, to AC with voltage (230 V) and frequency (50 Hz) to energise ODIH appliances. Besides, the hybrid inverter has its own Maximum Power Point Tracking (MPPT), and it is used to convert 48 VDC voltage to 230 V 50 Hz AC voltage. The solar inverter specifications are given in Table 3.9 (Pem Energy 2020).

The selected solar inverter is suitable for 10 kW rated power as well as 14.85 kW peak power at up to 900 V solar panel voltage. It is a 50 Hz grid compatible with a three-phase connection at the output side. The MPPT is suitable for 48 V.

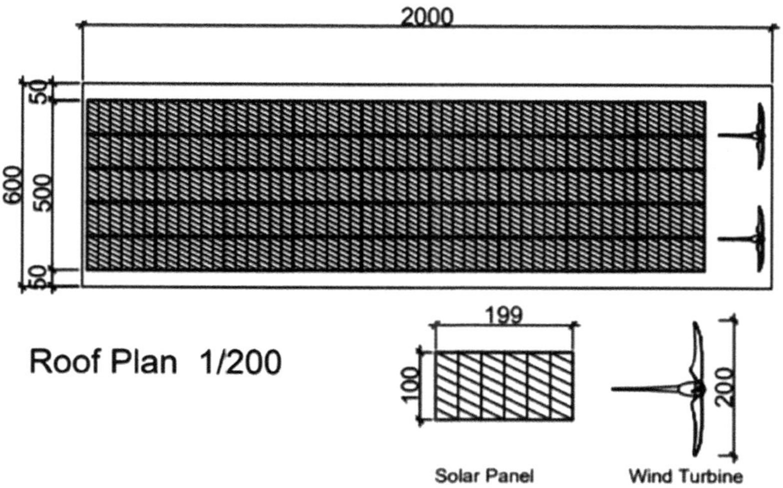

Fig. 3.12 ODIH roof plan

Table 3.9 Solar inverter electrical specifications (Pem Energy 2020)

Parameter	Value	Unit
Model	TRIGEN 10K3P	–
Rated power	10.000	W
Max. PV power	14,850	Wp
Max. DC voltage	720/900	V
Grid connection	Three-phase	–
Rated AC grid frequency	50/60	Hz
Nominal DC voltage	48	V
Number of MPPT	2	–
Max. input current per MPPT	18.6	A
Weight	45	Kg

3.3.2 Wind Turbine

Small wind turbines can be employed especially in urban areas, to foster local or even residential energy generation. The current trend in architecture and urban planning is to decrease energy consumption by increasing energy efficiency. Nevertheless, future urban settlements will require distributed and urban energy generation as well. Many technologies are utilised to face this challenge. One of the technologies is wind energy. Wind turbines are placed on both new and existing buildings in cities, both by businesses and domestic users. In cities, wind climate is highly related to building characters; thus, higher turbulences might be expected in urban areas, where high turbulence in hinders wind energy generation. However, some factors, such as willingness to minimize CO_2 emissions and the ability to generate electricity directly from households, encourage urban wind energy applications (Beller 2011). The energy output of a typical small wind turbine is highly depended on average wind speed and distribution[1] of wind speed. Average wind speed and direction can vary greatly depending on the seasons, geographical structures, buildings, and time of the day. The dominant wind direction in Istanbul is

[1]Classification of wind relative to speed and direction.

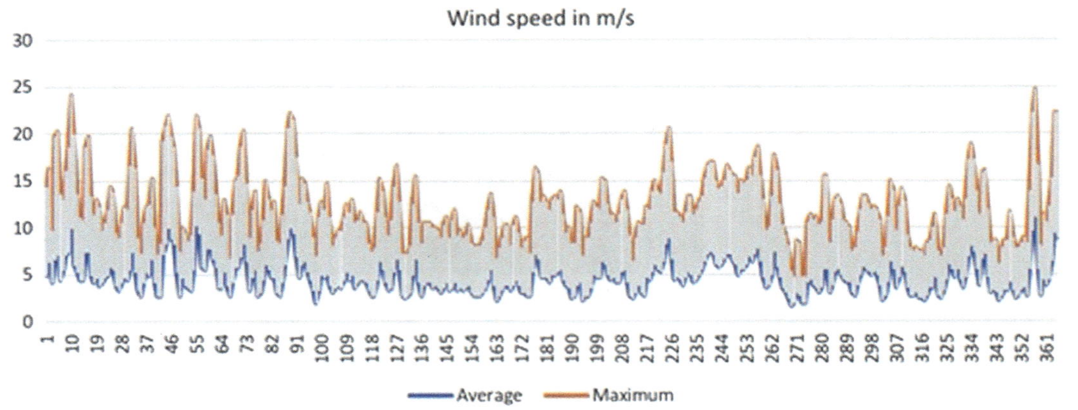

Fig. 3.13 2019 Istanbul bosphorus wind speed graph in m/s (Meteoroloji Genel Müdürlüğü 2020)

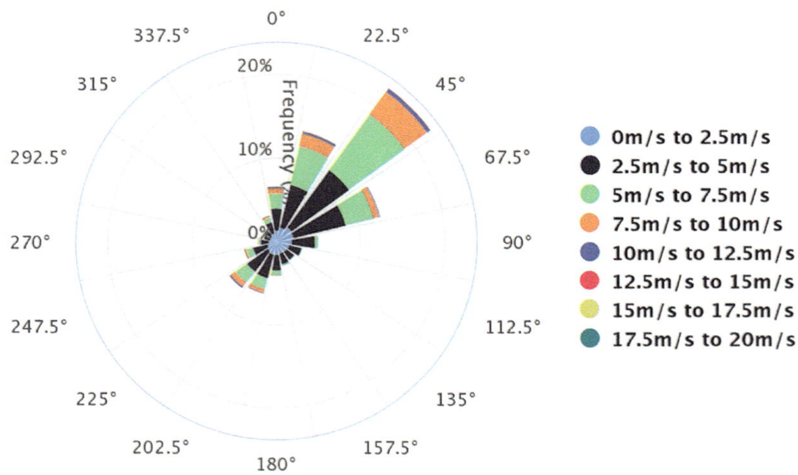

Fig. 3.14 Wind rose for Istanbul bosphorus (Meteoroloji Genel Müdürlüğü 2020)

NE (northeast), blowing Poyraz. All potential turbines will be selected in consideration of this information. We are planning to station ODIH on Bosphorus. Therefore, we need to consider specific wind information in that area. Figure 3.13 shows one-year average wind speeds in Istanbul, and in Fig. 3.14 demonstrates wind rose[2] for the Bosphorus area in Istanbul.

Considering the given information in Fig. 3.13, the monthly average and maximum wind speeds can be calculated. Table 3.10 summarises this information.

Below the wind turbine's cut-in speed, there is no significant power generation. Until the rated wind speed for the turbine, there is an increasing. Between rated wind speed and cut-out speed for the wind turbine, rated power is generated. After wind speed exceeds the cut-out speed, power generation is stopped for protecting the turbine in case of harm (Bilir et al. 2015). Figure 3.15 represents the power generation curve of the wind turbine.

In addition to the monthly average and maximum wind speeds, the type of turbine we will choose is of great importance in the energy we will obtain. There are two different wind turbines types in terms of built direction, and they are compared with respect to their advantages and disadvantages in Table 3.11.

[2]The wind rose is the time-honoured technique of graphically displaying wind conditions such as direction and velocity over a period at a specific location.

Table 3.10 Monthly average and maximum wind speed in Istanbul bosphorus 2019 (Meteoroloji Genel Müdürlüğü 2019)

	Jan	Feb	Mar	Apr	May	June	July	Aug	Sep	Oct	Nov	Dec
Average (m/s)	5.42	5.72	5.35	4.28	4.05	3.86	4.23	5.35	5.07	4.22	4.27	4.87
Max (m/s)	14.13	13.02	12.19	10.53	9.97	9.14	8.86	11.63	10.80	9.97	10.80	13.57

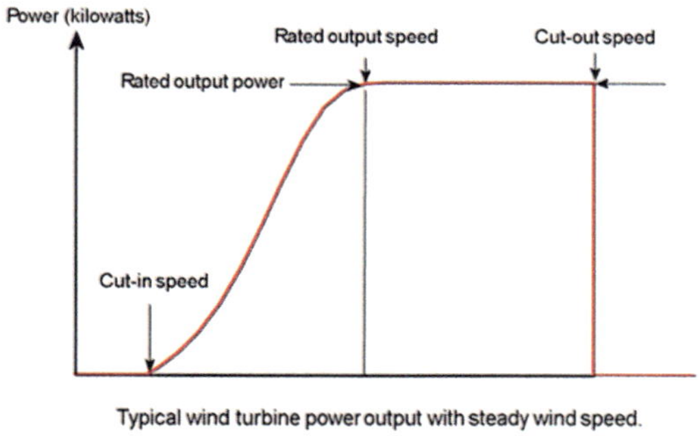

Fig. 3.15 Power generation curve of a wind turbine

The selection of wind turbines is important for an efficient energy generation system. Key factors are wind distribution and average wind speed. Table 3.12 presents three house scale wind turbines that are used frequently in the world.

Unlike Automax and Meigonju, technical specifications for Istabreeze turbines were available for public access. In Fig. 3.16 power curve for the Istabreeze turbine is given.

Istabreeze Wind Turbines are manufactured in Istanbul/Turkey, which makes them more reachable. Also, pricing is better than other options. Information given in Fig. 3.16 allows us to calculate potential power output and predict the percentage of energy needs that the wind side of this project will meet. Istabreeze Wind Turbine will be an efficient and reachable energy generation unit for this project.

3.3.2.1 Wind Inverter

Wind energy is harvested by mechanical forces from the wind and transferred into electrical energy by using generators within the wind turbine. Therefore, it requires power converters to use it effectively. Power converters maximise the harvested energy as well as ensures safe operation for the input and output of the system (Bianchi et al. 2007). The wind turbine is connected to MPPT charger that maximizes energy harvesting and supplies a DC voltage at the output. The charger circuit must be selected by considering the wind turbine output voltage. Therefore, a 48 V charger is selected for this application. The wind MPPT charger specifications are given in Table 3.13.

The charger is supplied from the wind turbine that has an electric generator with 48 V output

Table 3.11 House scale turbine types

Turbine type	Advantages	Disadvantages
Vertical axis (Whittlesey 2017)	Not dependent on direction of wind No need a lot of wind to generate power Easy to control and can be applied high structures Quiet Can be placed close to each other	Easily malfunctioning No self-starting system Needs more maintenance Does not work properly in strong winds Wind close to ground level and generated power is less
Horizontal axis (Saad 2014)	Maximum energy can be generated High efficiency Easy installation and easy maintenance Self-starting system	Difficulty to align properly and quickly Dependent on direction of wind Cannot be placed close to each other (bigger ones) More noisy

Table 3.12 Potential house scale wind turbines selected

Model	Istabreeze 2 kW (Istabreeze 2020)	Automax 1.5 kW (Automaxx 2020)	Meigonju 600 W (Meganju 2020)
Type	Horizontal	Horizontal	Vertical
Rated speed (m/s)	12	14	10
Rated power (W)	2000	1500	600
Voltage (V)	48	24	24
Rotor diameter (m)	2	1.7	0.6
Cut-in wind speed (m/s)	2	2.5	2.5
Survival wind speed (m/s)	–	50	40
Generator	3 phase	3 phase	3 phase
Weight (Kg)	24	15	21

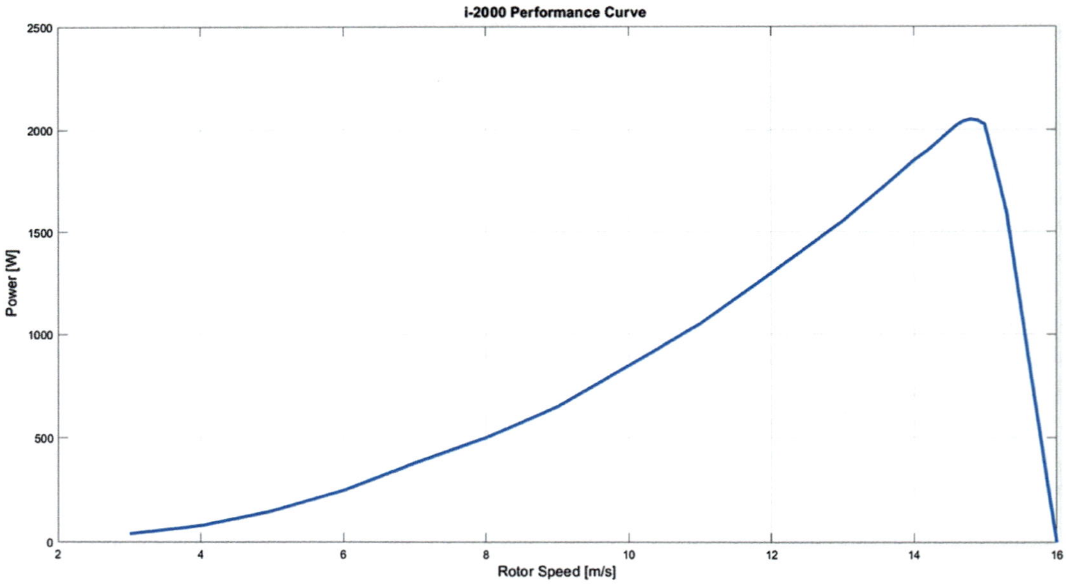

Fig. 3.16 Power curve for Istabreeze turbine (Istabreeze 2020)

voltage and regulates it by tracking maximum power. The other power conversion is done to keep the AC voltage and frequency constant to energise ODIH appliances. Thus, a power inverter is used to convert 48 VDC voltage to 230 V 50 Hz AC voltage. The power inverter is selected as per system requirements in this power flow. The wind inverter specifications are given in Table 3.14.

3.3.3 Biogas

The use of biogas on land takes place as follows. Substrate organic input materials such as food residues, sludge or fats can be fed into the biogas plant. The biogas, which mainly includes methane (CH4) and hydrogen sulphide, is deposited in the roof of the tank and is then burned in the combined heat and power plant

Table 3.13 Wind MPPT charger electrical specifications (Photonic Universe Ltd. 2020)

Parameter	Value	Unit
Model	HCON-2KW-48	–
Rated power	2000	W
Surge power	2700	W
Input voltage	50–65	V
Output voltage	48	V

Table 3.14 Wind inverter electrical specifications (RSComponents 2020)

Parameter	Value	Unit
Model	RS 179-3342	–
Rated power	3000	W
Surge power	6000	W
Output voltage	230	V
Output frequency	50 ± 3	Hz
Output waveform	Pure Sinewave	–
Input voltage	48	V
Weight	7.71	kg

(CHP) to produce electricity and heat. Electric power is fed immediately into the grid circuit. Gas supply takes place to the main grid or gas filling stations (Weltec Biopower Gmbh 2016). Reuse, reduction practices and recycling of waste have been identified as very important strategies for cities as an alternative to landfilling. Micro-Anaerobic Digesters for the production of urban biogas open up new opportunities for the on-site re-use of urban waste products for the generation of energy according to a circular green economy policy. (Pracucci and Zaffagnini 2019) Biogas production from animal waste is getting more attention from companies and entrepreneurs. Although, biogas has an electric generation potential, in this project, biogas is only used for recycling. Generating electricity from a biogas system is not preferred because that kind of system would be extremely costly and inefficient. The biogas system used in this project can generate 4 kWh energy per day on average which is quite low to take into consideration (Homebiogas 2020). Instead of generating electricity, this biogas system used only for recycling household waste to natural gas. This gas will be used for cooking on a gas stove. More detailed information is available in Chap. 2.

3.3.4 Storage

With the increasing intermittent power supply and development of energy storage technologies, batteries have become a viable component of modern energy systems. Batteries support flexibility in power systems, contribute to increasing energy savings and balance changes in supply and demand of electricity. In this section, we analysed numerous battery types and their properties.

3.3.4.1 Fundamental Terminology

Chemical energy storage systems contain different structures. Nonetheless, there are several common terms that are used frequently in the literature (Warner 2015).

Anode: Anode describes one major part of the battery cell, which stores the electrons.

Ampere-hour (Ah): Ampere-hour describes a capacity rate current multiplied by the time. A 10 Ah battery cell can deliver 10 A per hour or 1 A per 10 h in optimum conditions.

BMS: The BMS describes the control system of the battery system by monitoring several parameters and activates or deactivates different functions such as balancing, thermal management system, charging, discharging.

Beginning of Life (BOL): This term describes the initial condition of the battery cell or battery system when it is produced. Generally, the BOL condition is the condition that the user can get the maximum performance.

C-rate: C-rate describes the relative capacity to define a discharge or charge operation. For a 10 Ah battery cell, 2C discharge means a discharge rate of 20 A. Similarly, a battery cell that can deliver 500 W will deliver 1 kW at 2C operation.

Capacity: This term describes the energy content of the battery cell or battery system, which is mostly defined by Ah or Wh.

Cathode: Cathode describes the second major part of the battery cell, which contains active materials and stores ions.

Cycle: Cycle describes partial or complete charge or discharge operation of the battery cell or battery system. Generally, this term is used to measure the lifetime of a battery.

Depth of Discharge (DOD): This term describes a relative used energy value according to total system energy. In other words, DOD is the answer to how much energy is used out of the battery.

Electrode: Electrode is the general term for anode or cathode pairs. To increase the power or energy content of the battery cell, anodes or cathodes are collected.

Electrolyte: This term is used to refer to the intermediate layer of the battery cell. Electrolyte realizes ion flow between anode and cathode.

End of Life (EOL): This term describes the final condition of the battery cell or battery system. Generally, the EOL condition is the condition that the user can get the minimum performance. For battery systems that are utilized for EVs, EOL condition is described as the battery system that has 80% of its capacity.

Energy Density: This term describes the energy content of the battery relative to its weight or volume. Gravimetric energy density is calculated based on the energy content and weight (Wh/kg), while volumetric energy density is calculated based on the energy and volume (Wh/l).

3.3.4.2 Battery Selection

Batteries can be divided into two main categories according to their re-chargeable capabilities. Primary batteries are known as single-use, non-rechargeable batteries. Primary batteries cannot be recharged after they are used. However, secondary batteries are capable of being recharged many times. Especially for Plugged-in Hybrid Electrical Vehicles (PHEV) and Electric Vehicle (EV) applications, high power or high energy density, secondary batteries are preferred. Battery types and properties are detailed in Table 3.15.

Considering the advantages and challenges of different battery chemistries, Li-ion batteries promise the best performance in terms of power and energy as well as a good life cycle. Among several manufacturers and energy storage system products that are capable of being integrated with PV panels and wind turbines, Tesla Powerwall 2 has been preferred. It can be integrated into the power grid, and it has the capability to be connected in parallel based on the storage requirements. Therefore, Tesla Powerwall 2 will cover the energy storage part of this project. Table 3.16 represents the performance specifications of Tesla Powerwall 2 (Tesla 2018).

The electrical system consists of different types of power converters, cables, switch-off components as well as energy sources, and energy storage systems. Since the system will be off-grid most of the time, an AC main bus is designed within the Junction Box. DC side of the electrical system includes Solar Panels and Wind Turbines that delivers energy through Maximum Power Point Trackers (MPPT). Inverters are used to convert DC to AC which is suitable for grid compatible devices. Junction Box includes power

Table 3.15 General properties of battery types

Battery type	General usage area	Properties
ZnC (Warner 2015)	Flashlights Toys Clocks	Cheap Oldest battery technology available in the market
Na-ion (Warner 2015)	In short term, large scale storage application	Studied before 1980 but working model was produced in the 2000s Low cost Uniform distribution of Na 4th abundant element on earth High cycling efficiency Lots of improvement needed Non-toxic Non-flammable Non-corrosive Do not contain heavy metals Have long-life cycle
Li-ion (Warner 2015)	Mobility, consumer electronics and energy storage systems	Published 1991 by Sony High cost Un-uniform distribution of Li Few minerals of Li Already used frequent High volumetric density High charge–discharge efficiency Good power rates
Lead acid (May et al. 2018)	Vehicles, uninterruptible power supplies, and several back-up powers supplies	Low maintenance requirements, relatively low costs and ease of manufacturability Used as a 12 V supply system The lifetime is up to 5000 cycles and 15 years while the battery system costs are around \$150–\$200/kWh
Nickel-based (Warner 2015)	Aviation industry	It has been actively used since the 1800s Can endure as 1000 cycles NiCd batteries have energy densities between 45 Wh/kg and 80 Wh/kg where NiMH has 60–120 Wh/kg, NiFe has 50 Wh/kg, NiZn has 100 Wh/kg and NiH has 75 Wh/kg

switches to enable and disable any electrical source, fuses to protect the complete system in case of a short circuit, and an energy management system that analyses the energy flow and logs the complete data. Figure 3.17 shows the electrical diagram of ODIH.

3.3.5 Calculation Methods

Calculations on this chapter are based on a floating compound project which will be located on Bosphorus/Istanbul. Solar radiation and wind speed data are retrieved from online sources (Meteoroloji Genel Müdürlüğü 2019). According to these data, calculations have been made using the equations below.

3.3.5.1 PV Calculations

GI Hourly Global Irradiance $(\frac{W}{m^2})$
A_u Unit Area of PV Panel m^2
n Number of PV Panel
η Efficiency of PV Panel
μ Efficiency of PV System.
P_{h_i} Hourly Electricity Production.
W_{h_i} Hourly Consumption.
W_{h_t} Hourly Consumption.

Table 3.16 Tesla Powerwall 2 performance specifications (Tesla 2018)

AC voltage (Nominal)	230 V
Feed-in type	Single phase
Grid frequency	50 Hz
Usable energy	13.5 kWh
Grid standards (UK)	G98/G99/G100
Real power, max continuous	3.68 kW/5 kW (charge and discharge)
Apparent power, max continuous	3.68 kVA/5 kVA (charge and discharge)
Power factor output range	+ /– 1.0 adjustable
Power factor range	+/– 0.85
Dimensions	1150 × 755 × 155 mm
Weight	125 kg

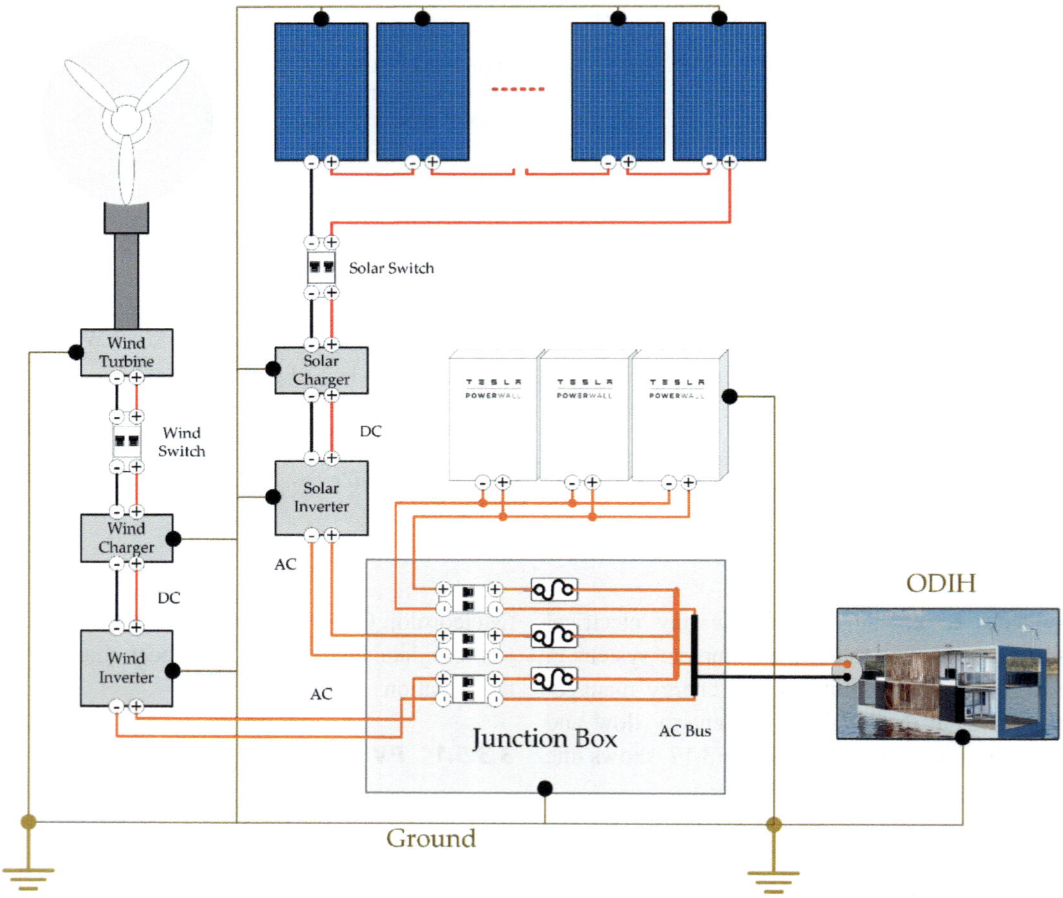

Fig. 3.17 Electrical diagram of ODIH

W_{n_i} Net Hourly Consumption.
W_{n_t} Net Total Daily Consumption.
S_i Hourly Storage.
S_t Total Daily Storage.

Power production of the system (kW) is calculated with Eq. 3.1:

$$P_{h_i} = \frac{GI.Au.n.\eta.\mu}{1000} \tag{3.1}$$

E Electric energy PV production (kWh/year)
H_i Global incident radiation $\left(\frac{\frac{kWh}{m^2}}{year}\right)$
P_{stc} Sum of peak power at STC conditions of photovoltaic solar panels (kWp)
PR Performance ratio of the solar PV system.

Annual Energy Output of Solar Photovoltaic:

$$E_{solar} = P_{stc} \times H_i \times \text{PR} \tag{3.2}$$

3.3.5.2 Wind Turbine Calculations

P_{wind} Power Output (W)
ρ: Air Density $\left(\frac{kg}{m^3}\right)$
r: Rotor Radius (m)
A: Rotor Swept Area (m^2)
C_p: Maximum Power Coefficient.
C_f: Capacity Factor.
v: Wind Speed $\left(\frac{m}{s}\right)$
η_g: Generator Efficiency.
η_b Gearbox Efficiency.

Power production of Wind Turbines is calculated with Eq. 3.3:

$$Pwind = \frac{1}{2} \times \times \pi \times r^2 \times Cp \times Cf \times v^2 \times g \times b \tag{3.3}$$

P_t Power generation of the wind turbine
P_r Rated power generation of the wind turbine
v_{ci} Cut-in wind speed of the wind turbine
v_r Rated wind speed of the wind turbine
v_{co} Cut-out wind speed of the wind turbine
v Wind speed.

$$P = \begin{cases} 0, < v_{ci} \\ 0 < P_t < P_r, v_{ci} \leq \leq v_r \\ P_r, v_r \leq \leq v_{co} \\ 0, > v_{co} \end{cases} \tag{3.4}$$

The wind turbine generates power between the interval of cut-in speed and cut-out speed. The total power generation of a wind turbine can be calculated by Eq. 3.5:

$$P = \int_{v_{ci}}^{v_r} P_t f()dv + P_r \int_{v_r}^{v_{co}} f()dv \tag{3.5}$$

$F(t)$: Annual time series function
P_{wind}: Wind turbine power relative to wind speed.

Annual Energy Output of Wind Turbine:

$$E_{wind} = \int_0^t P_{wind} f(t)dt \tag{3.6}$$

ODIH's total energy generation equation:

$$E_{total} = E_{wind} + E_{solar} \tag{3.7}$$

3.3.5.3 Battery Calculations

When choosing a battery, consumption and battery storage capacity should be calculated as well as electricity production.

W_{hi} Hourly Consumption
W_{ht} Total Daily Consumption
W_{ni} Net Hourly Consumption
W_{nt} Net Total Daily Consumption
S_i Hourly Storage
S_t Total Daily Storage.

Hourly net consumption is calculated by Eq. 3.8:

$$W_{ni} = W_{hi} - P_{hi} \tag{3.8}$$

Total net consumption is calculated with sum of W_{ni} values are shown in Eq. 3.9:

$$W_{nt} = \sum_{i=0}^{23} W_{ni} \tag{3.9}$$

If the W_n value is negative. It means the system must store the electricity. Therefore, it is necessary to calculate the electricity to be stored in the battery according to Eq. 3.10:

$$S_i = \begin{cases} |W_{n_i}|, W_{n_i} < 0 \\ 0, W_{n_i} \geq 0 \end{cases} \tag{3.10}$$

Net total storage is calculated by the formula below (Eq. 3.11):

$$S_t = \sum_{i=0}^{23} |S_i| \tag{3.11}$$

One of the most important factors affecting the size of the battery is the net total storage capacity. Customers should choose the right battery according to their own consumption and solar irradiation in their area.

3.4 Results

The main objective of this section is to illustrate all production and consumption data of ODIH, as explained in detail in the Methodology and Materials section. First, Fig. 3.18 shows monthly consumption values based on electrical devices in ODIH such as HVAC (Heating, Ventilation, and Air Conditioning), water devices, household appliances, and agriculture systems. Then, Fig. 3.19 represents the monthly energy generation from solar, wind, and biogas throughout the year. Figure 3.20 demonstrates PV and Wind energy generation on winter and summer day (21st of June and 21st of December). Figure 3.21 presents the energy balance between generation and consumption as well as the sufficiency and deficiency of the ODIH energy flow. Finally, Fig. 3.22 shows the energy flow through the battery over a year.

As indicated in Fig. 3.18, the energy consumption values and consumption types of

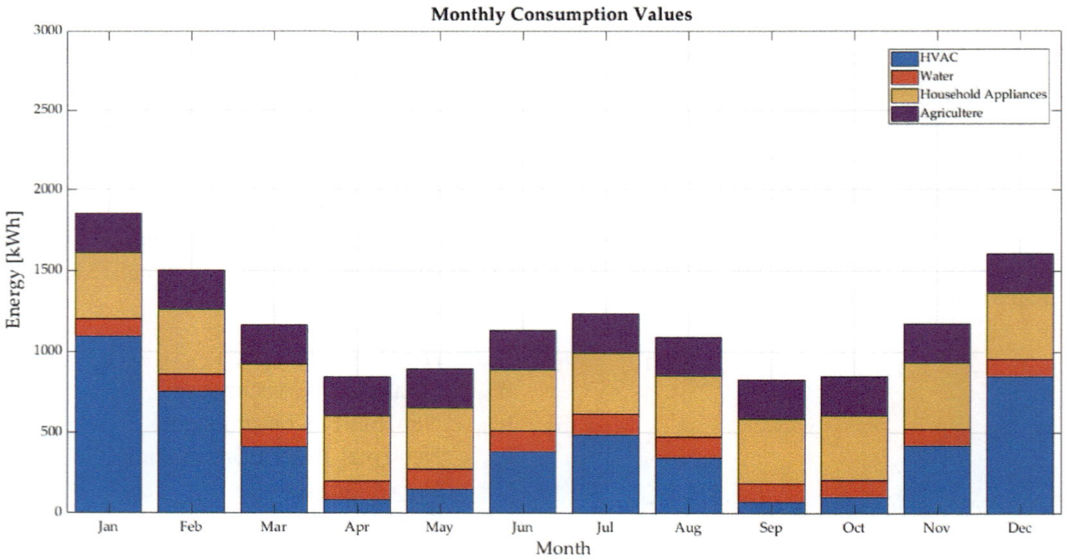

Fig. 3.18 Monthly energy consumption of ODIH

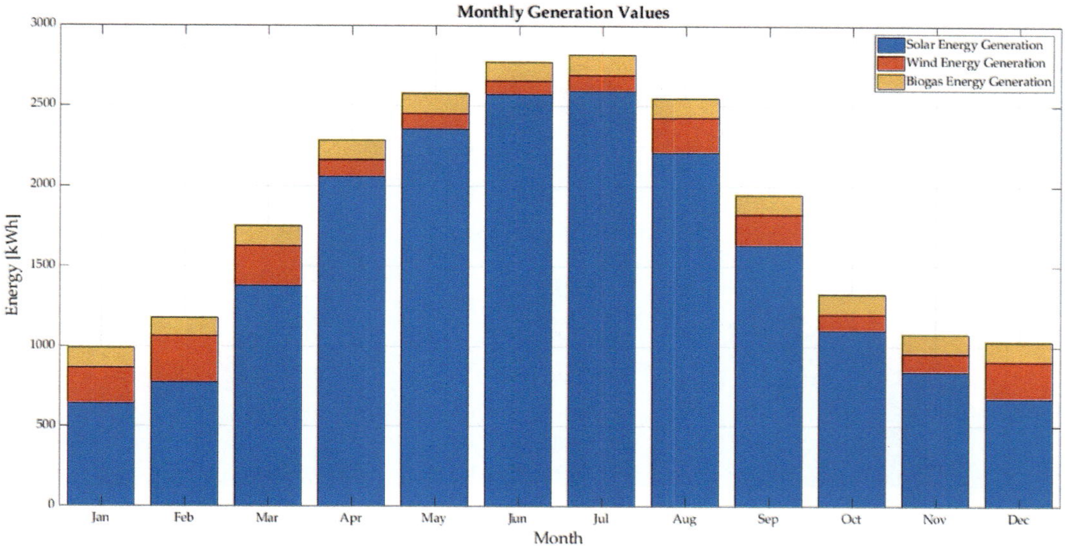

Fig. 3.19 Monthly energy generation of ODIH

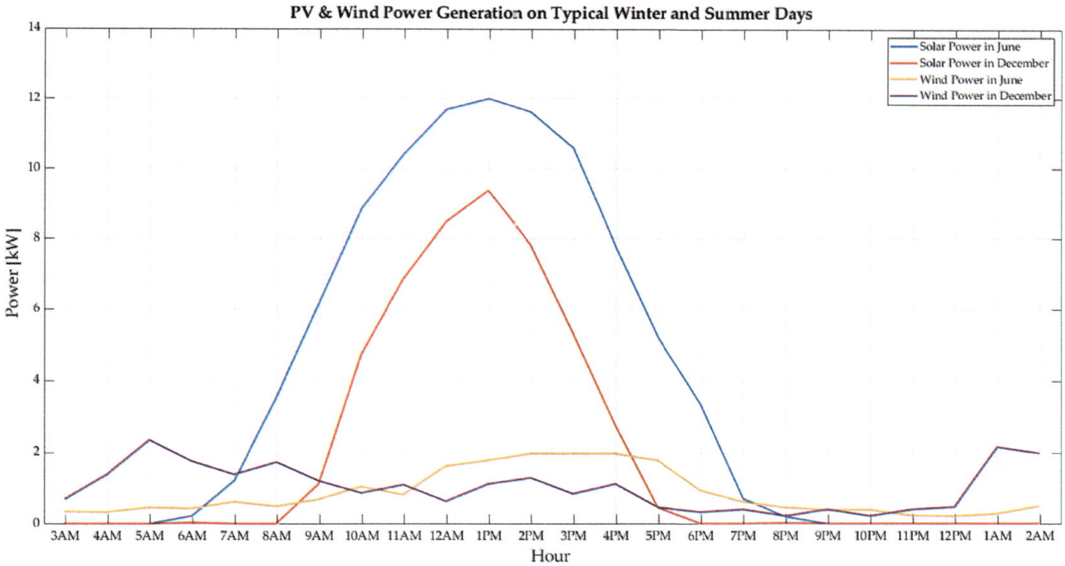

Fig. 3.20 PV and wind power generation throughout the day

ODIH vary by month. The minimum consumption is in May and September, and the highest consumption is in the winter months. According to calculations, HVAC consumption consists of the lion share in the energy consumption of ODIH. HVAC consumption is minimal in spring and autumn and increases significantly in winter. The reason for this is that the heating and cooling need is quite low during the spring and autumn months, whereas heating demand rises sharply during the winter months. There is another spike in HVAC energy demand during summer due to

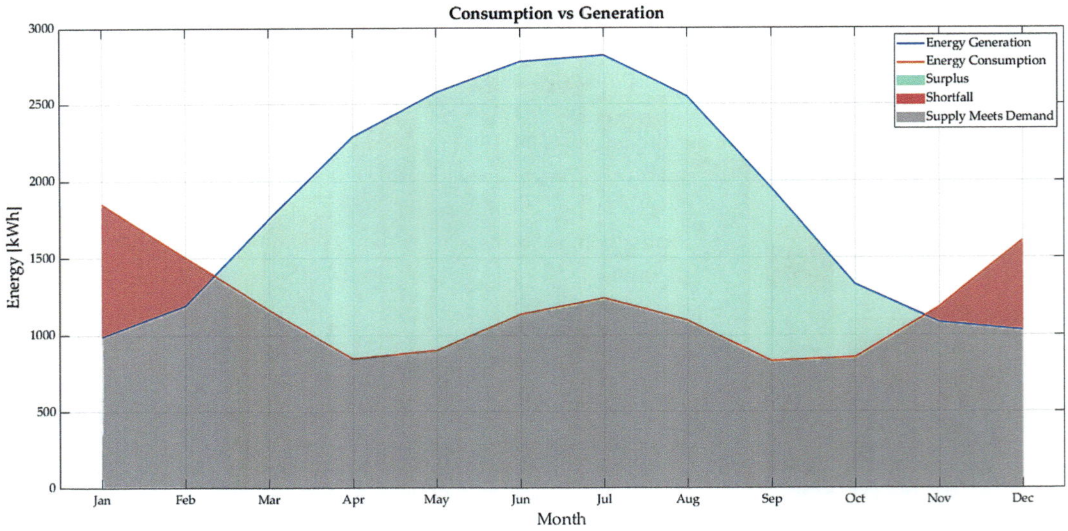

Fig. 3.21 Monthly energy generation and consumption comparison of ODIH

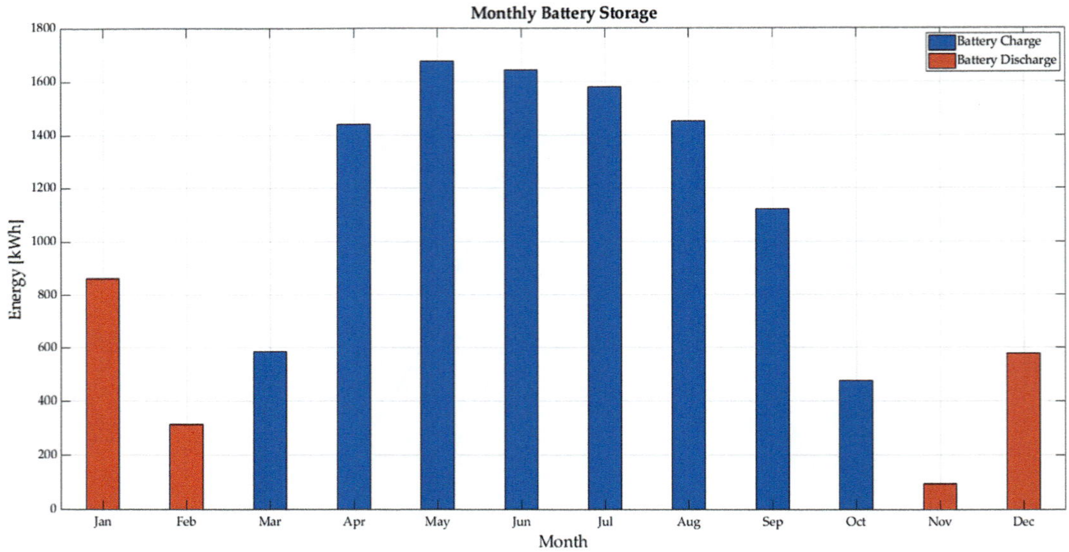

Fig. 3.22 Monthly battery storage of ODIH

cooling. Finally, the energy consumption of agriculture, water, and household appliances remain constant throughout the year.

Figure 3.19 shows the distribution of monthly energy generation of ODIH from solar, wind and biogas resources. As seen in the graph, ODIH generally meets most of its energy demand from rooftop solar energy. Energy generation by solar energy is higher in the spring and summer season compared to the winter and autumn season. However, during the term when the wind is intense along the Bosphorus, the energy generation by wind turbines is higher than in the other months. Energy generation by biogas is mostly produced in fixed amounts and not changing during the year.

As shown in Fig. 3.20, wind and solar power are not stable throughout the day. To illustrate this, we pick up two days, 21st of June and 21st of December, for being the longest daytime and nighttime days in a year. From the yearly data, we deduce that wind energy is not affected significantly by the calendar. However, since solar energy generation is directly related to solar radiation, the characteristics of solar energy generation vary in June and December. Solar power generation is 12 kW at its maximum point in June while it is 9 kW in December. Since energy generation is not stable, battery storage plays a crucial role in ODIH to avoid the intermittency nature of the energy supply.

Figure 3.21 indicates monthly energy consumption versus monthly energy generation. As seen in the graph, ODIH has an energy surplus from March to October. On the other hand, it has an energy shortfall from January to March and from October to December. Energy stored in batteries between March and October can be used to face this shortfall that is painted red in the graph. On the other hand, ODIH will be connected to the grid. Thus, surplus energy can be sold to the grid, or the shortfall can be met from it.

Figure 3.22 shows the distribution of battery storage by months. As we can see, the power flow of the storage will be bidirectional. If we wish to operate ODIH in an isolated mode, from March to October, we will charge our batteries whereas, in January, February, November and December will discharge it.

3.4.1 CO_2 Emission Calculations

According to Appendix A, the annual electricity consumption of ODIH is approximately 18,000 kWh. According to Eq. 3.7, the amount of electricity produced by ODIH using renewable energy resources in one year is 22303 kWh. This indicates that ODIH generates an additional 4303 kWh of energy per year. CO2 emissions from solar PV, biogas and wind generation is calculated as follows (World Nuclear Association 2020)

$$EmissionfromSolarPVGeneration(kgCo2) = 0.048 * GenerationValue(kWh) \quad (3.12)$$

$$EmissionfromBiogasGeneration(kgCo2) = 0.012 * GenerationValue(kWh) \quad (3.13)$$

$$EmissionfromWindGeneration(kgCo2) = 0.300 * GenerationValue(kWh) \quad (3.14)$$

The CO_2 emission from the Turkish energy mix in 2019 is calculated as follows:

$$EmissionfromGrid(kgCo2) = 0.481 * GenerationValue(kWh) \quad (3.15)$$

According to Eq. 3.4, if ODIH meets all the energy it needs from the grid, 8658 kg of CO_2 will be released. Equations 3.12, 3.13 and 3.14 yield that renewable energy sources employed within ODIH emit only 1360.74 kg of CO_2 per year. As a result of the calculations made, ODIH prevents 9367 kg of CO_2 emission every year. A tree can sequester roughly 25 kg of CO_2 per year for an expected lifetime of 40 years (Green Earth Appeal 2020). With this information, it can be said that ODIH saves 425 trees from cutting in a year. Figure 3.23 represents a comparison of the emission values of energy sources.

3.5 Discussion and Policy Recommendation

This section analyses the simulated energy generation results of the ODIH, discusses the simulation inputs and makes recommendations to the regulatory buddies. We can list several main points about the discussion;

- Although ODIH is completely sustainable when observed annually, it has to draw energy from the grid in certain months. The high energy consumption for heating and the

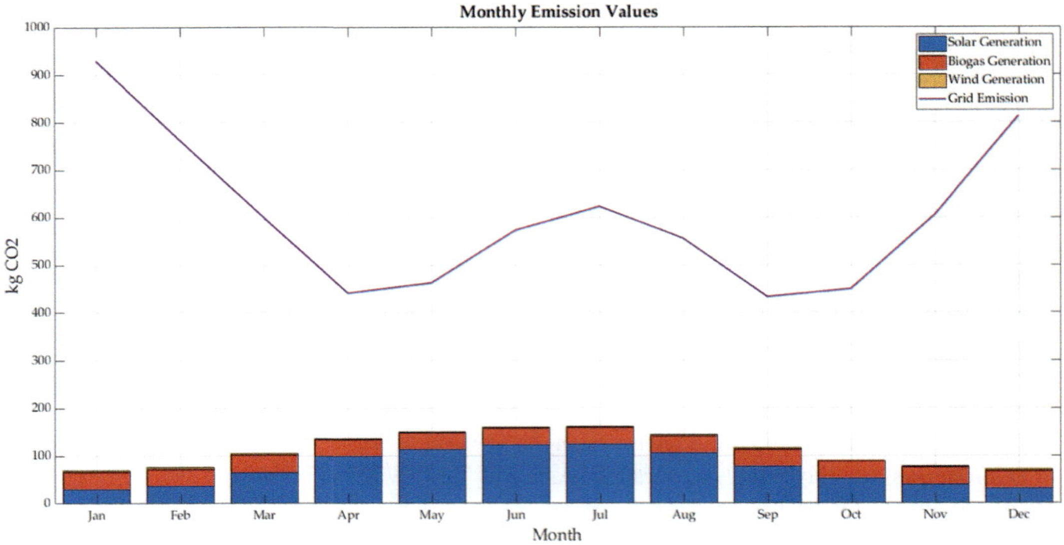

Fig. 3.23 Comparison of the emission values of energy sources

significant decline in energy production are the main reasons for this situation.

- The explicitly underlined issue for the battery development industry is the installation cost. Although Li-ion battery cell price reduced significantly to 150–120 kWh/$ and the expected price for 2024 is less than 100 kWh/$ (BATTERY 2030+, 2020), electrification in the automotive and aviation industries has started to be demanding in terms of energy storage systems and batteries. The expected global battery demand by application is given in Fig. 3.24. This intriguing change can definitely affect all the supply change of the battery production and can lead a challenge against smart grid applications (Fig. 3.24).

- Solar generation calculations were made by assuming albedo,[3] soiling[4] and shadowing[5] parameters.

- The main goal of ODIH is to be a self-sustaining compound by using natural resources and modern energy systems. With that being said, our team estimates that ODIH will prevent the usage of fossil fuels and increase access to modern energy systems for all. ODIH, by nature, serves the 1st Sustainable Development Goal of the United Nations "No Poverty".
- Instead of meeting the energy needs of the house from sources with high carbon emissions, ODIH produces energy from renewable energy sources. In this power generation model, the main energy flow comes from rooftop solar panels. Wind turbines and biomass reactors are used as auxiliary production. In this way, ODIH serves the United Nation's Sustainable Development Goals, "Affordable and Clean Energy".
- Under the scope of responsible consumption and production goal of UNSDG, the efficiency and health of the installed energy generation system are constantly monitored, and improvements are made in ODIH.
- Consuming 100% renewable energy over a year, thanks to ODIH will prevent 10 tons of CO_2 emissions per year, saving roughly 425 trees from cutting annually. Thus, ODIH

[3]Albedo is identified as the proportion of solar radiation's diffuse reflection to the overall received solar exposure.

[4]Soiling is the result of the accumulation of substances during a period in which no exterior cleaning is occurring.

[5]The electrical performance of the PV panel is negatively affected by the shading of the faceplate module glass due to particle build-up.

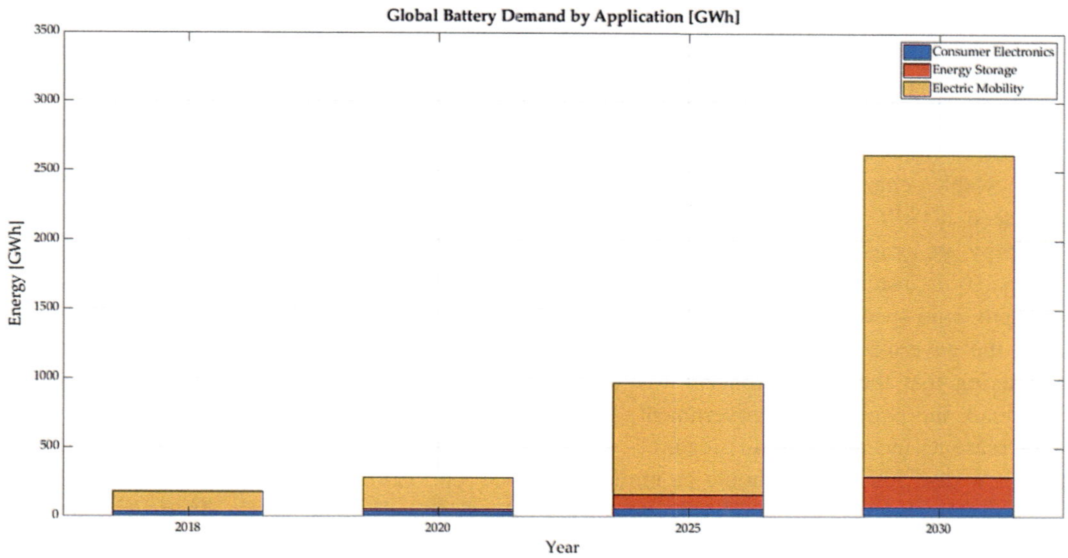

Fig. 3.24 Global battery demand by application

prevents climate change, which is one of the biggest problems of our age and is also the 13th goal of UNSDG.

- Third-party databases were used while performing energy generation simulations. However, these open-source databases could not be sufficient in more detail-based simulations, and this data type creates uncertainty in studies. Transparency and detailed sharing of meteorological data by government agencies will pave the way for projects such as ODIH.
- The feed-in tariff (FIT), which is shown in Table 3.17, under the renewable energy resource support mechanism should be made permanent. Thus, renewable energy will be a more reliable investment source.
- The rate of customs taxes made it difficult to access imported products. It increases the prices of equipment that is already expensive enough. This hurdle strains ODIH's economic budget. By decreasing the rates of these taxes, projects such as ODIH will be more attainable.
- In the current scheme of Turkish legislation, Peer-to-Peer (P2P) energy trade is not allowed. This legal hurdle troubles ODIH's business plan. Energy surplus should be sold with P2P for the sustainability of the system. This trade model decreases energy import by supporting local energy production. In this respect, this trade model adopts the 8th goal of UN Sustainable development goals.
- The ODIH utilizes a battery energy storage system to prevent intermittency on the electric power supply. The used product has a high percentage of the complete system cost. The significant development in the prevalence of sustainable storage systems can be achieved by government incentives. Additional research opportunities can help to have a different supplier in the local market. This will lead to reducing the total cost of ownership of

Table 3.17 Feed-in tariff table (T.C. Cumhurbaşkanlığı 2010)

Renewable energy source	Price (cent/kWh)
Solar	13.3
Wind	7.3
Biogas	13.3

batteries as well as contributing SDG9, UN's industry, innovation and infrastructure development goal.

- According to the legislation published by EMRA (EPDK 2018), in urban areas, installed renewable energy power capacity cannot exceed 10 kW in Turkey. Considering the energy generation potential of ODIH this capacity is insufficient.
- There is no special renewable energy subsidy by the government for floating houses. Considering that the ODIH concept will spread around the world, the government should increase its incentives in this regard.
- A new description class should be introduced for floating houses like ODIH. It is not correct to consider this concept as a house or boat. It must have its own legal regulations and legislation.

3.6 Conclusion

Presciently, UN sustainable development goals focus on 17 major topics, including zero hunger, clean water and sanitation, affordable and clean energy, and climate action. One can claim that climate change is one of the dominant causes of world nations' crisis, and it has been affecting the whole world irrepressibly. Thus, serious actions need to be taken for reducing greenhouse gas emissions to achieve a significant amount of sustainable development goals.

The fundamental ambition of this project is to create a sustainable ecosystem including water sanitation, food production and renewable energy generation by producing minimum pollution to the environment. Throughout different major work packages, this study supports the project by proposing the design of crucial components such as solar and wind energy generation systems, energy storage system and developing a strategy for power supply system for ODIH. Therefore, PV panels, solar charger, solar inverter, wind turbine, wind charger, wind inverter and battery energy storage systems that constitute the basis for this application are selected according to energy consumption characteristics.

To achieve continuous power supply, system needs are gathered, and a requirement list is created. Peak and continuous power characteristics of every device used in ODIH are considered as well as their energy requirements. Then, the energy generation system is structured by mechanically integrating renewable energy generation and storage systems in ODIH.

Considering the radiation values in Istanbul and the 120 m^2 roof area of ODIH, it was deemed appropriate to use 45 monocrystal PV panels. The panel we chose to use in ODIH can provide a maximum of 40.56 V, 9.62 A and 390 W of power with 19.61% efficiency. The energy generated only from 45 PV panels during the winter months cannot meet the consumption in those months. But it can meet the annual consumption. With 74% of the energy generated from PV panels, all annual consumption can be satisfied. To consume this generated electricity in AC form, one DC to AC inverter is needed. The chosen solar inverter is suitable for 10 kW of rated power and 14,85 kW of peak power at up to 900 V of solar panel voltage with 50 Hz frequency. Furthermore, due to solar energy generation being highly dependent on radiation, and it is only available during the daytime, wind energy generation is studied. We placed two 2 kW horizontally axed wind turbines with 2 m rotor diameter on the top of the ODIH platform as auxiliary power sources. We calculated the daily wind generation by using long term wind data (40 years). The prevailing wind direction is North-East for Bosphorus, Istanbul. Wind energy depends on wind conditions, so it often changes in the day time. Annual wind energy generation is about 1465 kWh; this energy generation is 8% of ODIH's annual consumption. According to site conditions, in summer, turbines produce higher at day time than night. However, in winter, turbines produce higher at night than day time. December, January and February are the peak months for wind energy generation. However, since these energy sources have intermittent nature, a battery-based energy storage system is designed. Different battery technologies are investigated, and Li-ion battery technology is decided to use due to its superiorities regarding cycle life, electrical performance and efficiency. The battery system which is chosen in ODIH is Tesla Powerwall 2 that can provide 230 VAC at 50 Hz with 13.5 kWh of usable energy capacity.

Considering the energy generation and consumption values of ODIH, the power flow shall be managed intelligently since the energy generation is higher than the consumption during 66.6% of the year. Although the system is designed as off-grid most of the time, ODIH will be connected to the grid when the energy demand is higher than the generation capacity, especially during the winter and the battery state of charge is low.

After the selection of the main electrical system components, we extracted the energy generation and storage strategy of ODIH. It is observed that the installed solar system is qualified to provide the required energy between mid-February to mid-November in a year. As a result of high radiation rates, especially during spring and summertime, the produced electricity can be sold if there is a grid connection. On the other hand, during the wintertime, it is not feasible to provide ODIH's energy without a grid connection, considering the investment and maintenance cost of the required battery energy storage capacity. Therefore, we can conclude that ODIH needs to be connected to the grid power supply to sell its excess energy generation during summers and buy electricity during winters. As an isolated, mini-grid, ODIH will boost the capital expenditure costs considerably. Thus, we prefer our compound to be a prosumer and engage in energy trading whenever possible.

Appendix 3.1 Yearly Consumption of Equipment and Household Appliances

See Table 3.18.

Table 3.18 Consumption of appliances in ODIH

Devices/Months	J	F	M	A	M	J	J	A	S	O	N	D	Yearly (kWh)
Fridge	53	53	53	53	53	53	53	53	53	53	53	53	636
Stove	135	135	135	135	135	135	135	135	135	135	135	135	1620
Oven	13.2	13.2	13.2	13.2	13.2	13.2	13.2	13.2	13.2	13.2	13.2	13.2	158.4
Range hood	0.88	0.88	0.88	0.88	0.88	0.88	0.88	0.88	0.88	0.88	0.88	0.88	10.56
Microwave oven	2	2	2	2	2	2	2	2	2	2	2	2	24
Dishwasher	12	12	12	12	12	12	12	12	12	12	12	12	144
Washer and dryer	43.52	43.52	43.52	43.52	43.52	43.52	43.52	43.52	43.52	43.52	43.52	43.52	522.24
Robot vacuum	1.8	1.8	1.8	1.8	1.8	1.8	1.8	1.8	1.8	1.8	1.8	1.8	21.6
Robot Mop	1.5	1.5	1.5	1.5	1.5	1.5	1.5	1.5	1.5	1.5	1.5	1,5	18
TV	4.2	4.2	4.2	4.2	4.2	4.2	4.2	4.2	4.2	4.2	4.2	4,2	50.4
Kettle	6.6	6.6	6.6	6.6	6.6	6.6	6.6	6.6	6.6	6.6	6.6	6.6	79.2
Iron	7.8	7.8	7.8	7.8	3466	3466	3466	3466	7.8	7.8	7.8	7.8	76,264
Hair dryer	11	11	11	11	0	0	0	0	11	11	11	11	88
Toaster	13	13	13	13	13	13	13	13	13	13	13	13	156
Coffee maker	5	5	5	5	2	2	2	2	5	5	5	5	48
Modem-router-switch-access points—computer	65.3	65.3	65.3	65.3	65.3	65.3	653	65.3	65.3	65.3	65.3	65.3	783.6
Phone charger	0.6	0.6	0.6	0.6	0.6	0.6	0.6	0.6	0.6	0.6	0.6	0.6	7.2
Laptop	48	48	48	48	48	48	48	48	48	48	48	48	576
Tablet	2.16	2.16	2.16	2.16	2.16	2.16	2.16	2.16	2.16	2.16	2.16	2.16	25.92
Desk lamp	1.2	1.2	1.2	1.2	1.2	1.2	1.2	1.2	1.2	1.2	1.2	1.2	14.4
Reoling cam outdoor	5.76	5.76	5.76	5.76	5.76	5.76	5.76	5.76	5.76	5.76	5.76	5.76	69.12
Burglar alarm	0	0	0	0	0	0	0	0	0	0	0	0	0
Smoke and CO alarm	0	0	0	0	0	0	0	0	0	0	0	0	0

(continued)

Table 3.18 (continued)

Devices/Months	J	F	M	A	M	J	J	A	S	O	N	D	Yearly (kWh)
Motion sensor (water leak, temperature)	0	0	0	0	0	0	0	0	0	0	0	0	0
Door lock	0	0	0	0	0	0	0	0	0	0	0	0	0
Thermostat	0	0	0	0	0	0	0	0	0	0	0	0	0
LED bulb	37.8	32.4	32.4	32.4	27	27	27	27	27	32.4	37.8	37.8	378
Smart plug	0	0	0	0	0	0	0	0	0	0	0	0	0
Smart meter	1.8	1.8	1.8	1.8	1.8	1.8	1.8	1.8	1.8	1.8	1.8	1.8	21.6
Daikin Altermo—heat pump	1095	753	415	84	152	383	487	346	72	101	420	851	5159
Watergen Genny—water production	52.5	52.5	52.5	52.5	52.5	52.5	52.5	52.5	52.5	52.5	52.5	52.5	630
Watermaker Nemo reverse osmosis	90	90	90	90	90	90	90	90	90	90	90	90	1080
Home biogas—biogas reactor	132	132	132	132	132	132	132	132	132	132	132	132	1584
Grey water treatment system	90	90	90	90	90	90	90	90	90	90	90	90	1080
Smart agriculture (Vahaa)	240	240	240	240	240	240	240	240	240	240	240	240	2880
Total (kWh)	2172.62	1825.22	1487.22	1156.22	1,200,486	1,431,486	1,535,486	1,394,486	1138.82	1173.22	1497.62	1928.62	17,941,504

References

Albrecht TR, Crootof A, Scott CA (2018) The water-energy-food nexus: a systematic review of methods for nexus assessment. Environ Res Lett 13(4):043002

Automaxx (2020). Amazon. [Online] Available at: https://www.amazon.com/WINDMILL-Generator-controller-automatic-installation/dp/B01ASNP062?ref_=ast_sto_dp. Accessed 21 Sept 2020

Balat H (2005) Solar energy potential in Turkey. Energy Explor Exploit 23(1):61–69

Batman A, Bagriyanik FG, Aygen ZE, Bagriyanik M (2012) A feasibility study off-grid connected photovoltaic systems in Istanbul Turkey. Elsevier 16 (8):5678–5686

BATTERY 2030+, 2020. Inventing the Sustainable Batteries of the Future. Available at: https://ec.europa.eu/digital-single-market/en/news/inventing-sustainable-batteries-future#:~:text=The%20ultimate%20objective%20of%20Battery,and%20Support%20Action%20(CSA). Accessed 10 Aug 2020

Baumann M et al (2019) A review of multi-criteria decision making approaches for evaluating energy storage systems for grid applications. Elsevier 107:516–534

Beller C (2011) Urban wind energy. Technical University of Denmark, Lyngby

Bianchi FD., de Battista H, Mantz RJ (2007) Wind turbine control systems: principles, modelling and gain scheduling design, 1st ed. Springer London Limited

Bilir L, Imir M, Devrim Y (2015) An investigation on wind energy potential and small scale wind turbine. Energy Convers Manage 103:910–923

Dehghani-Sanij A, Tharumalingam E, Dusseault M, Fraserbc R (2019) Study of energy storage systems and environmental challenges of batteries. Elsevier 104:192–208

Doymaci OA, Yilmaz Ulu E (2018) Wind energy in Turkey: potential and development. DergiPark 4:132–136

Eagles D (2020) Diamond Eagles Co. Ltd.. [Online] Available at: https://diamondeaglesgh.com/diamond-eagles-solar/. Accessed 15 Aug 2020

EIA (2019) Solar explained. [Online] Available at: https://www.eia.gov/energyexplained/solar/. Accessed 15 Aug 2020

Ekins-Daukes DN (2006) Solar energy for heat and electricity: the potential for mitigating climate change. Grantham Institute for Climate Change, London

Ekmekci I (2016) A study on solar radiation calculation for Istanbul with measured data. Int J Solar Energy Res (JSER) 1(2)

El-din AAH, Gabra CF, Ali AHH (2014) Effect of ambient temperature on performance of different types of PV cells at different location in Egypt. Cairo, Middle east power conference

Energy (2020) Energy. [Online] Available at: https://www.energy.gov/energysaver/save-electricity-and-fuel/buying-and-making-electricity/small-wind-electric-systems. Accessed 20 Aug 2020

Eni (2020) Eniscuola-Energy Knowledge. [Online] Available at: https://www.eniscuola.net/en/argomento/energy-knowledge/energy-sources/primary-and-secondary-sources/. Accessed 19 Aug 2020

EPDK (2018) EPDK T.C Enerji Piyasası Düzenleme Kurumu. [Online] Available at: https://www.epdk.gov.tr/Detay/Icerik/4-139/10-kwa-kadar-lisanssiz-elektrik-uretim-tesisleri. Accessed 03 Aug 2020

Fonash SJ (2020) Solar Cell. [Online] Available at: www.britannica.com/technology/solar-cell. Accessed 28 July 2020

Food and Agriculture Organization of the United Nations (2017) Water for sustainable food and agriculture. Food and Agriculture Organization of the United Nations, Rome

Green Earth Appeal (2020) Green Earth appeal. [Online] Available at: https://greenearthappeal.org/co2-verification/#:~:text=In%20summary%2C%20whilst%20the%20Carbon,KG%20per%20tree%20planted%20in. Accessed 04 Aug 2020

Hall PJ, Bain EJ (2008) Energy-Storage Technologies and Electricity Generation. Elsevier 36(12):4352–4355

Hepbasli A, Özdamar A, Ozalp N (2010) Present status and potential of renewable energy sources in Turkey. Taylor & Francis Online 23(7):631–648

Hoff H (2011) Understanding the Nexus. Stockholm, Bonn Nexus Conference

Homebiogas (2020) The home biogas. [Online] Available at: https://www.homebiogas.com/Products/HomeBiogas2. Accessed 07 Sept 2020

International Energy Agency (2014) How solar energy could be the largest source of electricity by mid-century. [Online] Available at: https://www.iea.org/news/how-solar-energy-could-be-the-largest-source-of-electricity-by-mid-century. Accessed 14 Aug 2020

Istabreeze (2020) Istabreeze. [Online] Available at: https://www.istabreeze.com.tr/online/Ruzgar-Turbini/i2000-48V-Ruzgar-Turbini-Windsafe-Korumali-iSTA-BREEZE. Accessed 21 Sept 2020

Markham D (2018) Treehugger [Online] Available at: https://www.treehugger.com/largest-vertical-axis-wind-turbine-installation-us-operating-texas-m-4857642. Accessed 20 Aug 2020

May GJ, Davidson A, Monahov B (2018) Lead batteries for utility energy storage: a review. J Energy Storage 15:145–157

Meganju (2020). Amazon. [Online] Available at: https://www.amazon.com/MEIGONGJU-Vertical-Permanent-Generator-Controller/dp/B089FFNFR6/ref=sr_1_3?crid=2UUFB9LQNKT11&dchild=1&keywords=wind+turbine+generator&qid=1598446615&sprefix=wind+tu%2Caps%2C279&sr=8-3. Accessed 21 Sept 2020

Meteoroloji Genel Müdürlüğü (2019) Resmi İstatistikler. [Online] Available at: https://mgm.gov.tr/veridegerlendirme/il-ve-ilceler-istatistik.aspx?k=A&m=ISTANBUL. Accessed 15 Sept 2020

Meteoroloji Genel Müdürlüğü (2020) Istanbul Icin Detaylı Hava Durumu. [Online] Available at: https://www.mgm.gov.tr/tahmin/il-ve-ilceler.aspx?il=ISTANBUL. Accessed 10 Sept 2020

Miao Y, Hynan P, Jouanne AV, Yokochi A (2019) Current li-ion battery technologies in electric vehicles and opportunities for advancements. MDPI Open Access J 12(6):1074–1094

Moseley PT, Garche J (2015) Electrochemical energy storage for renewable sources and grid balancing, 1st edn. Elsevier, s.l.

Newkirk M (2016a) Clean energy reviews. [Online] Available at: https://www.cleanenergyreviews.info/blog/2014/5/4/how-solar-works. Accessed 13 Aug 2020

Newkirk M (2016b) Clean energy reviews. [Online] Available at: https://www.cleanenergyreviews.info/blog/2014/5/4/how-solar-works. Accessed 20 Aug 2020

Ogden JM, Williams RH (1989) Solar hydrogen: moving beyond fossil fuels, 1st edn. World Resources Inst, United States

Ozturk I (1999) Anaerobik Biyoteknoloji ve Atık Arıtımındaki Uygulamaları, 1st edn. Su Vakfi, Istanbul

Pem Energy (2020) Abax Invertors, Istanbul: Pem Energy

Photonic Universe Ltd. (2020). Photonic universe. [Online] Available at: https://www.photonicuniverse.com/upload/file/Manuals/Wind/hybrid_controllers/HCON-1KW-48_and_HCON-2KW-48_user_manual.pdf. Accessed 21 Aug 2020

Pracucci A, Zaffagnini T (2019) Organic waste management through anaerobic digester technologies in urban areas. A urban strategies. Keywords: organic waste, predesign tool, socio-technical transition, urban strategy, biogas. [Online] Available at: https://www.researchgate.net/publication/338645946_Organic_waste_management_through_anaerobic_digester_technologies_in_urban_areas_A_multicriterial_predesign_tool_to_support_urban_strategies_Keywords_Organic_waste_Predesign_tool_Socio-technical_transi. Accessed 14 Aug 2020

RES4Africa Foundation (2015). Applying the water-energy-food nexus approach to catalyse transformational change in Africa, RES4Africa Foundation, Rome

RSComponents (2020). RS Components Ltd.. [Online] Available at: https://docs.rs-online.com/5b95/A700000006571393.pdf. Accessed 05 Aug 2020

Saad M (2014) Comparison of horizontal axis wind turbines and vertical axis wind turbines. IOSR J Eng 08, 4(8):27–30

Schmid Pekintas Energy (2018) Schmid&Pekintas energy. [Online] Available at: https://schmid-pekintas.com/mono380.pdf. Accessed 15 Aug 2020

Simpson GB, Jewitt GP (2019) The development of the water-energy-food nexus as a framework for achieving resource security: a review. Frontiers Environ Sci

Solars (2016) Solars. [Online] Available at: https://www.solars.com.tr/en/p/solar-radiation-map-of-turkey. Accessed 20 Aug 2020

Spirit Energy (2020). Spirit energy. [Online] Available at: https://www.spiritenergy.co.uk/kb-pv-solar-hybrid-inverters. Accessed 3 Oct 2020

Statista (2020) Statista. [Online] Available at: https://www.statista.com/statistics/892878/electricity-generation-by-fuel-turkey/. Accessed 7 Oct 2020

Stoker L (2015) Chaotic and unpredictable subsidy regime slammed as London rooftop market continues to falter. [Online] Available at: https://www.solarpowerportal.co.uk/news/chaotic_and_unpredictable_subsidy_regime_slammed_as_london_rooftop_3884. Accessed 13 Aug 2020

T.C. Cumhurbaşkanlığı (2010) Yenilenebilir Enerji Kaynaklarının Elektrik Enerjisi Üretimi Amaçlı Kullanımına İlişkin Kanun. [Online] Available at: https://www.mevzuat.gov.tr/MevzuatMetin/1.5.5346.pdf. Accessed 09 Sept 2020

Tesla (2018) Tesla Powerwall 2 AC datasheet—United Kingdom. [Online] Available at: https://www.tesla.com/sites/default/files/pdfs/powerwall/Powerwall%202_AC_Datasheet_en_GB.pdf. Accessed 20 Sept 2020

The Clean Energy Regulator (2019) Defining small scale and large scale solar systems. [Online] Available at: https://www.cleanenergyregulator.gov.au/RET/How-to-participate-in-the-Renewable-Energy-Target/eligibility-for-the-renewable-energy-target/defining-small-scale-and-large-scale-solar-systems. Accessed 29 July 2020

The Solar Nerd (2019) How do you heat a house with solar power? [Online] Available at: https://www.thesolarnerd.com/blog/heating-house-with-solar-power/#:~:text=Basically%2C%20there%20are%20two%20types,to%20power%20your%20heating%20system. Accessed 15 Aug 2020

The World Bank (2019) The World Bank data—population. [Online] Available at: https://data.worldbank.org/indicator/SP.POP.TOTL. Accessed 22 Sept 2020

Top Alternative Energy Sources (2020) Top alternative energy sources. [Online] Available at: https://www.top-alternative-energy-sources.com/horizontal-axis-wind-turbine.html. Accessed 20 Aug 2020

U.S. Energy Information Administration (2019) EIA. [Online] Available at: https://www.eia.gov/energyexplained/wind/types-of-wind-turbines.php. Accessed 01 Sept 2020

U.S. Energy Information Administration (2020) U.S. Energy Information Administration. [Online] Available at: https://www.eia.gov/energyexplained/what-is-energy/sources-of-energy.php. Accessed 07 Sept 2020

United States Environmental Protection Agency (2020) United States Environmental Protection Agency. [Online] Available at: https://www.epa.gov/agstar/how-does-anaerobic-digestion-work. Accessed 10 Sept 2020

Warner J (2015) The handbook of lithium-ion battery pack design: chemistry, components, types and terminology. Elsevier Inc., Grand Blanc, MI, USA

Weltec Biopower Gmbh (2016) Weltec Biopower. [Online] Available at: https://www.weltec-biopower.com/info-center/biogas/how-does-a-biogas-plant-work.html#: ~ :text=Organic%20input%20materials%20such%20as,the%20biogas%20plant%20as%20substrate.&text=The%20final%20product%20of%20this,also%20contained%20in%20the%20biogas. Accessed 14 Sept 2020

Whittlesey R (2017) Vertical axis wind turbines: farm and turbine design. In: Letcher TM (ed) Wind energy engineering: a handbook for onshore and offshore wind turbines. Academic Press, Los Angeles, pp 185–202

World Nuclear Association (2020) How can nuclear combat climate change. [Online] Available at: https://www.world-nuclear.org/nuclear-essentials/how-can-nuclear-combat-climate-change.aspx. Accessed 09 Sept 2020

YEGM (2020) Enerji Isleri Genel Mudurlugu. [Online] Available at: https://www.yegm.gov.tr/MyCalculator/pages/34.aspx. Accessed 15 Sept 2020

YEGM (2020) Yenilenebilir Enerji Genel Müdürlüğü. [Online] Available at: https://www.yegm.gov.tr/yenilenebilir/ruzgar-ruzgar_enerjisi.aspx. Accessed 19 Aug 2020

4 Food

Abstract

This chapter is built upon the argument that conventional farming is not sufficient to meet the rising demand for nutrition for humankind. On the other hand, with a futuristic perspective, meeting the food demand of the human population on earth seems unsustainable in terms of water, energy, and land use. To meet the ever-increasing demands, we must find a better way that provides higher yield and higher product quality. We studied the average need of a person in terms of nutrition and ingredients such as carbohydrate, protein and fat. We inspected the nutrients needed for human-beings. We compared soilless agriculture systems in terms of applicability, yield, consumed water, consumed energy and needed space. As our next step, we calculated all the necessary resources to nourish two adult humans and decided to use the nutrient film technique. We show a water-energy-food nexus comparison between traditional and soilless farming. At the Open Digital Innovation Hub (ODIH), our available space for farming is too small for plant diversity. Therefore, we carry out our calculations with potato farming and explain the reasons in detail

The author would like to acknowledge the help and contributions of Murat Can Özbaşaran, Aybüke Zeynep Cengiz, Celal Bayraktar, Hasan Basri Yüksel, and Sedef Güraydın in completing of this chapter.

4.1 Introduction

Climate change damages agriculture, food security, and the incomes of millions of people all over the globe. There are lots of organizations and foundations whose aim is to avoid the negative impacts of climate change on food production and to find a permanent solution. One of the leading organizations is the United Nations (UN). The Sustainable Development Agenda 2030 was formally adopted by the UN in 2015. The promise 'nobody is left behind' has sponsored Seventeen Sustainable Development Goals (SDGs). It guarantees fair and balanced global growth (Hellin and Fisher 2019). Although, there are problems caused by climate change which could be separated into three main areas; economic, environmental, and social, and they may trigger hunger. One of the problems that the world has been encountering for a long time is hunger. To achieve food security in SDGs, under and overuse food highlighted. In addition, producing more food is the most important step. About 800 million people are impacted by chronic undernourishment and micronutrient shortages, primarily in low-income countries (Whitfield et al. 2018). This fact stresses the immense difficulty and also the necessity of meeting the goal of zero hunger by 2030. The condition in Africa is particularly troubling, where the Prevalence of Undernourishment (PoU) reveals minor and gradual growth in almost all sub-regions since 2015. The trouble in Western Africa is rapidly growing in the last

S. Küfeoğlu, *The Home of the Future*, Sustainable Development Goals Series,
https://doi.org/10.1007/978-3-030-75093-0_4

years. In addition, the Middle African PoU level is 26.5%; also, the Eastern African level is 30.8% (FAO, IFAD, UNICEF, WFP, WHO 2019). And a quarter of babies under the age of five in the world is stunted. This ensures that 165 million children who are too undernourished can never achieve their maximum capacity, both physical and emotional (Food and Agriculture Organization of the United Nations 2013a). And also, the economic effect of this problem cannot be underestimated. The expense of poverty to the global economy might reach up to 5% of global income, which equals to US$3.5 trillion per year or US$500 per person (Food and Agriculture Organization of the United Nations 2013a). The agricultural sector plays a vital role in ending hunger, but it faces a different challenge: to increase food production while decreasing greenhouse gas (GHG) emissions due to food production. Agriculture, forestry, and land usage add up to 25% of the world's anthropogenic GHG pollution, including additional food-processing and food-circulation pollution. With the adoption of the Paris Agreement in 2016, political awareness and acknowledgement of the increasing agricultural footprint have gained importance. Limiting heating to 2 or 1.5 ° C involves not only increasing efficiency and durability without increasing temperature but also reducing pollution by significant quantities. (Whitfield et al. 2018). To create a solution for these problems from a systematic point of view, Climate-Smart Agriculture, WEF Nexus, and Sustainable Food concepts should be highlighted.

4.1.1 Climate-Smart Agriculture (CSA)

4.1.1.1 What Is Smart Agriculture?

The definition of Climate-Smart Agriculture (CSA) was presented by the Food and Agriculture Organization of the United Nations (FAO) in 2010 (Leslie Lipper 2018). CSA aims to achieve SDGs, decrease GHG emissions, establish systems that are less affected by the changes in parameters, and increase agricultural productivity and incomes. To achieve sustainable agricultural development, CSA is a method to establish technical, political, and economic requirements (Food and Agriculture Organization of the United Nations 2013b).

Agriculture and food systems must enhance and maintain food security as a response to climate change by reducing the effect of climate change. These objections are interrelated. For these purposes, we build systems that are more efficient and resilient. There are lots of parameters that affect systems efficiency. To create more efficient systems, resource efficiency must be increased, integrated systems must be used, and food loss and waste must be reduced. Adaptive capacity is defined as resilience, and it means a response to changes. Lots of parameters seem to risk agriculture, such as stress combinations, the effect of climate change on agriculture, pernicious bugs, etc. However, the prediction of these setbacks is impossible. The "No regret" approach is the most effective one. This approach aims to reduce instability and enhance durability. To achieve both efficiency and resilience in systems, agriculture investments must be increased in developing countries (FAO, OECD 2012).

4.1.1.2 Why Do We Need Smart Agriculture?

There is a lot of problems now, and there will have been many by 2050. These problems can be categorized into three parts: economic, social, ecological. For example, in 2050, FAO clarified that agricultural production must be increased by 60% if we want to feed all the people in the world (Food and Agriculture Organization of the United Nations 2013b) (Asfaw and Branca 2018). Therefore, a transformation of agriculture is inevitable for feeding the global population and for reducing poverty. The difficulty arises due to climate change. As a consequence of the increasing incidence of natural disasters, the frequency of excessive incident, and the unpredictability of weather conditions, climate change has a vital role in agriculture and food security. In addition, prices of foods are responses to these issues. To tackle climate change, we have to reduce GHG emissions per unit of land (Long et al. 2015).

4.1.1.3 The Importance of Managing Landscapes for CSA

All societies are focused on structures of human-managed cultivation, forestry, and fisheries. Converting land from woodland into fields has positive and adverse impacts on the environment. When woodland is converted into a field, there will be more area to plant. On the other hand, this action leads to environmental degradation, loss of some important environmental services, and the loss of biodiversity. Many of the existing agricultural production systems are focused on large monocultures, so systems depend on a few plants and diversity. There are several manufacturing processes which are not sustainable regardless of their socially harmful activities in land conservation and their inefficient usage of water resource. These systems also use fossil fuels and increase GHG emissions. In fact, there has been a major difference between the theoretical yield potential and real yields in many agricultural production systems. In short, the approach to a landscape has a significant duty to take part in the transformation to CSA. This is an innovative approach that aims to control natural and human-managed systems in the ecosystem sustainably (Food and Agriculture Organization of the United Nations 2013b).

4.1.1.4 Water Management

Ecology has affected climate change in means of water problems, for instance, droughts and floods. Crops, livestock, trees, fish, rural communities, and physical infrastructure are affected by the changes in water cycling. More strategic programs are needed because old ones are not enough to improve in a socio-economical way in rural areas. To build resilient systems, water loss should be fought systematically (Food and Agriculture Organization of the United Nations 2013b).

4.1.2 Sustainable Food Production

Sustainable food production is described as the protection and control of natural resource infrastructure, and the direction of the technological change is on its way to guarantee human requirements continuously for present and upcoming generations. Sustainable agriculture does not damage the environment, and maintains soil, water, and the genetic codes of plants and animals, and is technically convenient, economically reasonable, and socially agreeable (FAO 1988). Unfortunately, today's food production system is not succeeding in terms of supplying enough food and nutrients for all people to have a healthy diet, and it will not be possible to make it real with the present approach, and agriculture will face extraordinary pressure. For example, by 2050, the world's population will have increased by 34%, almost all of which will be in developing countries and will be 9.1 billion. Urban population will be around 70% of the world's population due to an increase in urbanization, and people will be richer. Food production has to increase by 70% to provide enough food for a richer, larger, and more urban population. Yearly cereal production and yearly meat production will require 3 billion tonnes and 470 million tonnes, respectively (FAO 2009) and also increase from the current 8.4 billion tonnes to around 13.5 billion tonnes in food production will be needed (FAO 2014a, b). In addition, agriculture has a significant impact on climate change which is one of the most critical issues that the world faces. 70% of freshwater resources is used by the crop and livestock industries. When forestry is added to them, they cover 60% of the earth's surface. 80% of grassland and the total crop is used by only livestock (Food and Agriculture Organization of the United Nations 2013a). It is predicted that crop and animal production and forestry, particularly deforestation, lead to 25% of overall GHG emissions (IPCC 2014) and losses of soil organic carbon as many as 20 to 80 tonnes per hectare, most of it emitted into the atmosphere results from the transformation of natural to agricultural ecosystems (Lal 2004). When the considerable level of necessity of more food and the impacts of agriculture on climate change are taken into account, feeding the growing population and meeting the increasing demand for food will not be possible without making our food system sustainable and shorter this transformation process from traditional to sustainable agriculture which ensures the food security of the world as much as possible. In addition, it is not enough to improve only one parameter or to generate benefits for only one dimension, which can be social, economic, and environmental, to consider food production sustainable. All three dimensions have to be enhanced simultaneously and have to have benefited from what is done in the name of sustainability. There are some actions to make our food system more sustainable such as digitalization for global sustainability and using clean and efficient technology. For instance, ways to diminish the imbalance between poor and richer countries can be improved. Excessive consumption is a huge problem for sustainability (Living Sustainably 2020). Thus, we have to develop a policy to reduce over-consumption. As a result, if we demand better and higher-quality life, sustainable food production is an indispensable solution for our world.

4.1.3 The Water-Energy-Food (WEF) Nexus

Water, energy, and food are crucial for human well-being, destitution decrease, and sustainable growth (FAO 2014a, b). According to global projections, the request for freshwater, energy, and food will rise dramatically over the upcoming decades because the world population and urbanization are increasing. Cultural and technological changes and climate change create pressure on them (Hoff 2011). Agriculture consumes 70% of the total fresh water on earth. Water is utilized for entire food production processes, fishery, forestry, and livestock farming,

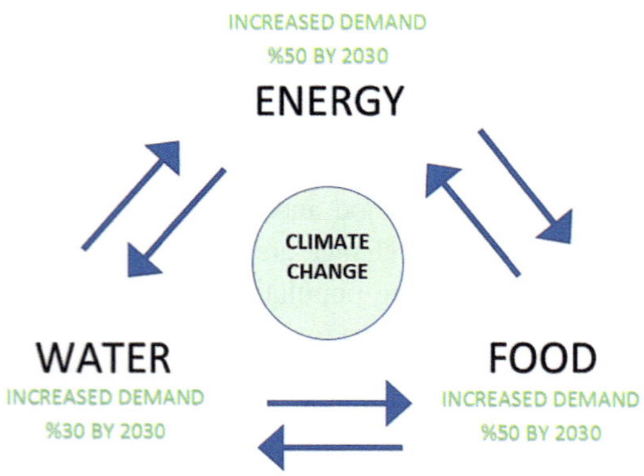

Fig. 4.1 Prediction of water, energy, food demands (Aboelnga et al. 2018)

and it is utilized to create or transport energy in various forms (FAO 2011a). Besides, food production and supply chain expend approximately 30% of all of the energy used around the world (FAO 2011b). Energy is needed to extract, hoist, collect, and transport water. Cities, industry, and other users are also increasingly demanding water, energy, and land resources; therefore, environmental degradation problems and, in many situations, resource scarcity arise in recent years. Figure 4.1 shows us the prediction of water, energy, food demands.

In this context, the WEF Nexus has come out as a beneficial approach for defining and addressing the complicated and relevant nature of our global resources necessary to achieve many economic, environmental, social, and technological goals (FAO 2014a, b). In terms of application in the real business world, the nexus provides a notional approach to systematically analyze the interplay between nature and human action and to use natural resources. It presents a perspective to understand sync management across sectors in a better way. On the part of agricultural of the WEF security nexus, Goal 2 is easily recognized. Goal 2 seeks to end hunger, improved nutrition and succeed in food security, and support sustainable agriculture and has four associated targets and three means of implementation. These are summarized in Table 4.1.

4.1.4 Future Problems

4.1.4.1 Food

The diet of the people changes in time, and the required calorie amount changes with it. In 1964, the average kcal of a person consumed in a day was around 2000. It increased to around 2600 in 1997, and it is predicted that it will increase to 2800 in 2030 (FAO 2009). Hunger is the situation of not getting at least 1800 calories per day (Action Against Hunger 2020). Even today, there are 690 million people who suffer from hunger, although there is sufficient food production (FAO 2009). If there is no change in this trend, this number will exceed 840 million by 2030 (United Nations 2018). Hunger is caused mostly by poverty. By growing population predicted at 9

Table 4.1 SDG 2 Zero hunger

Goal	2030 Targets	Implementation means
Zero hunger	By the year of 2030, end hunger and guarantee access by everyone, especially both the unguarded and poor people, baby, to safe, nutritious, and adequate food all the time	1. Prevent trade distortions in global markets, by the parallel removal of all types of agricultural export subsidies 2. Increase the number of investment, which is done by enhanced international cooperation, agricultural study and extension services, technology enhancement, and farming gene banks to develope the productive capacity of agriculture 3. Make sure that the food markets run smoothly without sharp price increases and facilitate market information for all
	By the year of 2030, aiming to end every type of malnutrition, on stunting and emaceration in children between 0–5, and address the nutritional requirements of young girls, pregnant women, and old people	
	By the year of 2030, increase the gains of small and medium-scale agricultural producers by 100% percent and productivity, including other knowledge, opportunities, and markets for value addition and non-farm employment	
	By the year of 2030, ensure a sustainable food production system and apply agricultural practices that helps the rise of production, that support sustain natural systems, enhance the ability for adaptation to climate change, regularly develop soil quality	

billion in 2050 and changing diet in time requires 70% more food production according to FAO (2009). If these requirements cannot be satisfied, then food prices will rise even more. A study shows that even if these requirements satisfied, food prices will rise by 11%. In the worst-case scenario, it may rise by 53.9% (Goswami 2010). It will lead to a worse hunger than nowadays. Even in the best-case scenario, there is still an 11% increase in food price. This shows that increasing food production is not enough for itself. There should be some changes in policies and developments in food trading (FAO 2009). UN works on this problem, especially in the policy change part. UN has seventeen sustainability goals, and one of them is "Zero Hunger". It has five targets, and the first two targets are about, as the name suggests, ending hunger, starting from children under five and young, pregnant or lactating women by 2030 (United Nations 2018).

4.1.4.2 Agricultural Land

Agricultural land is one of the limited resources, and it is reaching its limits. Productive land scarcity is already a critical issue. In the world, there is 13.2 billion ha ice-free land. Its usage distribution is 12% cultivatable area, 28% forest, 35% grasslands, and woodland (United Nations Convesion to Combat Desertification 2017). It is estimated that agricultural land will decline by 11.5% by 2050 (Abd EL-kawy et al. 2019). Not just lands, but the soil is also a non-renewable type of resource (Rounsevell et al. 1999). While the need for food production and naturally for productive soils to produce food is increasing, available agricultural lands are decreasing because of climate change and urbanization. Besides, due to climate change efficiency of the soil is decreasing. Unfortunately, in some regions, the diversity of the crops also decreases owing to the changes in the climate since some crops do not to grow in that region (United Nations Convesion to Combat Desertification 2017). Zero Hunger sustainability goal, as mentioned in the previous part, has targets to ensure sustainable food production that is protected against climate change and to save the diversity of the seed all over the world (United Nations 2018).

4.1.4.3 Uncontrolled Urbanization

The demographic definition of urbanization is the increase in the urban living portion of countries' population over the rural living portion. Urbanization occurs in two ways: the natural way, which results in birth and migration (Satterthwaite, et al. 2010). In today's world, people tend to migrate from rural to urban places. In 2007, the urban population passed the rural population first time in history. After 2007, while the urban population grew by 780 million, rural population growth was only 70 million. In 2050, the world population will be 9.8 billion, and it is predicted that 6.7 billion of them will be living in urban places (Ritchie 2018). More than sixty per cent of the irrigated agricultural lands are placed close to urban areas (United Nations Convesion to Combat Desertification 2017). Because land is a scarce resource, to fit people in urban areas, urban areas need to be expanded over rural areas or forests. It is estimated that in the period between 2000 and 2030, urbanization cause the loss of prime agricultural farmlands in a range of 1.6 and 3.3 million hectares (United Nations Convesion to Combat Desertification 2017). Some countries have policies to prevent this productive land loss. For example, in Turkey, an individual cannot get a building license in a productive farming area. However, the lack of control over the rural areas causes people to build illegal buildings on productive lands. It is hard to stop urban expansion, but to eliminate the negative effects of urbanization on agriculture, urban farming applications can be encouraged. Urban farming is a wide spectrum of food production projects and activities in urban areas and has a commercial value. Green roof farming is one of the example applications of urban farming. The main idea of green roof farming is utilizing the unused area on the rooftops of the buildings. There are a few methods to grow food on rooftops. Although most of the rooftop farming applications use soil-based methods, there are some other methods like hydroponics and aquaponics (Thomaier et al. 2015). The green roof systems are environmentally, socially, and economically sustainable (Hui 2011).

4.2 Aim of the Study

This study is a part of designing and building a self-sustaining compound. By self-sustaining, we mean supplying food in a sustainable manner. Sustainable food production, food security, nexus, and CSA will be the basis of this paper and urban farming. We aim to provide:

- Calculation of daily nourishment (calorie) of an average human being.
- Essential conditions for food production.
- Question and analyze the interrelation between food, water, and energy.
- Investigating smart and sustainable food production techniques and technologies.
- Proposing the most applicable and efficient solution for food production in the self-sustaining compound.

4.3 Methodology

4.3.1 Recommended Ratios of Macronutrients for Energy Intake

People need macro and micronutrients to maintain their vital functions. Micronutrients consist of vitamins and minerals. Macronutrients such as protein, fat, and carbohydrate should be taken in higher amounts. Proteins are responsible for regulating metabolic activity by participating in the structure of enzymes and hormones. Also, they provide the preservation of tissue structure. Fats have the duty to store energy, also provide insulation and protection for vital organs and transportation of fat-soluble vitamins. Carbohydrates are the main energy sources of the body, but other macronutrients can supply energy as well (British Nutrition Foundation 2020). It is clear that about half of the energy requirement is supplied from carbohydrates (45–65%), and fat (20–35%) and proteins (10–35%) constitute the other half (Institute of Medicine 2000). They should contribute to the energy intake at close rates. These rates vary according to a person's lifestyle, such as energy requirements. Total energy requirement estimated with Eqs. 4.1 and 4.2 depending on these factors: person's energy intake, energy expenditure, gender, age, weight, height, and physical activity (Gerrior, et al. 2006).

$$TEE(w) = 387 - 7.31 * A + PA * [(10.9 * W) + (660.7 * H)] \quad (4.1)$$

$$TEE(m) = 864 - 9.72 * A + PA * [(14.2 * W) + (503 * H)] \quad (4.2)$$

TEE(w) = Total energy expenditure of woman
TEE(m) = Total energy expenditure of man
A = Age (year)
W = Weight (kg)
H = Height (m)
PA = Physical activity coefficient. (It depends on the duration and intensity of the physical activity performed and the efficiency of performance.)

In addition to macronutrients, micronutrients are essential for living healthy. Most of them are not synthesized in the body; thus, they must be taken with foods. Vitamins work as regulators and participate in the structure of enzymes. Minerals; form the structure of cells in tissues such as bones, teeth, and nails and act as a catalyst in enzymatic reactions. Vitamins B and C cannot be stored in the body, so regular intake is very important every day (Samur 2008).

Apart from macro and micronutrient intake, diet food must be selected according to their water consumption, GHG emissions, and biodiversity effects in terms of sustainability (Friel et al. 2014). Figures 4.2, 4.3, and 4.4 show water consumption, GHG emission, and land use of various food types, respectively.

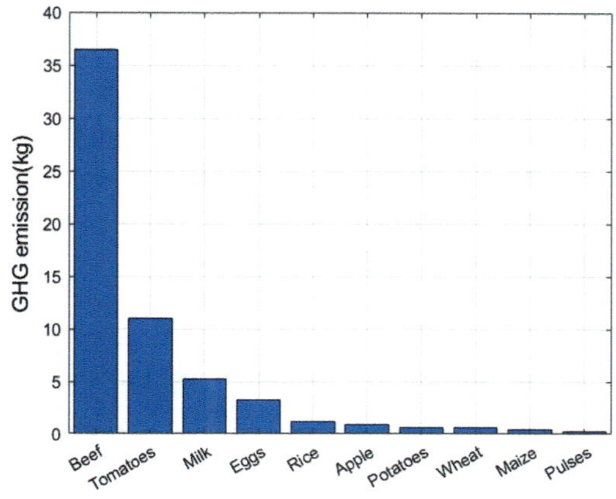

Fig. 4.3 GHG emission of per 1000 kcal (Nemecek and Poore 2018; Our World in Data 2020)

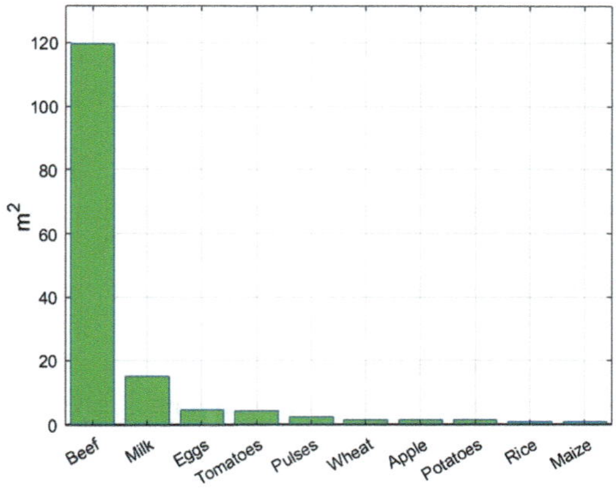

Fig. 4.4 Land usage for 1000 kcal (Clarck and Tilman 2017; Our World in Data 2018)

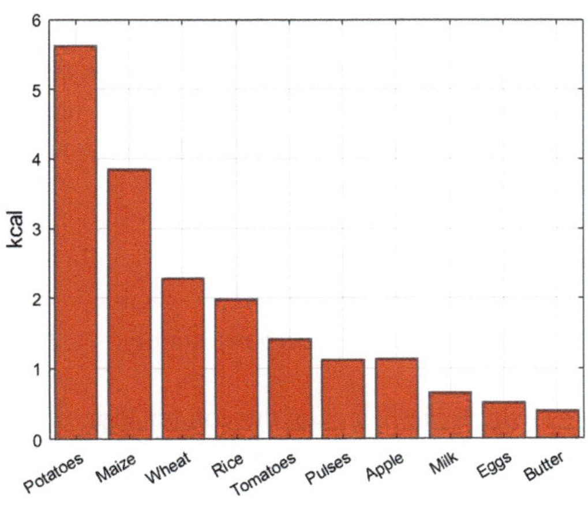

Fig. 4.2 Energy provided per litre of water (Renault and Wallander 2000)

It is clear that plant-derived foods require less water than animal-derived food when considering the energy that foods provide, and potato is the most water-friendly food.

GHG emissions of food production 1000 kcal are summarized in Fig. 4.3. It appears that the production of animal-derived foods has more negative effects on the environment by emitting considerably more GHG.

In terms of biodiversity impact of food production, animal-derived foods cover more area and have more negative effects on flora. Thus, a large part of the diet must consist of plant-derived food in terms of sustainability. The article "Towards Healthy and Sustainable Food Consumption: An Australian Case Study" provides more and detailed information on foods that can be chosen instead of the foods that consume more water and emit more GHG to reduce the negative ecological impact (Friel et al. 2014). So as far as we are concerned, potato is the most sensible one among our research.

4.3.2 Why Potato?

A diet consisting only of potatoes is not feasible practically, like other foods. If we still had to consume a single variety of food, that food would be potato which is the 4th most important food in the globe, where the first three are rice, wheat, and corn (FAO 2008a, b, c). We choose potato with the following grounds;

- The least water-consuming food per kilocalorie (Renault and Wallander 2000).
- Compared to other stable crops, it supplies more protein per unit growing area (International Potato Center 2020).
- It contains three macronutrients, especially carbohydrates (FAO 2008a, b, c).
- Proteins in potatoes consist of amino acids that are essential for people (FAO 2008a, b, c).
- It is good for zinc, iron, protein, and potassium intake (International Potato Center 2020).
- It contains a sufficient amount of vitamins C, E, and several B vitamins (International Potato Center 2020).
- The carotene it contains can be converted into vitamin A in the body (International Potato Center 2020).

It is clear that potatoes contain most of the nutrients people need. Denali, which is one of the varieties of potato, was chosen for calculation in this study because of Dearborn (1979), Canadian Food Inspection Agency (2020);

- Resistance to temperature change
- High dry matter content
- High yield
- High specific gravity.

4.3.3 Nutrient Film Technique (NFT)

It becomes more important to find a sustainable and efficient food production method in terms of many aspects. One promising solution is soilless farming. The basic distinction between conventional and soilless farming is how plants take the nutrients in them (Abdullah 2016). Soilless farming has numerous advantages over traditional farming. First of all, there are some concerns in traditional agriculture regarding the soil-borne disease, unfavourable soil composition, soil pollution, and poor water drainage (Kürklü et al. 2018). Various extensive unfavourable effects on the environment may arise from soil-based farming (Shrouf 2017). Unlike traditional farming, it is possible to grow crops in soilless farming, where there is no appropriate soil or where the soil is polluted by disease (Mason 2011). Also, plants can be put nearer to each other since soilless farming eliminates the need for roots to reach the nutrients, and time is not wasted to construct large-scale root systems, and lots of nutrients are available, and these lead to a larger yield (Elkazzaz 2017). Thanks to the advantages soilless farming provides us, we are going to use a soilless farming method in the "Open Innovative Digital House (ODIH)" study. As seen from Fig. 4.5, soilless farming can be divided into 2 sub-topics which are hydroponics and aquaponics.

Aquaponics is a food production system that incorporates hydroponics with aquaculture (Verner et al. 2017) and the system uses nutrient-enriched water from the fish tanks for the growth of plants (Goddek et al. 2015). There are certain drawbacks in aquaponics, and the main reasons why we do not utilize this technology are as follows (Verner et al. 2017):

- Great risk of system failure
- Huge energy demand
- Need for reliable electricity supply.

Hydroponic, which is the combination of two Greek words, "hydro-ponos" meaning "water-work" (Lakkireddy et al. 2012), is an agricultural technology that contains crop cultivation in water-based solution (Gumisiriza et al. 2020). The use of nutrient solution has the need for

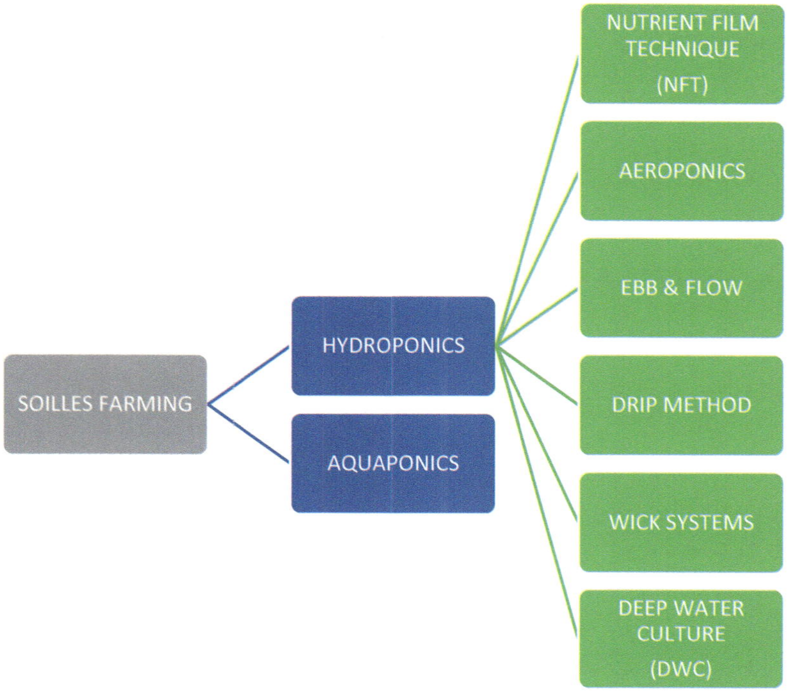

Fig. 4.5 Classification of soilless farming

soil's place in hydroponic (Verner et al. 2017). As a soilless farming method, hydroponic also has many benefits. It diminishes or removes the presence of pesticides and toxicity resulted from it, since the contamination of the crops with the soil-borne disease, insects or pests are not possible in the hydroponic environment (Sharma et al. 2019). Also, weeds in the hydroponic system nearly do not exist (Mason 2011) and the cultivation of the land in preparation for planting is not necessary (Payne 2020). Moreover, hydroponic systems can be built up almost anyplace because these systems do not rely on external circumstances (Verner et al. 2017), for example, indoor farming in a controlled environment makes farming possible in areas that weather and soil condition is not appropriate for conventional farming (Green Our Planet 2020).

In addition, hydroponic has subsystems; NFT, aeroponics, ebb & flow, drip method, wick systems, and DWC. We chose the NFT, shown in Fig. 4.6, as the method to employ in ODIH.

In NFT, plants are placed in long plastic grow trays, and there is circling water through the system. A water level is adjusted in a way that the roots can reach the water (Verner et al. 2017). The main placement of the equipment used in the system and the plants in trays can be seen in Fig. 4.6. Via irrigators at the highest point of each sloping pipe or tray, the solution is drained by holding tank and the run-off from the lowest point of the channels coming back to the tank. Therefore, the nutrient solution is frequently recycled (Elkazzaz 2017). Although aeroponics, which is one of the sub-systems of hydroponic, has many advantages, it has a huge vulnerability about power cut-offs, the equipment is highly dependent on automatic systems, and it can disturb indoors because of its sounds (Miller 2020). All the other sub-systems of hydroponic, except aeroponics, are less water-efficient than NFT (Verner et al. 2017). Figure 4.7 shows the water usage, number of growing seasons, and production comparison between conventional farming and NFT.

From Fig. 4.7, we can see that the amount of water consumed by NFT is almost eight times less than conventional farming. The number of growing seasons in NFT is twice as much as conventional farming and production are higher in NFT. Some other advantages of NFT are the flow of abundant oxygen and being space-sufficient systems (Verner et al. 2017). Due to

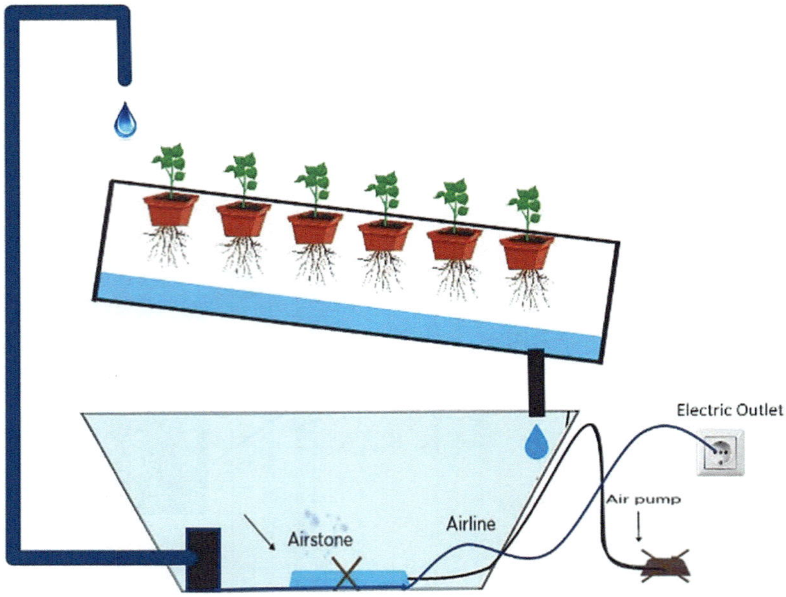

Fig. 4.6 The basic configuration of NFT

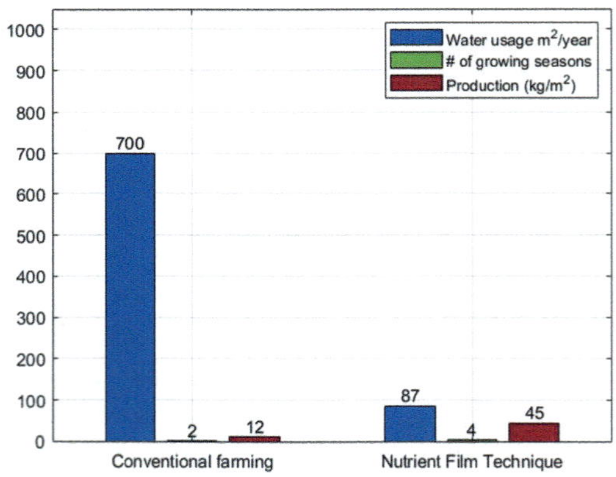

Fig. 4.7 Comparison between conventional farming and NFT (Verner et al. 2017)

all of the mentioned drawbacks of other hydroponic systems, we have decided that the NFT is the most suitable soilless farming method for ODIH.

4.3.4 Required Quantity of Potato for One Average Human in a Year

The amount of potato to provide daily energy requirement is calculated with Eq. 4.3:

$$\mathcal{X}d = \frac{\eta}{\mu} \tag{4.3}$$

where,

η = Daily required kcal [kcal]

μ = kcal per gram in baked potato [kcal g^{-1}]

$\mathcal{X}_d$ = Required amount of potato for providing nutritional requirements for one average human in a year [kg].

The total required potato amount is the same, and it is calculated with Eq. 4.4:

$$\mathcal{X} = 360 * \frac{\eta}{\mu} \tag{4.4}$$

where,

η = Daily required kcal [kcal]

μ = kcal per gram in baked potato [kcal g^{-1}]

$\mathcal{X}$ = Required amount of potato to provide nutritional requirements for one average human in a year [kg].

Total produced calorie calculated with Eq. 4.5:

$$\eta total = 1000 * \mathcal{X} * \mu \tag{4.5}$$

where,

η_{total} = Total kcal need for nourished one average person in a year[kcal]

μ = kcal per gram in baked potato [kcal g^{-1}]

$\mathcal{X}$ = Required amount of potato to provide nutritional requirements for one average human in a year [kg].

4.3.5 Calculations of Conventional Agriculture

4.3.5.1 Area Needed to Provide Nutritional Requirements

In Conventional Agriculture, FAO calculated to approximate yield per hectare of Turkey (FAO 2020a, b). Yield per hectare datum is calculated with respect to detailed area and production data combined in hectares and tonnes. These data were used for calculating approximate production for 1 m^2. Required space for providing enough nutrient for 1 average person is calculated with Eq. 4.6:

$$\alpha = 360 * \beta * \frac{\eta}{\mu * k} \quad (4.6)$$

where,

α = Required space for providing enough nutrient for 1 average person [m^2]

β = Approximate production per m^2 [kg m^{-2}]

η = Daily required kcal [kcal day^{-1}]

k = Number of harvests in a year []

μ = kcal per gram in baked potato [kcal g^{-1}].

4.3.5.2 Water Consumption of Conventional Farming

Water consumption of conventional agriculture is taken from FAO (2020a, b). Conversions from mm to litre and from hectare to m^2 are made, and daily water consumption is calculated with Eq. 4.7:

$$W_{dc} = \frac{W_{gc}}{360} \quad (4.7)$$

where,

W_{gc} = Given water consumption [L]

W_{dc} = Daily water consumption [L d^{-1}]

Total water consumption is calculated with Eq. 4.8:

$$W_{tc} = \alpha * W_{dc} * k * t \quad (4.8)$$

where,

k = Number of harvests in a year []

α = Required space for providing enough nutrient for 1 average person [m^2]

W_{tc} = Total water consumption in 1 year [L]

t = Harvest time of potato [day].

4.3.5.3 Energy Consumption of Conventional Farming

Mechanization in agriculture has reduced the use of muscle power in agricultural production considerably. This opened the way to the excessive usage of fossil fuels as an energy source to replace human labour. In countries, which are developing, huge amounts of fossil fuels are utilized in agricultural production, especially in fertilizer production and the usage of machinery. Energy use in agriculture is examined in two categories:

(1) Direct energy use: Electricity, fuel, petrol, coal, petrol derivative, natural gas, biomass, etc.
(2) Indirect energy use: Human and animal workforce, agriculture device/machinery, fertilizer, pesticides, amount of consumed energy for irrigation.

A. Fuel Energy for Agriculture Equipment/ Machinery:

The most extensive mechanical power resources are agricultural trucks. Direct energy consumption of one tractor is measured in L/h and kg/h. L/h shows litres of fuel expended per hectare. Specific fuel consumption (g/kWh) demonstrate the efficiency of a tractor motor. This is calculated based on the power and fuel consumption of the tractor. Fuel consumption of tractor changes depending on motor weight and work (Yaşar et al. 2011).

In energy analysis for the production of particular agricultural produce, fuel consumption is normally considered as the amount of fuel consumed per unit area (L/ha). The use of direct fuel energy per unit area (ha) during agricultural production processes is calculated with Eq. 4.9 (Yaşar et al. 2011):

$$FE = WEA * FC * LCVF * AN \quad (4.9)$$

where,

FE = Fuel energy [MJ/ha]
WEA = Work efficiency per area [h/ha]
FC = Fuel consumption [L/h]
LCVF = Lower calorific value of fuel [MJ/L]
AN = Application number [].

The fuel consumption of the tractor, which is mainly used as a power source in field works, varies depending on the power usage and the process.

B. Direct Energy Consumption in Irrigation Applications

The energy consumption of irrigation systems is determined similarly to the method used in calculating the energy consumption of agricultural equipment/machinery. Direct energy consumption per irrigated area is determined with Eq. 4.10 (Yaşar et al. 2011):

$$EI_D = \frac{Hm * Q * \rho * g}{\partial p * \partial m} \quad (4.10)$$

where,

EI_D = Direct energy consumption in irrigation [J/ha]
H_m = Total monomeric height [m]
Q = Irrigation water flow [m^3/h]
g = Gravitational acceleration [9.81 m/s^2]
ρ = Density of irrigation water [kg/m^3]
∂p = Efficiency of irrigation pump [%]
∂m = Efficiency of the electric motor [%].

C. Indirect Total Energy Consumption

The energy that is not directly consumed is categorized as indirect energy consumption. Figure 4.8 shows the entire energy consumption items.

Indirect energy inputs are seeds, fertilizers, phytosanitary, pesticides, human power, and maintenance activities (Hatırlı et al. 2005). Our calculation and studies in the literature show that indirect energy consumption is more than direct energy consumption in traditional farming methods.

4.3.5.4 Total Energy Consumption of Conventional Farming

To complete the calculations, it is necessary to add direct and indirect energy consumptions. Total energy consumption is calculated with Eq. 4.11:

$$Pc = \beta + \Gamma t \quad (4.11)$$

where,

Pc = Total energy consumption of conventional farming in 1 year [kWh]
β = Direct energy consumption [kWh]
Γ_t = Indirect energy consumption [kWh].

4.3.5.5 Calculations for WEF Nexus Phenomenon for Conventional Farming

To calculate of WEF nexus phenomenon, the required space for producing 100 kcal in conventional farming calculated with Eq. 4.12:

$$\alpha_{100} = 100 * \frac{\alpha}{total} \quad (4.12)$$

where,

α_{100} = Required Space for producing 100 kcal [m^2 $kcal^{-1}$]
η_{total} = Total kcal need for nourished one average person in a year[kcal]
α = Required Space for providing enough nutrient for one average person [m^2].

Water consumption for producing 100 kcal in conventional farming is calculated with Eq. 4.13:

$$W_{100con} = 100 * \frac{W_{tcon}}{total} \quad (4.13)$$

where,

W_{tcon} = Total water consumption in 1 year [L]
W_{100con} = Water consumption for producing 100 kcal [L]

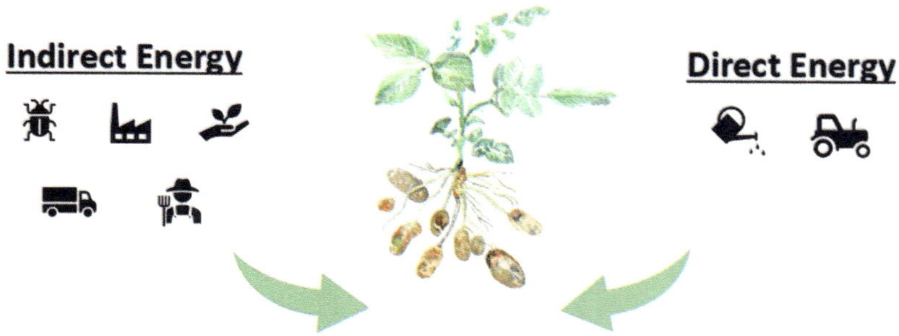

Fig. 4.8 Energy consumption in conventional agriculture

η_{total}=Total kcal need for nourished one average person in a year [kcal].

Energy Consumption for producing 100 kcal in conventional farming is calculated with Eq. 4.14:

$$P_{100} = 100 * \frac{\text{Pc}}{total} \tag{4.14}$$

where,

Pc = Total Energy Consumption of conventional farming in 1 year [kWh]

η_{total} = Total kcal need for nourished one average person in a year [kcal]

P_{100} = Energy Consumption for producing 100 kcal in conventional farming [kcal].

4.3.6 Soilless Agriculture (NFT) System

Calculations of soilless agriculture that adopt NFT are based on NASA CELSS (Control Ecological Life Support Systems) project (Wheeler et al. 1990). The food production part of this project aims to determining the optimal plant and providing nutritional requirements. For these purposes, minimum space and minimum power were used (Wallace and Powers 1990). NASA chose harvest time to be 112 days (Wheeler et al. 1990). However, for the sake of simplification of calculations, we chose harvest time to be 120. Therefore, we assumed we would be harvesting three times a year at ODIH.

4.3.6.1 Area Needed to Provide Nutritional Requirements

In the CELSS project, two of Denali Potato plants were planted in each tray (every tray is 0.2 m^2), and the total dry weight of Denali tubers is 2834 g (Wheeler et al. 1990). According to daily nutritional requirements, the required space for potato production is calculated with Eq. 4.15:

$$\Delta = 360 * \frac{\eta}{\mu * W * T} \tag{4.15}$$

where,

Δ = Required space for providing enough nutrient for one average person [m^2]

η = Daily required kcal [kcal]

μ = kcal per gram in Baked Potato [kcal g^{-1}]

T = Harvest time of Denali potato in NFT system [d]

W = Number of harvests in a year [].

The required amount of potato to provide nutritional requirements for one average human in a season is calculated with Eq. 4.16:

$$X season = \frac{X}{3} \tag{4.16}$$

X = Required amount of potato to provide nutritional requirements for one average human in a year [kg]

X_{season} = Required amount of potato to provide nutritional requirements for 1 average human in a season [kg].

4.3.6.2 Water Consumption of NFT System

Water consumption of the NFT system is taken from the CELSS research (Wheeler et al. 1990) and consumption of the system is approximately 2 $Lm^{-2}d^{-1}$. Daily water consumption is calculated with Eq. 4.17:

$$W_d = \Delta * 2 \quad (4.17)$$

where,

Δ = Required space for providing enough nutrient for one average person [m^2]

W_d = Daily water consumption [L d^{-1}].

Total water consumption is calculated with Eq. 4.18:

$$W_t = W_d * t * k \quad (4.18)$$

where,

W_t = Total water consumption [L]

t = Harvest time [d]

k = Number of harvests in a year []

W_d = Daily water consumption [L d^{-1}].

4.3.6.3 Energy Consumption of NFT System

Energy consumption is considered in four parts: consumption for climatization, consumption of water pump, consumption for lighting, and consumption of water maker. Climatization and lighting data are calculated month by month because other parts do not depend on natural changes. Total energy consumption of NFT system is calculated with Eq. 4.19:

$$\Psi = \phi_{\text{total}} + \Sigma_{\text{total}} + \Pi_{\text{total}} + \Gamma_{\text{total}} \quad (4.19)$$

where,

Ψ = Total energy consumption in a year [kWh]

Σ_{total} = Total energy consumption of water pump [kWh]

Π_{total} = Total energy consumption of lighting [kWh]

Γ_{total} = Total energy consumption of producing enough amount of water [kWh].

A. Energy Consumption of Climatization

Data of monthly energy consumption of climatization are taken from Daikin (2016). Monthly energy consumption of climatization for the total area is calculated with Eq. 4.20:

$$\phi_n = \Delta * \phi_{mpn} \quad (4.20)$$

where,

Δ = Required Space for providing enough nutrient for 1 average person [m^2]

ϕ_n = Monthly energy consumption of climatization [kWh]1

ϕ_{mpn} = Monthly energy consumption of climatization per m^2 [kWh].

Total energy consumption for climatization in a year is calculated with Eq. 4.21:

$$\phi\text{total} = \sum_{n=0}^{n=12} \phi n \quad (4.21)$$

where,

ϕ_{mp} = Monthly energy consumption of climatization per m^2 [kWh]

ϕ_{total} = Total energy consumption of climatization in a year [kWh].

B. Energy Consumption for Lighting

In the CELSS study, (Wheeler et al. 1990), PPFD value is approximately 244/μmol m^{-2} s^{-1} (±26 μmol m^{-2} s^{-1}). For providing enough PPFD, MIGRO ARAY 2 with a Full-spectrum Samsung LM301B lamp (Migro Lights 2020) was selected due to providing the necessary photosynthetic photon flux density (PPFD) level. Consumption of lighting in 1 m^2 is designed to involve all trays area. Data of approximate hours of sunshine of a day was taken from the Turkish State Meteorological Service (T.C. Ministry of Agriculture and Forestry 2020). The daily requirement for lighting time is 24 h in the first 28 days and 12 h later on because 24 h of lighting expedite the shoot growth in the first 28 days of potato growth (Wheeler et al. 1990). However, in this research, we assumed 30 days of continuous lighting for the beginning phase just to simplify the calculations. The total energy consumption of lighting is calculated with Eq. 4.22:

$$\Pi_t = \sum_{n=0}^{n=12}((\Lambda r - \Lambda sn) * 30 * \Pi m) \quad (4.22)$$

where,

Λ_{sn} = Approximate hours of sunshine per day [hour]

Λ_r = Required lighting time per day [hour]

Λ_t = Required lighting time per month [hour]

Π_m = Energy consumption of lighting per m^2 [kWh]

Π_t = Total energy consumption of lighting [kWh].

C. Energy Consumption of Water Pump

The nutrient solution is continuously pumped for 12 months and 24 h (Wheeler et al. 1990). We aimed to pick up the most efficient water pump to reduce the overall consumption of energy. The energy consumption of the water pump is calculated with Eq. 4.23:

$$\Sigma\text{total} = (\text{w}*12 * 24 * 30) \quad (4.23)$$

where,

w = Electrical power of the water pump [kWh]

Σ_{total} = Total energy consumption of water pump [kWh].

D. Energy consumption of making irrigation water

Irrigation water will be produced compound water supplying system. As shown in Chap. 2, to produce water, the system requires 0.008 kWh per litre. The energy consumption of producing irrigation water is calculated with Eq. 4.24:

$$\Gamma\ \text{total} = (W_t * \Gamma) \quad (4.24)$$

where,

Γ = Energy consumption of producing 1 L water from water maker [kWh]

Γ_{total} = Total energy consumption of producing enough amount of water [kWh]

W_t = Total water consumption [L].

4.3.6.4 Calculations for WEF Nexus Phenomenon

To calculate of WEF nexus phenomenon, Water consumption for producing 100 kcal in the NFT system is calculated with Eq. 4.25:

$$W_{100nft} = 100 * \frac{W_{tnft}}{total} \quad (4.25)$$

where,

W_{tnft} = Total water consumption in a year [L]

W_{100nft} = Water consumption for producing 100 kcal [L $kcal^{-1}$]

η_{total} = Total kcal need for nourished one average person in a year[kcal].

Energy Consumption for producing 100 kcal in NFT system is calculated with Eq. 4.26:

$$\Psi_{100} = 100 * \frac{\Psi}{total} \quad (4.26)$$

where,

Ψ = Total energy consumption in a year [kWh]

Ψ_{100} = Energy consumption for producing 100 kcal [kWh]

η_{total} = Total kcal need for nourished one average person in a year [kcal].

To calculate of WEF nexus phenomenon, the required space for producing 100 kcal in the NFT system calculated with Eq. 4.27:

$$\Delta_{100} = 100 * \frac{\Delta}{total} \quad (4.27)$$

where,

Δ = Required space for providing enough nutrient for 1 average person [m^2]

Δ_{100} = Required space for producing 100 kcal [m^2 $kcal^{-1}$]

η_{total} = Total kcal need for nourished one average person in a year [kcal].

4.4 Materials

Due to the space limitation of ODIH, instead of supplying a sufficient amount of nutrients for human beings, in the construction of ODIH we will only include soilless agriculture for demonstration purposes. Vahaa is a smart farming start-up based in Istanbul, Turkey. Their area of expertise is urban indoor farming using vertical farming and IoT technologies (Vahaa 2018). For the ODIH, Vahaa is chosen as a farming application supplier. Among several companies supplying similar products, Vahaa was the most suitable option due to the following reasons. Firstly, they are open to incorporation to develop their product in line with ODIH's needs. As a result of this incorporation, the product can easily be connected to Home Management System (HMS). The design of the product provides a monthly self-sustainable yield production after the first harvest. In addition, Vahaa also supplies all additional requirements, such as seeding materials, including the seed itself on a monthly basis. Water usage of the given product is one of the superior options compared to the other products in the market. Besides all, the location of the company provides constant support for repair and maintenance. Technical specifications of the offered product are as follows:

- Physical dimensions are 100 cm * 130 cm * 18 cm (width * height * depth)
- Monthly electric consumption is 60 kWh
- Water tank capacity is 10 L
- Weekly water consumption is 1 L (Vahaa 2018).

Figure 4.9 shows a soilless vertical farming system of Vahaa.

Fig. 4.9 Soilless vertical farming system of Vahaa (Vahaa 2018)

The company suggests green-leafy vegetables grow at their product. Because the interior temperature of ODIH will be kept around 20 °C, and most of the green-leafy vegetables grow in 18–23 °C temperature interval, we decided to grow leafy vegetables. As mentioned before, the company supplies the seeds, and the variety of seeds is limited. They provide seeds for lettuce, arugula, basil, parsley, thyme, coriander, rosemary, and mint. Among these, the most commonly used in salads and directly eaten ones are lettuce, arugula, basil, and parsley. Table 4.2 shows macronutrient values in grams per 100 g, amount of kcal per 100 g, total kcal supplied in two harvests, estimated maximum harvest time for two harvests this execution, energy and water consumed for two harvests for chosen vegetables.

As seen in Table 4.2, most kcal gathered by using the least water and energy is via lettuce farming. Furthermore, lettuce has the quickest growth rate. Thus, lettuce is the most reasonable option to grow in the ODIH's farming rooms which are numbered 1–4 in Figs. 4.10 and 4.11.

Table 4.2 Macronutrient values in grams per 100 g (Tūrkomp 2013)

	Protein (g)	Fat (g)	Carb. (g)	Kcal (Total)	Kcal/kWh	Kcal/L	Harvest Time (day)	Energy Cons. (kWh)	Water Cons. (L)
Lettuce	0.87	0.23	1.69	6800	18.9	269.8	45	359.6	25.2
Arugula	2.57	0.27	2.38	4160	6.76	69.3	77	615.12	43.2
Basil	3.15	0.64	2.65	920	1.37	19.5	84	671.2	47.2
Parsley	3.39	0.56	2.82	3552	4.11	58.8	108	863.2	60.4

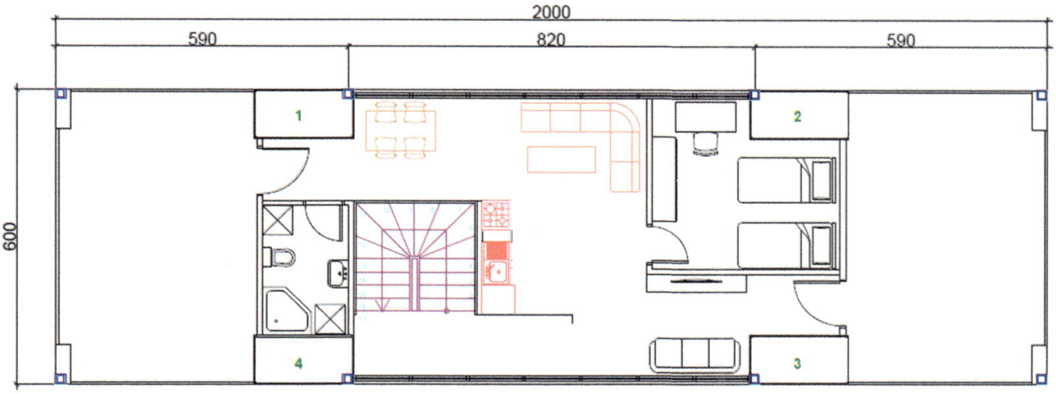

Fig. 4.10 Sketch of ODIH 2nd floor

Fig. 4.11 Illustration of farming room in ODIH

The farming rooms have dimensions of 2 × 1 × 3 m in length, width, and height. We will be growing lettuce to demonstrate sustainable Hydroponic agriculture at ODIH.

4.5 Results

4.5.1 Healthy Diet

According to the appendix "An Easy Approach to Calculating Estimated Energy Requirements", the physical activity coefficient (PA) value of an average woman (A = 30, W = 54, H = 1.61) and an average man (A = 30, W = 70, H = 1.77) is 1.27. According to this value and Eqs. (4.1) and (4.2), the average person needs 2624 kcal energy for her/his normal life (Gerrior et al. 2006). The annual energy requirement of an average person is calculated with Eq. (4.5) and found as as 9,44,444 kcal.

In this research, it is assumed that potatoes do not lose weight when it is baked and Solanum tuberosum L. cv. Denali's calories equal to white potatoes since there is no information about this topic. 108 g baked potato provides with 100 kcal (Kolasa 1993). 2834 g baked potatoes (Eq. 4.3) can supply most of the nutrition and calorie needs for one average person to have a healthy diet. Depending on this knowledge, one person should consume 1020 kg (Eq. 4.4) baked potatoes for one year.

4.5.2 Conventional Agriculture

In this research, the harvest time of potatoes is accepted to be 120–150 days (FAO 2020a, b). The required area of the conventional agriculture method is calculated with Eq. 4.6, and the result is 155 m^2 for providing healthy nourishment for an average person. Water consumption of conventional agriculture method calculated with Eq. 4.7 and daily water consumption found to be 215.2 L. Also, 77 500 L of irrigation water need for producing 1020 kg of potato (Eq. 4.8). Calculations were made to get a harvest twice a year. The quantity of used energy in irrigation for 1 kg of potato production is calculated with Eq. 4.10. The most efficient irrigation method is the drip irrigation method in conventional agriculture (Ökten 2011). For this reason, electric motors used in drip irrigation systems have been selected. Most of the potato production takes place in Central Anatolia in Turkey (Yegül 2020) and sump pumps are preferred to benefit from underground water resources in this region. The efficiency of irrigation pump value has selected with the optimal value of the sump pump (Yürdem et al. 2012). Also, the total manometric height value has been determined by the average distance between the underground water resources in the specified region and the surface (Yılmaz 2010). Altogether, these data are applied in the formula, and the direct energy consumption per unit area turned out to be 0.3630 kWh/m^2 (Eq. 4.10).

The total energy consumption caused by a tractor while planting is calculated by using Eq. 4.9. For work efficiency (h/ha) Agromaster OPD 4 potato planter's specifications are taken because the result of conducted market research shows that it is an efficient option in the market. In the catalogue given work efficiency is 1.11 h/ha. (AgroMaster 2018). For fuel consumption, (L/h) Deutz-Fahr 6160P model was chosen because of 21.3 l/h fuel consumption. According to the DLG PowerMix test, this model was chosen as the most fuel-efficient model in the market. (Traction Maganize 2014). The lower calorific value of the fuel is 36.0 MJ/kg for diesel fuel (Engineering ToolBox 2003). Diesel fuel is chosen because a given tractor model uses diesel as fuel (Traction Maganize 2014). Lastly, because there will be two planting sessions in a year application number is taken as two. As a result, these data are applied to Eq. 4.9 give 1702.3 MJ/ha. It is converted to 0.047 kWh/m2 for convenience. The indirect energy consumption of conventional farming is 1.2708 kWh/m^2 (Allali, et al. 2017).

4.5.3 Soilless Agriculture

In our study, the NFT, which is one of the soilless agriculture methods, is used to calculate the needed area, water, and energy to grow 1020 kg potatoes (Eq. 4.8). Since optimum 120 days are required to harvest potatoes in NFT (Wheeler et al. 1990), a year is divided into three growing seasons, which can provide the most efficient energy consumption for lighting and air conditioning. Table 4.3 demonstrates the three growing seasons.

340 kg potatoes (Eq. 4.16) can be harvested in each season, and 49 m^2 area (Eq. 4.15) is needed to produce that much of potatoes. The

Table 4.3 Growing seasons

1st season	2nd season	3rd season
February	June	October
March	July	November
April	August	December
May	September	January

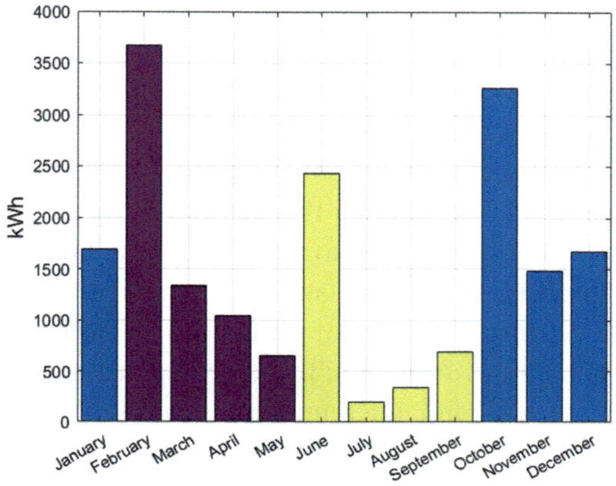

Fig. 4.12 Energy consumption (kWh) for lighting by month

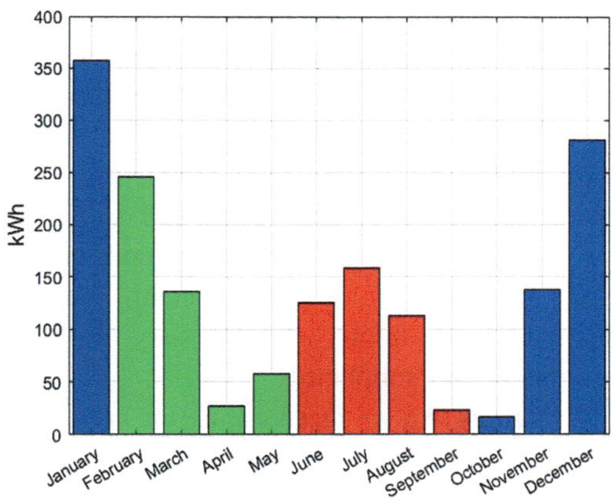

Fig. 4.13 Energy consumption (kWh) for air conditioning by month

daily water consumption of the NFT system is 98 L per day (Eq. 4.17). The water consumption of 49 m^2 in one year is fixed to 32,928 L (Eq. 4.18). Energy requirements for lighting 49 m^2 area may vary by months depending on daily sunlight exposure time. Figure 4.12 shows the energy consumption for lighting each month.

The total energy consumption of lighting is 18 469.08 kWh (Eq. 4.22). Like lighting, the energy need for air conditioning may change by month. Figure 4.13 shows the energy consumption (kWh) for air conditioning by month (Eq. 4.20).

The total energy consumption of climatization is 1679.23 kWh (Eq. 4.21). Table 4.4 shows the energy consumption in each season for lighting and air conditioning as follows.

The required energy per year for the water pump is fixed to 207.36 kWh (Eq. 4.23), and 26.34 kWh (Eq. 4.24) is consumed by the Compound water supplying system a year. When all of the energy consumption data are taken into account together, the overall required energy to grow 1020 kg potatoes hydroponically in one year is 20,382.01 kWh (Eq. 4.19). Water, energy, and food are essential elements of human well-being. Figure 4.14 (Eqs. 4.12–4.14, 4.25–4.27) shows how much water, energy, and land area required for 100 kcal dietary energy provided by potato via comparing conventional farming with NFT.

Table 4.4 Energy consumption of lighting and climatization season by season

	Energy consumption of lighting (kWh)	Energy consumption for climatization (kWh)
1st season	6 703.20	466.48
2nd season	3 651.48	419.93
3rd season	8 114.40	792.82

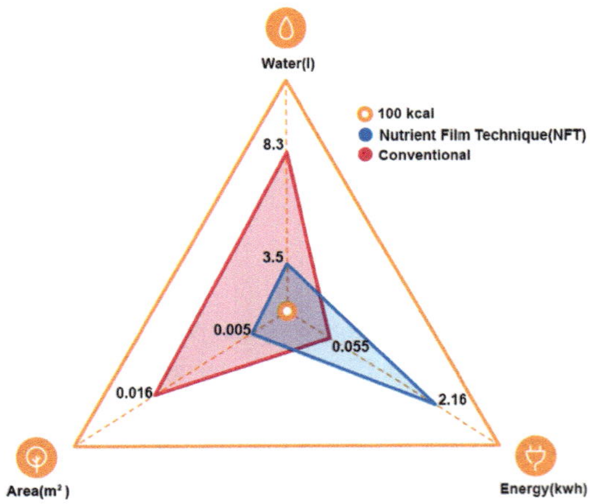

Fig. 4.14 WEF Nexus for 100 kcal

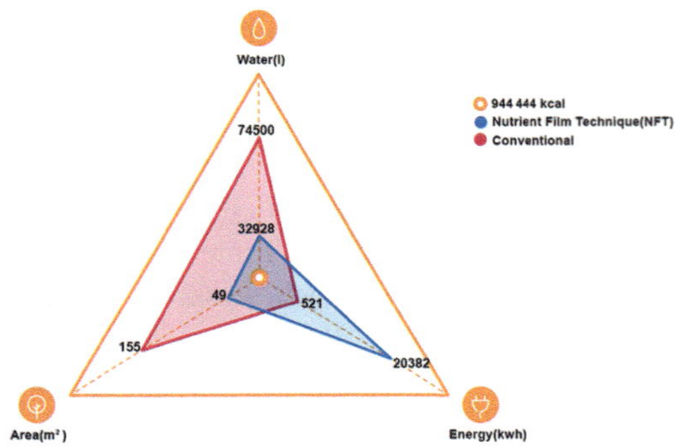

Fig. 4.15 WEF Nexus to nourish one average person per year

NFT consumes less water and demands less area than conventional farming. However, conventional farming consumes dramatically low energy compared to NFT. One average person needs to take 9,44,444 kcal dietary energy, which equals 1020 kg potatoes, to live healthy for one year. Necessary potatoes produced by NFT are harvested three times a year by consuming 32,928 L water and 20,382 kWh energy in a 49 m^2 area. Figure 4.15 shows the WEF and Land nexus on a yearly basis to nourish an average human being.

Although; energy consumption in NFT is considerably higher than in conventional farming, the amount of water and required area are remarkably less in NFT compared to conventional agriculture.

4.6 Discussions and Policy Recommendation

4.6.1 Discussion

In this chapter, we calculated the energy requirement for an average person. It changes depending on the physical activity level, gender, or age of people. Also, the average daily energy expenditure increases every year. Thus, we need more nutrients in upcoming times to meet our energy requirements.

Calculation of the amount of food to be taken was made considering the energy need and on a potato-only diet. Due to the lack of baked data of Denali type, the calculation was made according to the nutritional data of white potatoes instead of Denali type. Although; this diet contains most of the macro and micronutrients necessary for a healthy diet, some of them are taken more than they need due to too much consumption.

NFT is neither the most efficient technique in terms of water and energy consumption nor the best method for producing potatoes. The potato was produced with NFT in a special experimental setup. On the other hand, the aeroponic technique needs less water and energy compared to NFT. Thus, aeroponic is the best soilless agriculture method for potato production. However, the system is dependent, and it requires experience. To sum up, producing food with the NFT technique is more convenient, even for beginners, in terms of necessary technical equipment and simplicity.

Obstacles that we faced during the calculation of WEF Nexus Phenomenon for NFT system and Conventional agriculture:

- It was very difficult to acquire the data of yield in NFT system.
- The calculations of conventional agriculture are made concerning average values.
- The data of yield in NFT system are taken from an old study (1990).
- The calculations of energy usage in NFT system are higher than expected.

- The energy consumption of sensors was neglected.
- The vertical farming technique was neglected for calculations of the needed area.
- The gas requirement from the air was neglected, and it was assumed that optimal conditions were achieved.

Obstacles that we faced during the selection of NFT system (Vahaa) for ODIH:

- The lighting system of Vahaa has not controlled by the feedback system
- Vahaa is not appropriate for growing potatoes.
- Vahaa does not have a climatization system.
- Vahaa is not designed for an area that is inside ODIH. So, the given space can be used more appropriately.

No research has been done on storage conditions and the required storage area. We assume that produced nutrients will be consumed without spoiling and without any nutrient loss.

4.6.2 Policy Recommendations

Multiplicities can conduct their soilless farming greenhouses to encourage local farmers to imply soilless farming techniques and to expand the usage of soilless farming. As an example, Istanbul Metropolitan Multiplicity conducted a soilless farming greenhouse project with the purpose of developing a new business model and increase agricultural production by expanding it throughout Istanbul (Istanbul Metropolitan Multiplicity 2019).

Despite the following of a low GHG pathway, the rise in global sea level from the level in 2000 will have been at least 0.3 m by 2100, and in case of the worst scenario, this rise will be at least 2.5 m (Sweet et al. 2017). As a result, considerable farmland losses are inevitable in the near future. Vertical farming is a solution to use the remaining agricultural lands effectively and increase the yield per m^2. Therefore, vertical farming initiatives should be supported.

Water quality is expected to deteriorate in many regions due to increased pollutant activity, salinization caused by rising sea levels, and water resource changes. 70% of water consumption worldwide is because of agricultural irrigation (OECD 2020a). Soilless agriculture is recommended so as to use water in the most effective way for agriculture. Also, foods with a high calorie/water ratio should be preferred more.

Agriculture accounts for a quarter of GHG emissions. Half of the gas emissions from agriculture stem from deforestation to create agricultural land (OECD 2020b). In addition, deforestation changes the climatic conditions visibly, such as temperature, humidity, amount of precipitation. Because of climate change, the expected yield may not be obtained, or even food losses may occur (Lawrence and Vandecar 2015). We aim to produce nutrients effectively in existing areas with soilless and vertical farming rather than traditional techniques. If soilless and vertical farming is assimilated, essential steps will be taken to achieve SDG #13 and #15.

Big cities are almost entirely dependent on outside for food requirements. Also, extended supply chains and prolonged logistics cause food losses (Girgin 2020). Balconies, roofs, or terraces can be used for agriculture in cities. Soilless agriculture is quite suitable for urban agriculture. If it is done together with vertical agriculture, this will increase the usable farm area up to 2, 3, or more folds. When necessary conditions are created with plant growing lights and air climatization, food can be produced even in basements. The sustainability of cities (SDG 11) can increase when urban agriculture is supported by governments. If it becomes widespread enough with these supports, everyone can produce at least some of the food they will consume. Thus, access to food becomes easier (SDG 2), and poverty decreases (SDG 1). We recommend that urban agriculture be supported not only with greenhouses but also in homes.

Technological developments reduce the demand for manual labour. IoT and AI; have brought a new perspective to agriculture, and smart agriculture concepts are created. For the first time in 2017, plants could be grown without a human's touch (OECD 2020c). Smart agriculture can support the purpose of prospering more resilient, productive, and sustainable agriculture and food systems that better fulfil the needs of consumers. We may claim that agriculture can be integrated with technology to make agriculture more manageable and improve nutrients' quality.

Agricultural data are fundamental in testing the appropriateness of policies implemented (OECD 2020d). Collecting the data and its analysis is effortless with technological developments. The collected data should be evaluated by experts, and policy improvements should be made on how to achieve the highest efficiency with the least energy and water consumption.

Turkey has the Mediterranean region's 3rd largest cultivation area. Turkey has a 52,000 ha greenhouse area in 2016 (TUIK 2016), but all of the areas were covered by Low Technology greenhouses (LTG) that have a primitive structure. Besides, its average surface area is 0.1–0.2 ha, LTG does not have an advanced heating system. Heating is commonly used against frost. To ensure food security and resilience, the area covered by high technology greenhouses (HTG) that have a central climatization system, and the average surface area is 2–4 ha. HTG commonly use heating resources that are geothermal and coal (Gül 2018). HTG must be increased systematically. Also, investment costs are an issue. The initial investment cost is around 7 € per m^{-2} for LTG and 55–70 € per m^{-2} for HTG (Gül 2018). We recommended the policymakers increase incentives for HTG's. Also, policymakers add tax breaks (tax deductions) for soilless cultivation greenhouses to increase the soilless cultivated area.

4.7 Conclusion

Climate change has negative impacts on many areas, and agriculture is one of the many. As a result, the food security and incomes of individuals are affected. The United Nations has a set of goals named SDGs to eliminate or minimize the negative effects of climate change globally by 2030. Hunger is one of the most crucial causes of the lack of food security. To solve the hunger problem, the UN determined one of the SDGs as Zero Hunger. Increasing food production and decreasing the overuse of food is vital to achieving Zero Hunger. While increasing food production, doing it in a sustainable way is essential. CSA concept emerges at this point. CSA is an agricultural production method to decrease the negative impacts of agricultural production on the environment like unnecessary land, water, and energy usage. Another important concept to provide sufficient resources to increasing demand is WEF Nexus. WEF Nexus shows the deep relationship among three different concepts where actions in any of the three affect the others. In the future, scarce resources, increasing population, and boosting demand on the resources make the situation that is already difficult to handle more challenging. Thanks to WEF Nexus, SDG of UN, and CSA applications, it seems possible to overcome those challenges.

The amount and types of nutrients a healthy adult should take daily are clear. We researched from which foods humans can get these nutrients and the required amounts. Many animal and plant products can meet the necessary nutritional values. However, energy and water consumption and GHG emission should be a very low level in their production so that sustainable production can be made. Because of these reasons, the potato was preferred as the most appropriate nutrient. The method to be used in its cultivation should not be traditional agriculture because when we make comparisons with soilless farming methods, according to the results of our calculations, the yield per unit area is low. Water and energy consumption are very high, and also pesticides are used. Soilless agriculture is much more suitable for this project for every criterion. We preferred the hydroponic NFT method among soilless farming methods because the efficiency per water consumption is higher than all other approaches.

Because of the limited space of ODIH, agricultural applications will only be for demonstration purposes instead of nourishing human beings. Vahaa will be suppliers for farming applications in ODIH, which is a start-up in smart indoor farming applications. System Vahaa will set will be connected to the Home Management System of the ODIH. Leafy green vegetables can be grown in ODIH because of the suitability of the both Vahaa module and the interior temperatures. However, for the sake of efficiency and supplied kcal amount, lettuce is the most reasonable choice among the other leafy green vegetables. In ODIH there will be four farming rooms, each having dimensions of 2*1*3 m (Length*Width*Height).

An average person needs to consume 9,44,444 kcal of energy in a year. 1020 kg of baked potatoes meet this energy requirement. According to our calculations, by using conventional agriculture with two harvests a year, 155 m^2 of land use, 74,500 L of water, and 521 kWh of energy are needed to produce this potato. On the other hand, using the NFT, a soilless farming mechanism, with three harvests a year, 49 m^2 of land use, 32,928 L of water, and 20,382 kWh of energy are needed.

At ODIH, our available space for farming is too small for plant diversity. We need to create more space to increase plant diversity and the number of plants. The chosen Vahaa product was not very suitable for our field, but it was the best alternative. We could find more appropriate a system and entirely automatic device. Climate-control systems are integrated into the smart home management portal with IoT applications. Growing processes can be improved with the data collected from plants using artificial intelligence and machine learning technologies. The data obtained can be shared as open data. The status of application methods can be examined within the scope of city farms and sustainable cities. Innovation profiles like water management, microgrids, energy-water nexus, and sustainability, and COP21 (2015 United Nations Climate Change Conference) experience a high priority in the smart city framework discussion around cities' climate change charters. Initiatives by the World Economic Forum seek to rebuild society after COVID-19 with a "Great Reset" facilitated through the Fourth Industrial Revolution in urban centres and a project that sustainability and business development, especially in cities, will increase (World Economic Forum 2020a, b). Especially in Europe, regional prioritization (for example, the European Green Deal) will provide extensions to the circular urban environment approach. In this long span, conditions of the smart city project related to, for example, the city government's policy or finances may change (World Economic Forum 2020a, b). Therefore, it is important to remain flexible when implementing the project to cope with changes. According to 'Hype Cycle for Smart

Life Stage	Group	Male	Female	Average	Potato (100 g)	Potato (2852 g)
Carbohydrate	(g/d)	130	130	130	21.08	601.2016
Total fiber	(g/d)	38	25	36.5	2.1	59.892
Fat	(g/d)	ND	ND	15,85	0.15	4.278
Protein	(g/d)	56	46	51	2.1	59.892
Energy	(kcal)	2965	2282	2624	92	2623.84
Calcium	(mg/d)	1000	1000	1000	10	285.2
Fluoride	(mg/d)	4	3	3.5	0	0
Iron	(mg/d)	8	18	13	0.64	18.2528
Magnesium	(mg/d)	400	320	360	27	770.04
Manganese	(mg/d)	2.3	1.8	2.05	0.189	5.39028
Phosphorus	(mg/d)	700	700	700	75	2139
Zinc	(mg/d)	11	8	9.5	0.35	9.982
Potassium	(g/d)	4.7	4.7	4.7	0.544	15.51488

(continued)

Life Stage	Group	Male	Female	Average	Potato (100 g)	Potato (2852 g)
Sodium	(g/d)	1.5	1.5	1.5	0.007	0.19964
Chloride	(g/d)	2.3	2.3	2.3	0	0
Vitamin C	(mg/d)	90	75	82.5	12.6	359.352
Vitamin E	(mg/d)	15	15	15	0.04	1.1408
Thiamine	(mg/d)	1.1	1.2	1.15	0.048	1.36896
Riboflavin	(mg/d)	1.3	1.1	1.2	0.043	1.22636
Niacin	(mg/d)	16	14	15	1.528	43.5786
VitaminB6	(mg/d)	1.3	1.3	1.3	0.211	6.01772
Pantothenic Acid	(mg/d)	5	5	5	0.383	10.92316
Choline	(mg/d)	550	425	487.5	14.4	410.688
Chromium	(μg/d)	35	25	30	0	0
Copper	(μg/d)	900	900	900	127	3622.04
Iodine	(μg/d)	150	150	150	0	0
Molybdenum	(μg/d)	45	45	45	0	0
Selenium	(μg/d)	55	55	55	0.5	14.26
Vitamin A	(μg/d)	900	700	800	7	199.64
Vitamin D	(μg/d)	15	15	15	0	0
Vitamin K	(μg/d)	120	90	105	2.7	77.004
Folate	(μg/d)	400	400	400	38	1083.76
Vitamin B12	(μg/d)	1.3	1.3	1.3	0	0
Biotin	(μg/d)	30	30	30	0	0

City Technologies and Solutions'report of Gartner, digital ethics and WEF nexus will be the peak of inflated expectations in 5–10 years (Gartner 2020). Therefore, we are going to focus on these issues.

Appendix 4.1

This appendix (U.S. Department of Agriculture 2018; Ross et al. 2011).

References

Abd EL-kawy OR, Ismail HA, Yehia HM, Allam MA (2019) Temporal detection and prediction of agricultural land consumption by urbanization using remote sensing. Egypt J Rem Sens Space Sci 22(3):237–246

Abdullah N-O (2016) Vertical-horizontal regulated soilless farming via advanced hydroponics for domestic food production in Doha, Qatar. Res Ideas Outcomes 2(2):4–5

Aboelnga HT et al (2018). The water-energy-food security nexus. Bonn: Nexus Regional Dialogue Programme (NRD)

Action Against Hunger (2020). World hunger: key facts and statistics 2020. [Online] Available at: https://www.actionagainsthunger.org/world-hunger-facts-statistics#: ~ :text=What%20is%20hunger%3F,1. Accessed 17 Aug 2020

AgroMaster (2018) Potato planter. [Online] Available at: https://www.agromaster.com/machinedetail/2/3/48/Seed-Sowing-Equipment/Potato-Equipment/Potato-Planter. Accessed 22 Oct 2020

Allali K, Dhehibi B, Kassam SN, Aw-Hassan A (2017) Energy consumption in onion and potato production within the province of El Hajeb (Morocco): towards energy use efficiency in commercialized vegetable production. J Agric Sci 9(1):118–127

Asfaw S, Branca G (2018) Introduction and Overview. Climate smart agriculture building resilience to climate change. FAO, Rome, pp 3–4

British Nutrition Foundation (2020) British nutrition foundation. [Online] Available at: https://www.nutrition.org.uk/healthyliving/basics/exploring-nutrients.html. Accessed 19 Oct 2020

Canadian Food Inspection Agency (2020) Government of Canada. [Online] Available at: https://www.

inspection.gc.ca/plant-varieties/potatoes/potato-varieties/denali/eng/1312587385697/1312587385698. Accessed 19 Oct 2020

Clarck M, Tilman D (2017) Comparative analysis of environmental impacts of agricultural production systems, agricultural input efficiency, and food choice. Environ Res Lett 12(6):1–11

Daikin (2016) Daikin Altherma heat pump. [Online] Available at: https://altherma.daikin.com.tr/assets/images/urunler/dusuk-sicaklikli-daikin-altherma/pdf/teknik-ozellikler.pdf. Accessed 24 Oct 2020

Dearborn CH (1979) Denali: a high dry matter potato with wide adaptation. Am Potato J 56(8):373

Elkazzaz A (2017) Soilless agriculture a new and advanced method for agriculture development: an introduction. Agric Res Technol Open Access J 3 (2):2–3

Engineering ToolBox (2003) The engineering toolbox. [Online] Available at: https://www.engineeringtoolbox.com/fuels-higher-calorific-values-d_169.html. Accessed 22 Oct 2020

FAO, IFAD, UNICEF, WFP, WHO (2019) The State of Food Security and Nutrition in the World 2019. FAO, Rome

FAO, OECD (2012) Building resilience for adaptation to climate change in the agriculture sector. In: Proceedings of a joint FAO & OECD workshop. FAO, Rome

FAO (1988) Report of the FAO council. FAO, Rome

FAO (2008a) International year of potato/hidden tresure/why potato. [Online] Available at: http://www.fao.org/potato-2008/en/aboutiyp/index.html. Accessed 18 Oct 2020

FAO (2008b) International year of potato/potatoes, nutrition and diet. [Online] Available at: http://www.fao.org/potato-2008/en/potato/factsheets.html. Accessed 17 Oct 2020

FAO (2008c) International year of potato/the plant. [Online] Available at: http://www.fao.org/potato-2008/en/potato/index.html. Accessed 17 Oct 2020

FAO (2009) How to feed the World in 2050. FAO, Rome

FAO (2011a) The state of the world's land and water resources for food and agriculture(SOLAW)-Managing system at risk. Food and Agriculture Organization of the United Nations, Rome

FAO (2011b) Energy-smart food for people and climate. Food and Agriculture Organization of the United Nations, Rome

FAO (2014a) Building a common vision for sustainable food and agriculture: principles and approaches. FAO, Rome

FAO (2014) The water-energy-food nexus. A new approach in support of food security. Food and Agriculture Organization of the United Nations, Rome

FAO (2020a) FAOSTAT. [Online] Available at: http://www.fao.org/faostat/en/#data/QC. Accessed 24 Oct 2020

FAO (2020b) Land & water—potato. [Online] Available at: http://www.fao.org/land-water/databases-and-software/crop-information/potato/en/. Accessed 24 Oct 2020

Food and Agriculture Organization of the United Nations (2013a) Healthy people depend on healthy food systems. FAO, Rome

Food and Agriculture Organization of the United Nations (2013b) Climate-smart agriculture sourcebook. FAO, Rome

Friel S, Barosh L, Lawrence M (2014) Towards healthy and sustainable food consumption: an Australian case study. Public Health Nutrit 17(5):1156–1166

Gartner (2020) Hype cycle for smart city technologies and solutions 2020. Gartner, Stamford

Gerrior S, Juan W, Peter B (2006) An easy approach to calculating estimated energy requirements. Prevent Chronic Dis 3(4):A129

Girgin D (2020) Kentsel Tarımın İstanbul'da Uygulanabilirliği. AURA İstanbul, İstanbul

Goddek S, Delaide B, Mankasingh U, Ragnarsdottir KV (2015) Challenges of sustainable and commercial aquaponics. Sustainability 7(4):4199–4224

Goswami UA (2010) World food prices may rise from 31–101% by 2050. [Online] Available at: https://economictimes.indiatimes.com/news/economy/agriculture/world-food-prices-may-rise-from-31-101-by-2050/articleshow/7026749.cms. Accessed 17 Aug 2020

Green Our Planet (2020) Benefits of hydroponics. [Online] Available at: https://greenourplanet.org/benefits-of-hydroponics/. Accessed 24 Oct 2020

Gumisiriza MS, Ndakidemi PA, Mbega ER (2020) Memoir and farming structures under soil-less culture (hydroponic farming) and the applicability for Africa: a review. Agric Rev 41(2):139–145

Gül A (2018) Soilless cultivation in Turkey. İstanbul, ISHS

Hatırlı SA, Ozkan B, Fert C (2005) Energy inputs and crop yield relationship in greenhouse tomato production. Elsevier, Antalya

Hellin J, Fisher E (2019) The Achilles heel of climate-smart agriculture. Nature Clim Change 7(9):493–494

Hoff H (2011) Understanding the nexus. Background paper for the Bonn2011. Stockholm, Stockholm Environment Institute

Hui SC (2011) Green roof urban farming for buildings in high-density urban cities. In: Hainan, Hainan China world green roof conference

Institute of Medicine (2000) Dietary reference intakes. The National Academic Press, Washington, DC

International Potato Center (2020) Case for investment: nutrition. [Online] Available at: https://cipotato.org/nutrition/. Accessed 17 Oct 2020

IPCC (2014) Climate Change 2014: Mitigation of climate change. Contribution of working group III to the fifth assessment report of the intergovernmental panel on climate change. Cambridge University Press, Cambridge and New York

Istanbul Metropolitan Multiplicity (2019) İBB TOPRAKSIZ TARIM SERASI KURDU. [Online] Available at: https://www.ibb.istanbul/News/Detail/35196. Accessed 19 Nov 2020

Kolasa KM (1993) The potato and human nutrition. Am Potato J 70(5):375–384

Kürklü A, Ghafoor A, Khan FA, Ali Q (2018) A review on hydroponic greenhouse cultivation for sustainable agriculture. Int J Agric Environ Food Sci 2(2):59–66

Lakkireddy K, Kondapalli K, Rao KS (2012) Role of hydroponics and aeroponics in soilless culture in commercial food production. Res Rev J Agric Sci Technol 1(1):26–35

Lal R (2004) Soil carbon sequestration impacts on global climate change and food security: soils: the final frontier. Science (Washington, D.C.) 304 (5677):1623–1627, 304(5677):1623

Lawrence D, Vandecar K (2015) Effects of tropical deforestation on climate and agriculture. Nature Clim Change 2(5):27–36

Leslie Lipper DZ (2018) A short history of the evolution of the climate smart agriculture approach and its links to climate change and sustainable agriculture debates, 52nd edn. Springer Nature, Cham

Living Sustainably (2020) 6. What can we do?. [Online] Available at: https://www.sustainability-yes.ch/living-sustainably-chapter-6/. Accessed 7 Dec 2020

Long TB, Blok V, Coninx I (2015) Barriers to the adoption and diffusion of technological innovations for climate-smart agriculture in Europe: evidence from the Netherlands, France, Switzerland and Italy. J Cleaner Prod 112(1):9–21

Mason JL (2011) Commercial hydroponics, 3rd edn. ACS Distance Education, London

Migro Lights (2020) MIGRO ARAY 2 | 120W. [Online] Available at: https://www.migrolight.com/product/migro-aray-120w-ii-bar-light/. Accessed 24 Oct 2020

Miller B (2020) 21 Big advantages and disadvantages of aeroponics. [Online] Available at: https://greengarageblog.org/21-big-advantages-and-disadvantages-of-aeroponics. Accessed 12 Dec 2020.

Nemecek T, Poore J (2018) Reducing food's environmental impacts through producers and consumers. Science 360(6392):987–992

OECD (2020a) Organisation for economic co-operation and development. [Online] Available at: https://www.oecd.org/agriculture/topics/water-and-agriculture/. Accessed 21 Nov 2020

OECD (2020b) Organisation for economic co-operation and development. [Online] Available at: https://www.oecd.org/agriculture/topics/climate-change-and-food-systems/. Accessed 21 Nov 2020

OECD (2020c) Organisation for economic co-operation and development. [Online] Available at: https://www.oecd.org/agriculture/topics/technology-and-digital-agriculture/. Accessed 21 Nov 2020

OECD (2020d) Organisation for economic co-operation and development. [Online] Available at: https://www.oecd.org/agriculture/topics/farm-level-analysis-network/. Accessed 21 Nov 2020=

Our World in Data (2018) Land use per 100 kilocalories by food and production type. [Online] Available at: https://ourworldindata.org/grapher/land-use-per-kilocalorie-by-food. Accessed 13 Dec 2020

Our World in Data (2020) Our world in data. [Online] Available at: https://ourworldindata.org/grapher/ghg-kcal-poore. Accessed 18 Oct 2020

Ökten S (2011) Konya Havzasında Su Yönetimi Politikalarının Yol Açtığı Çevre Sorunları ve Genel Çözüme Yönelik Çalışmalar. Süleyman Demirel Üniversitesi Vizyoner Dergisi 3(5):124–147

Payne M (2020) 20 Advantages & disadvantages of hydroponics that you should know. [Online] Available at: https://www.trees.com/advantages-disadvantages-of-hydroponics. Accessed 8 Nov 2020

Renault D, Wallander WW (2000) Nutritional water productivity and diets. Agricultural water management 45(3):275–296

Ritchie H (2018) Urbanization. [Online] Available at: https://ourworldindata.org/urbanization. Accessed 17 Aug 2020

Ross CA, Taylor CL, Yaktine AL, Valle HB (2011) Dietary reference intakes for Calcium and Vitamin D, 1st edn. The National Academies Press, Washington D.C.

Rounsevell M, Evans S, Bullock C (1999) Climate change and agricultural soils: impacts and adaptation. Clim Change Interdisc Int J Devoted Description Causes Implications Clim Change 43(4):683–709

Samur G (2008) Vitaminler mineraller ve sağlığımız. Klasmat Matbaacılık, Ankara

Satterthwaite D, McGranahan G, Tacoli C (2010) Urbanization and its implications for food and farming. Philos Trans R Soc London Issue 365:2809–2820

Sharma N, Acharya S, Kumar K (2019) Hydroponics as an advanced technique for vegetable production: an overview. J Soil Water Conserv 17(4):364–371

Shrouf A (2017) Hydroponics, aeroponic and aquaponic as compared with conventional farming. Am Sci Res J Eng Technol Sci (ASRJETS) 27(1):247–255

Sweet WV et al (2017) Global and regional sea level rise scenarios for the United States. National Oceanic and Atmospheric Administration, Maryland

T.C. Ministry of Agriculture and Forestry (2020) Resmi İstatistikler. [Online] Available at: https://www.mgm.gov.tr/veridegerlendirme/il-ve-ilceler-istatistik.aspx?k=undefined&m=ISTANBUL. Accessed 24 Oct 2020

Thomaier S et al (2015) Farming in and on urban buildings: Present practice and specific novelties of Zero-Acreage Farming (ZFarming). Renew Agric Food Syst 1(30):43–54

Traction Maganize (2014) DLG PowerMix Test | Deutz-Fahr 6160P. Traction, 14 3. Issue Mai/Juni 2014:30–38

TUIK (2016) Bitkisel Üretim İstatistikleri, 2016. [Online] Available at: https://www.tuik.gov.tr. Accessed 23 Nov 2020

Türkomp (2013) Ulusal Gıda Kompozisyon Veri Tabanı. [Online] Available at: http://www.turkomp.gov.tr/database. Accessed 2020 Oct 2020

U.S. Department of Agriculture (2018) Potatoes, white, flesh and skin, baked. [Online] Available at: https://ndb.nal.usda.gov/fdc-app.html#/food-details/170434/nutrients. Accessed 23 Oct 2020

United Nations Conversion to Combat Desertification (2017) Global Land Outlook first edition (2017). [Online] Available at: https://knowledge.unccd.int/glo/GLO_first_edition. Accessed 17 Aug 2020

United Nations (2018) Goal 2: zero hunger. [Online] Available at: https://www.un.org/sustainabledevelopment/hunger/. Accessed 17 Aug 2020

Vahaa (2018) Vahaa. [Online] Available at: https://www.vahaa.co/. Accessed 22 Oct 2020

Verner D, Vellani S, Klausen A-L, Tebaldi E (2017) Middle East and North Africa refugee and host communities & frontier agriculture: climate smart and water saving agriculture technologies for livelihoods: unleashing climate-smart & water-saving agriculture technologies in mena. The World Bank, Washington, DC

Wallace JS, Powers JV (1990) Publications of the NASA controlled ecological life support system (CELSS) program, 1979–1989. National Aeronautics and Space Administration, Washington, D.C.

Wheeler RM et al (1990) Potato growth and yield using nutrient film technique (NFT). Am Potato J 67 (3):177–187

Whitfield S, Challinor A, Rees B (2018) Frontiers in climate smart food systems: outlining the research space. Front Sustain Food Syst 2(2):1–5

World Economic Forum (2020a) The global risks report 2020. World Economic Forum, Switzerland

World Economic Forum (2020b) The great reset: dynamic briefing. World Economic Forum, Switzerland

Yaşar B, Eren Ö, Öztürk H (2011) TARIMDA ENERJİ KULLANIMI VE YENİLENEBİLİR ENERJİ KAYNAKLARI. Çukurova University Faculty of Agriculture, Adana

Yegül Z (2020) Ürün Raporu Patates. Agricultural Economy and Policy Development Institute, Ankara

Yılmaz M (2010) Karapınar Çevresinde Yeraltı Suyu Seviye Değişimlerinin Yaratmış Olduğu Çevre Sorunları. Ankara Üniversitesi Çevrebilimleri Dergisi 2 (2):145–163

Yürdem H, Demir V, Çalışır S, Günhan T (2012) Tarımsal Sulamada Kullanılan Bazı Dalgıç Pompaların Sistem Etkinliği Açısından Değerlendirilmesi. Tarım Makinaları Bilimi Dergisi 8(2):117–126

5 The Enabling Technology: Internet of Things (IoT)

Abstract

The study's primary purpose is to investigate the Internet of things (IoT) structure as the enabling technology in smart homes and create the Open Digital Innovation Hub (ODIH) as a real-life example. Smart homes pose many benefits with the integration of IoT-based systems, such as energy, water and food consumption control and monitoring. Besides, it can recognise the waste and provide an opportunity to take customised measures according to the user's behaviour. In this study, we aimed to establish an automated network of smart devices for ODIH, which focuses on high efficiency in energy. When we examine the IoT's working principles, we first look into the compatibility between sensors, such as temperature and motion sensors and home appliances. Secondly, we inspect the connection of sensors to a gateway device, and finally, the processes that occur during device management. Then, we deal with efficient and innovative systems for the IoT ecosystem's connection, such as ZigBee and WIFI. We continue with architectures such as Home Assistant, together with the modules' user interface preferences where the latest information is transferred and processed with the middleware such as hub or gateway used during communication in IoT.

The author would like to acknowledge the help and contributions of Hakan Demirer, Furkan Albayrak, Kerem Şen, Mehmet Ali Sarsıl, Meliha Çağla Kara, and Raşit Yiğit Kaya in completing of this chapter.

5.1 Introduction

The Internet of Things (IoT) is the bridge between the virtual world and real-life physical activities. The main concept of IoT is to provide autonomy and safe connection and to enable data sharing between physical devices and real applications (Chopra et al. 2019a). The word "IoT" is 21 years old; however, connected devices' concepts have been around for a longer time period in the form of "embedded internet" or "pervasive computing" since the 70s. In 1999, Kevin Ashton mentioned the actual term "Internet of Things" during his work at Procter & Gamble due to the popularity of the Internet at that time (Knud and Lueth 2015). The Internet we use today expands into the real world, and it embraces everyday objects that define the vision of the Internet of Things concept. Steady improvements that have been happening in information technologies, microelectronics and communication technologies, as well as future developments that we may expect from these fields, all contribute to the vision of IoT (Mattern and Floerkemeier 2010). Smart cars and mobility, smart regions (cities, etc.), smart industries, smart homes and assisted living, public safety, agriculture, tourism, and energy & environmental protection as

S. Küfeoğlu, *The Home of the Future*, Sustainable Development Goals Series,
https://doi.org/10.1007/978-3-030-75093-0_5

parts of a vast IoT ecosystem are getting more attention as a number of IoT applications increase rapidly. Further potential applications are estimated by combining related technologies and technological approaches and concepts (Vermesan and Friess 2013). Thus, IoT is intended to be an ecosystem that evolves for the integration of environments and the services required to meet human life expectations in a better way (Lutui et al. 2018). With the rise of IoT, by linking multiple artefacts to the internet, internet traffic will rise significantly. This will inevitably lead to higher data storage needs with regards to a data centre. Such a large network would also trigger privacy and security concerns. There should be an appropriate architecture to meet all these demands. The IoT architecture is yet to be developed at present, but it is distinguished by a range of consistencies. This architecture essentially consists of five different layers: perception layer, network layer, middleware layer, application layer and business layer. Figure 5.1 illustrates the architecture of IoT.

The layer of perception includes artefacts of the physical world and the instruments of sensors. Such sensors may be radio-frequency identification (RFID) tags, barcodes, or even infrared sensors depending on the identification system used by the object. The key function of this layer is to classify the objects and gather their information with the assistance of sensor tools. This information may be dependent on the type of sensor regarding the position, inclination, humidity, motion, and orientation of the object.

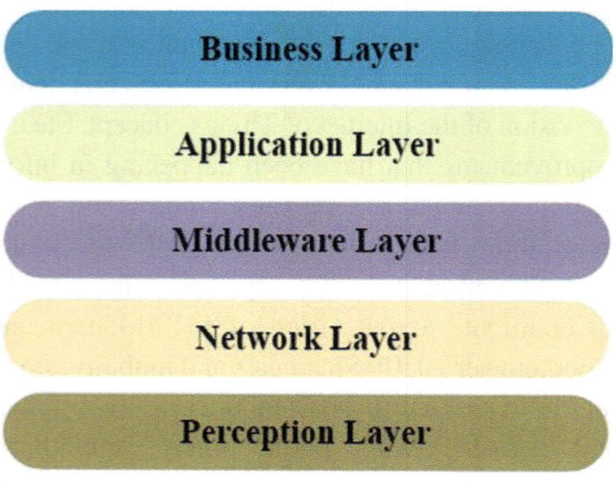

Fig. 5.1 Architecture of IoT

The network layer has the key function to safely relay the information that is obtained from sensor devices to the system's processing unit. Such knowledge can be transmitted either wired or wirelessly. The middleware layer is primarily responsible for the administration of service between IoT objects and applications. It can interact with the database as well. It gathers the information coming from the network layer and then stores the same information within the database. This layer processes the information and performs periodic calculations. Ultimately, by evaluating the tests, it makes automated decisions. The application layer provides global administration of the entire application depending on the object information processed in the middleware layer. Different applications can be introduced using IoT, such as smart homes, intelligent manufacturing, smart healthcare services, smart transportation, and more. The business layer communicates with the entire IoT network, which involves specific applications and services. This layer produces diagrams, flowcharts and business models based on the data from the application layer (Chopra et al. 2019b).

An object is considered a smart object if it has some of the following functions (Fig. 5.2).

Smart objects are not only things that make the IoT applications intelligent but also a network that connects and helps to communicate with each other. Some of the properties of a smart network are defined in Fig. 5.3.

The IoT applications exist in a large range of contexts. Five of these applications are summarised here as they provide essential background.

Firstly, smart homes, or also known as home automation or domotics, is a prominent subject of research in IoT. Home automation cannot be considered separate from IoT, as IoT offers ways of performing daily tasks. As indicated by Raj and Raman (2017), all passive house electronics are getting digitized, connected wirelessly and able to communicate. The first step to making the home environment smarter is more capable and smarter appliances. Energy costs can be significantly reduced by instant monitoring of the energy use of all smart home appliances, autonomous lighting, temperature and environmental

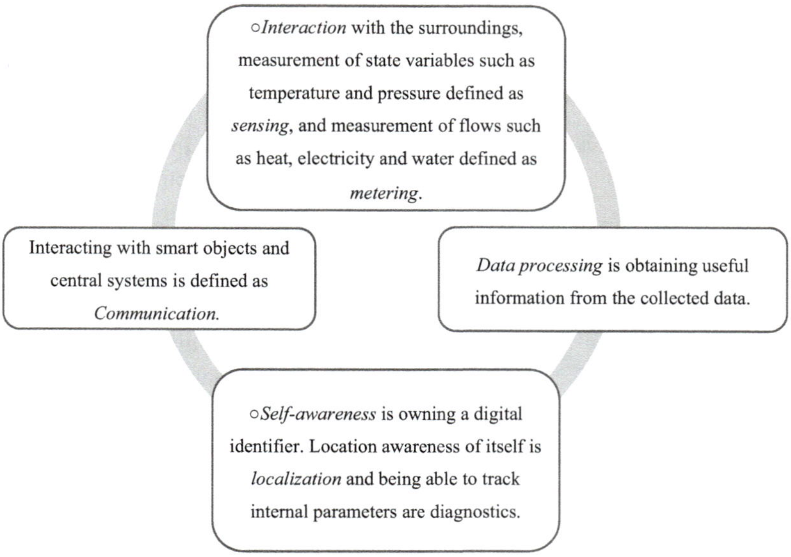

Fig. 5.2 Smart object functions (Miragliotta et al. 2012)

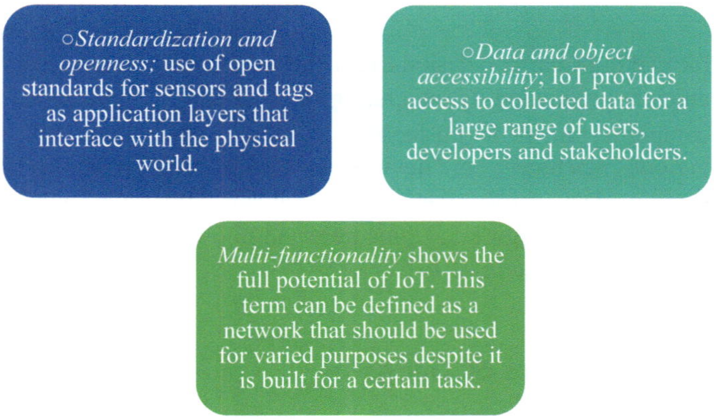

Fig. 5.3 Smart network properties (Miragliotta et al. 2012)

sensors. In terms of smart security, police, fire department, and homeowners can be automatically notified in collaboration with a network of alarm and sensors surrounding the house.

Secondly, the use of IoT in manufacturing can be examined under two categories; manufacturing and user experience. Manufacturers use IoT to automate their processes, track their machinery, and decide if machines need maintenance. Furthermore, on the user-experience side, IoT devices that are linked to a network are used to gather data from clients. Then, examine the gathered data to improve the product and maybe fix problems (DeNisco-Rayome 2017).

Thirdly, IoT supported Remote Patient Monitoring (RPM) could allow patients to be monitored while they are at their homes. With smart wearable devices gathering data about residents' vital statistics (e.g. blood pressure), caregivers can provide the first-step medical examination, regardless of distance (Greengard 2015; Sagahyroon et al. 2018).

Fourthly, IoT is the first step to introduce smart city applications with a large number of sensors in city infrastructure producing real-time data (Latre et al. 2016). Modern smart cities focus on sustainable and effective solutions specifically for energy management, transportation, healthcare, governance, and more to fulfil the demands of urbanization (Ejaz et al. 2017).

Lastly, agriculture is another application of IoT, which is worth mentioning. Farmers and the producers will benefit by gaining more options

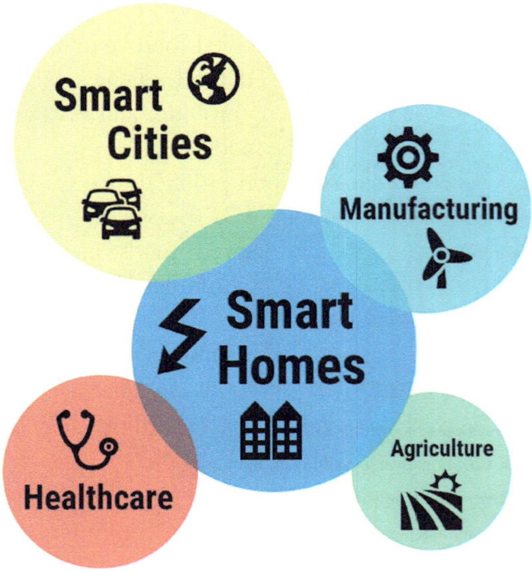

Fig. 5.4 Applications of IoT

for making smarter decisions, reducing costs, and efficient production by integrating IoT and cloud computing in their businesses, together with various machinery or devices gathering and sharing data across the cloud (Namani and Gonen 2020). Figure 5.4 illustrates the applications and ecosystem of IoT.

5.1.1 Internet of Things and Efficiency

There is a growing need for efficient systems for controlling and optimising the energy and resource consumption of households. This section suggests a monitoring and management framework for energy, water and food consumption efficiency, with the help of the newly emerging Internet of Things technology. Potential changes for decreasing resource consumption and increasing efficiency can be established with the implementation of the IoT (Chen et al. 2018).

Through the use of sensor devices and internet-based information systems, IoT technology has created an exciting opportunity in recent years to develop powerful tools for energy monitoring and management. With an effective power management system, energy hotspots can be detected, and the energy output can be enhanced in real-time in relation to their process (Chen et al. 2018).

Chen (2018) introduced a framework based on energy monitoring and management for a machining workshop. The unused auxiliary equipment can shut down depending on the tasks of the machinery. Real-time energy-focused planning can be applied to boost energy efficiency. Furthermore, by tracking the energy consumption frequency in real-time, it is possible to minimise energy consumption at peak time by making an effective load balance. IoT aims to improve energy usage visibility and understanding thanks to smart sensors and smart meters in the system. As a result, data from complex systems on real-time energy consumption can be easily obtained and then analysed to enhance energy-aware decision-making. These kinds of solutions produce an ample amount of knowledge by capturing large amounts of energy-related data, almost in real-time. For this reason, it is critical to plan in advance how to integrate these IoT energy monitoring solutions into the whole of the energy management of the environment (Shrouf and Miragliotta 2015).

Furthermore, IoT is regarded as a viable approach to be a technical solution for the tracking, evaluation and rapid decision of infrastructure provision before failure. A smart water management network could only be envisaged across the entire infrastructure arrangement by implementing IoT. Once IoT is completely integrated, the water monitoring system functions as a smart entity, in which problems are immediately discovered and treated without incessant observation by third parties (Ramakala et al. 2018). The key goals of online monitoring of water quality with IoT include assessing important water quality parameters such as microbial, physical, and chemical properties, detecting anomalies in parameters, and providing early warning hazard recognition. The control program can also offer an overview of the recorded data in real-time and recommends necessary remedial steps (Geetha and Gouthami 2016).

For food nexus, the biggest challenge in agriculture is resource management. For sustainable food production, traditional methodologies can be combined with state-of-the-art technologies such as the Internet of Things and Wireless Sensor Networks (WSNs) to enable different applications in the Modern Agriculture domain. New revolutionary IoT technologies tackle agricultural problems by increasing the efficiency, quantity, productivity and cost-effectiveness of farm production. IoT sensors capable of delivering producers with data on crop yields, irrigation, insect infestation, and soil fertility are useful for sustainable production, and they provide accurate data that can be used over time to enhance farming techniques. With this vigorous build-up of IoT in Agriculture, smart farming technologies are gaining grounds intending to have 24/7 insight into soil and crop wellbeing, equipment in service, storage conditions, livestock behaviour and energy and water consumption rates. Farmers, ranchers and conservationists now, more than ever, need a system for using and conserving resources more effectively. The most successful way to achieve this is by actionable data, and using connectivity from machine to the machine makes it easy and inexpensive to start gathering these data. To summarize, the Internet of Things in agriculture helps raise crop production by monitoring and regulating the operations (Satish et al. 2017).

Figure 5.5 illustrates these three components of the nexus mentioned above.

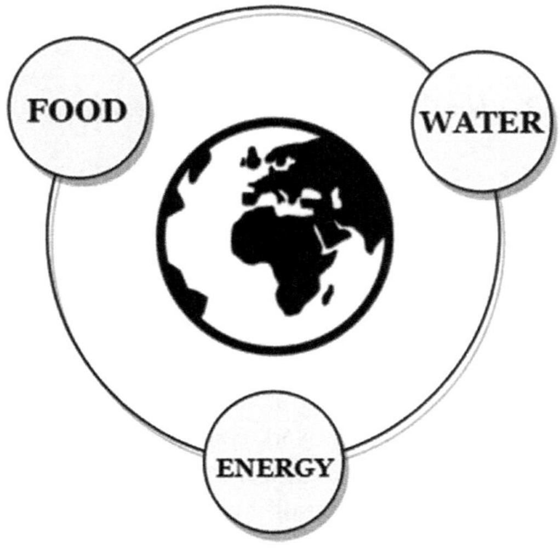

Fig. 5.5 Water-energy-food (WEF) nexus

5.1.2 The Place of Demand Response, Machine Learning and Artificial Intelligence in Internet of Things

In an intelligent environment, machines operate in concert to help users by leveraging knowledge and data that are embedded in network-connected devices in a simple way to carry out their daily activities. Artificial Intelligence applies to computer systems adaptive and attentive to people's presence (Madakam et al. 2015). At the same time, given the large consumer base and the growing proliferation of smart appliances and IoT devices in residential homes, the demand response for residential and Small and Midsize Business (SMB) sectors are very desirable for utility providers to have greater flexibility to manage energy and supply balance for more reliable and productive power grid service (Mashima and Chen 2016). Moreover, seeing that IoT will be one of the most important sources of new data, data science will make a major contribution to making IoT systems more intelligent (Mahdavinejad et al. 2018). IoT, along with machine learning, offers a way for publisher/subscriber interactions where small sensors generate knowledge that is validated by machine learning algorithms. In addition, machine learning algorithms produce sophisticated ideas that are delivered to IoT platforms for device adjustment and elevation (Din et al. 2019).

5.1.3 Capabilities and Future

The idea of obtaining and storing every relevant data, processing them and providing valuable outcomes allows IoT to emerge itself as a strong modifier of contemporary urban and rural life. After the first industrial revolution, we may claim that machines and humans will get more related because of gradually blurring borders between them (Greengard 2015). The disappearance of these borders gives rise to building stronger direct or indirect bonds between nature and humans. A direct way to interact with nature is air, and its pollution is a significant problem today. Toxic gases and detrimental particles such as carbon dioxide and soot are emitted from both factories and farmlands could be monitored by air quality data analysis (Patel et al. 2016). Thereafter, the required precautions could be taken immediately. In this project, air analysis and air conditioning work cooperatively to idealise the breathed air inside the self-sustaining compound thanks to the Home Management System (HMS). The most comprehensive field of human interaction is the energy drawn from nature. Especially electricity appears in every aspect of an individual's daily life. Its measurement, management, pricing, and tracking are highly complicated because of the large amount of data and traditional residential power network. Open Digital Innovation Hub project provides a residential solution for energy, water, food and air quality management. Every appliance in the compound will be coupled with IoT to collect and transfer data to the HMS to constitute usage behaviour data of the resident. The HMS will then analyse these data and take several actions such as warning the resident about spare usage of lightning or unnecessary power consumption and shifting energy consumption of various appliances (load shift). IoT appliances could occupy a vast field of residential appliances independently, but in an inclusive network, with loads of data and uninterrupted processing, the system gets smarter gradually and develops its own intelligence (Greengard 2015). In the ODIH project, we aim to minimise human decision making and shift the compound's management to the intelligent HMS. Greengard (2015) states that the spread of area and depth of the internet above human interaction will redefine how individuals live in the future, and Open Digital Innovation Hub Project will be a playground to test the interaction between humans and intelligent in-home management systems, the HMS.

The vision of IoT paves the way to enhance methods for efficient and sustainable energy, water and food management. By the dynamics of layers in architecture, the collected and processed data will be delivered to the user with an appropriate interface such as the display of HMS, websites or smartphone applications. In this project, instruments that were mentioned before are going to be fully integrated with Open Digital Innovation Hub to inspire the urban life of the future.

5.2 Aim of the Study

The study aims to research and build an automated network for Open Digital Innovation Hub among household appliances to minimise the decision process of residents upon chores, providing efficiency in utilized resources and consumables and analyse each data obtained by monitoring the environment with the Internet of Things.

The main objectives of the research are:

1. Making a load shift process possible by tracking electricity costs through smart counters and arranging working hours of the household appliances.
2. To avoid possible faulty and inefficient human decisions by diminishing human intervention and authorizing the Home Management System.
3. Remote controlling through the Internet of Things based on cloud systems and various interface protocols.
4. Setting up a self-sustaining house.
5. To design an ecosystem compatible with WEF Nexus.
6. Determine optimal utilization techniques of IoT compatible devices in order to digitalise a regular house in the smartest way.

5.3 Methodology and Materials

5.3.1 Setting an Intelligent Home System

In general, an intelligent home control system includes an illuminated management system, a room temperature control system, a kitchen process control, depending on the mode. Typically, a house automation network consists of a gateway system node, a home control subnetwork, a home gateway, and a control hierarchy. The home control sub-network via the communication module finishes the interconnection between different home's internal equipment nodes. The user interface is the contact module by which a typical household appliance which converted to an appliance management network. In order to ensure proper contact between different household appliances, household appliance networks, and gateways, a series of full communication protocols must be created (Zhao 2010).

5.3.2 Working Steps of IoT

5.3.2.1 Sensors/Devices

Sensors or devices gather data from their surroundings. This may be as easy as a temperature reading or as complicated as a full video stream. Several sensors can be clustered together, or sensors can be part of a system. However, whether it is a stand-alone sensor or a full device, the raw data is collected from the field or from the surroundings in this first step.

A. Gateway

The central unit of the intelligent home communication system is the gateway (or so-called home server). Gateway is a global function device that acts as a network coordinator to lead network setup and track its regular operation. It has more storage spaces to finish initializing the network, gathering data, and managing devices (Zhao 2010).

B. Device Management Platform

These platforms are focused on the management of remote devices and the optimization of network resources. By providing tools that enable data distribution, system identification, and network diagnostics, they accumulate network capabilities and optimize network resources. If a particular gateway is overloaded in the network or has a low battery, the platform should realize and take appropriate action. Plug and play are the other parts of this type of platform, so a simple user setup is needed when new devices join the network or get repositioned. It is important to remember that system management typically requires additional software installed on the unit (Da Cruz et al. 2018).

5.3.2.2 Connectivity

Sensors and devices can connect to the network via various methods, including cellular, satellite, Wi-Fi, Bluetooth, Low Power Wide Area Networks (LPWAN), or direct internet connectivity through ethernet. Additional protocols can be seen in Fig. 5.6; however, only the specific ones are covered in this chapter.

A. Wireless Sensor Networks

The Wireless Sensor Network refers to a collection of spatially distributed and dedicated sensors configured to track and document physical environmental conditions and coordinate the data gathered at a central location.

B. LPWAN (Low Power Wide Area Network)

LPWAN technology performs best in cases where computers need to relay small data over a wide area while retaining battery life for many years. It operates on lightweight with inexpensive batteries lasting for years, and it has an operating range that usually exceeds 2 km in urban settings for long-term implementation and monitoring of smart lighting, smart grid, and asset tracking sensors and meters (e.g., water metering, gas detectors, smart agriculture, and remote door locks).

C. Mesh Networks

Wireless Fidelity (Wi-Fi) (IEEE 802.11 a/b/g/n/ad/ac) consumes too much energy, does not cover a large area, and does not accommodate a large number of end devices. For this reason,

Fig. 5.6 Commonly used communication protocols for IoT

alternatives are being implemented in IoT solutions, such as Bluetooth 5 and IEEE 802.15.4. Bluetooth 5 is the newest iteration of the Bluetooth Popular standard. Bluetooth 5 also supports IP networks, as with Bluetooth 4.2. IEEE 802.15.4 is a low-rate wireless personal area network specification defining the Open Systems Interconnection (OSI) model's PHY and MAC layers. The most popular IEEE 802.15.4 implementations are IPv6 over low power personal area wireless networks (6LoWPANs) and ZigBee. The ZigBee Partnership was established and maintained by ZigBee. It is renowned for its mesh topology but also supports other topologies, including star and tree. 6LoWPAN's most notable advantage is that it supports the native IP networks. When using ZigBee or standard Bluetooth, communicating with the Internet requires a gateway, which increases the network traffic. In order to connect to the Internet, all technologies which do not support IP natively use a similar definition. ZigBee acknowledged the importance of IP networks and released ZigBee IP using several 6LoWPAN concepts, especially the fragmentation of headers and compression schemes. For this reason, using ZigBee in ODIH will be preferable.

D. Wi-Fi

Compared to other approaches, Wi-Fi-based technology has three benefits. Firstly, as a ubiquitous system, it is optimized for practical implementation. With the rapid development of Wi-Fi technology, the indoor coverage of wireless networks has become increasingly widespread. So, we do not have to mount complex sensors or RF connections. Second, Wi-Fi-based methods employ pattern extraction in the physical layer of wireless signals and, therefore, do not pose privacy concerns. Third, it can provide reasonably reliable sensing with limited infrastructure specifications, which is a reasonable trade-off between accessibility and efficiency. For coarse sensing applications such as occupancy detection, most devices use Channel State Information (CSI) curve similarity to achieve the objective. It is implemented entirely on IoT machines.

In wireless technology, state information on channels is a representation of a communication link's channel properties. They are affected by the physical environment during the transmission of wireless signals, which leads to reflections, diffractions, and dispersal. CSI may be able to explain these phenomena. On the physical layer, modern Wi-Fi networks follow Orthogonal Frequency Division Multiplexing (OFDM) and comply with the IEEE 802.11 n/ac protocol that requires multiple transmitting and reception with antennas of multiple input, multiple outputs (MIMO) communication. Therefore, CSI shows fine-grained wireless signals features on each communication subcarrier, combining time delay, amplitude attenuation, and phase change of several paths. Since the Received Signal Strength Indicator (RSSI) is just a multi-path signal superimposition, CSI inherently possesses higher resolution (Yang et al. 2018). In a nutshell, Wi-Fi can also be used as an alternative to ZigBee in ODIH.

5.3.2.3 Data Processing

When the data enter the cloud, installed software analyse it. This may be fairly basic, such as ensuring that climate measurements are within a reasonable range. However, identifying objects (for example, intruders in your home) using computer vision on video can often be very challenging. However, what happens when the temperature is high, or if there is a visitor in the house? That's when the user steps in.

5.3.2.4 User Interface

A message from the user can be sent as a warning to the system, such as an e-mail, post, or notification. A text message, for example, when the air temperature in the cold storage space of the house is too high. An app can also have a system that helps them to control the system proactively. A user can choose to use a mobile app or a browser to search for live footage in their household. According to the IoT software, the user would also operate and affect the system. The user can adjust the temperature remotely in cold storage with an app on his/her smartphone.

Moreover, the system can automatically perform those actions. Instead of waiting for the consumer to adjust the temperature, the scheme might do it automatically from predefined laws (or AI). The IoT system can alert the local authority immediately, rather than only calling the resident to warn of an intruder. In the next steps, as the Message Queuing Telemetry Transport (MQTT) broker, the cloud server is responsible for the machine learning system and message delivery. Finally, the sensing results are displayed on the user terminal, such as mobile phones and laptops (Yang et al. 2018).

A. Middleware

Middleware in IoT is serving as a translator. Devices (APP) communicate through their APIs (application programming interface), so each APP has its API. Every APP must understand any other API without a middleware.

Application enablement platforms concentrate on implementing and integrating external applications. They include data management and visualization means that accelerate the creation of applications and encourage enterprise systems integration, including customer relationship management (CRM) and enterprise resource planning (ERP). These platforms often protect user data and allow the sharing of information between different devices/applications. This type of platform is also known as the platform for IoT middleware. For instance, the Artik cloud is a platform that Samsung is developing. It provides various applications as well as control of devices. It supports communications with its server between MQTT, Representational state transfer (REST), Websockets, and Constrained Application Protocol (CoAP). One of the benefits of Artik Cloud is that it can be easily incorporated with common IoT apps and devices such as the Amazon Echo and Google Home. Their business model is the Platform as a Service (PaaS). PaaS solutions benefit because they are stored in the cloud, and authorized users can access the server data from anywhere in the world without thinking about installing or maintaining the infrastructure.

B. Modules of Middleware

i. Interoperability Module

The IoT is a diverse framework, and the middleware program is the integrator. It can also have an "API" that allows the software to report functionality to other programs without disclosing the individual code. API requests made by things/applications can be carried out with any protocol, so the middleware should at least support the most common IoT application protocols, such as CoAP, MQTT.

ii. Context Module

IoT environments must adapt to the environment, and context may have an essential role in this regard. Artificial intelligence is the best to run in this module resulting in IoT that is imagined, but external APIs can be utilised in the middleware platform to accomplish this aim. Some middleware suggestions, such as Linksmart and Open-IoT, currently rely on ontology to supply semantic interoperability between sensed data (Da Cruz et al. 2018).

5.3.3 Cloud-Based IoT System and Its Implementation

The storage, computation of the sensor data, and the applications to be implemented fed by these data will be enabled by the cloud platform. The usage of cloud-based IoT platforms eases application growth and development, vanishes the need for physical infrastructure, simplifies the management of things, and decreases maintenance expenses. The provision of unique capabilities to easily control devices, direct communication with things, storage of the data coming from devices, and transfer events are the significant advantages of the cloud platform. This platform has three considerable issues to analyse: storage, data processing, and communication issues that work with an application programming interface to achieve interoperability. Figure 5.7 illustrates these issues regarding cloud-based IoT platform.

5.3.3.1 Storage Issues

The primary issues associated with cloud data storage are mainly sharing, storing, visualizing, searching, and collecting the data. IoT integrated cloud is a solution that enables the system to acquire virtually limitless capacity coming at a low cost.

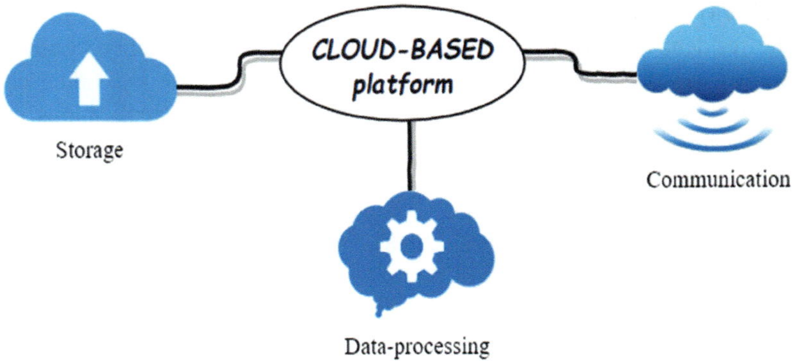

Fig. 5.7 Issues relating to cloud-based platform

In cloud-based IoT systems, data can be handled homogeneously by standard APIs, secured by the use of top-level protection, and directly accessed and visualized from anywhere. Besides, this platform offers reliability and high accessibility, easy deployment, high data backup security, archiving, and disaster recovery, which finally enables the smart management system to handle the issues regarding storage and access to IoT data (Pourqasem 2019).

The data storage framework should handle various types of data collected from many different devices, such as RFID readers, monitors, thermometers, etc. These data can be different in terms of volume, accessing methods, data structures, and many other aspects, so that utilizing a single method cannot be adequate to store and access these data efficiently. What is more, the volume of the data may rise rapidly, and thereby the processing of the data with a high throughput becomes a requirement for the storage framework.

Several modules are included in the data storage framework; file repository, database module, service module, and resource configuration module. Figure 5.8 demonstrates the modules included in the data-storage framework.

A. File Repository

Unstructured files in a distributed environment can be stored by file repository utilizing the Hadoop Distributed File System (HDFS). To enforce the management of the versioned model files and the isolation of tenants' data, there is a version manager and a multitenant manager. The file repository's ability to handle small files is enhanced by a file processor.

B. Database Module

This module incorporates many databases and uses both NoSQL and relational databases to manage structured data. To hide differences in databases during implementing and interfacing, the database module comprises unified APIs and object-entity mapping characteristics, which brings about simplification regarding the development of data access and migration of applications of databases.

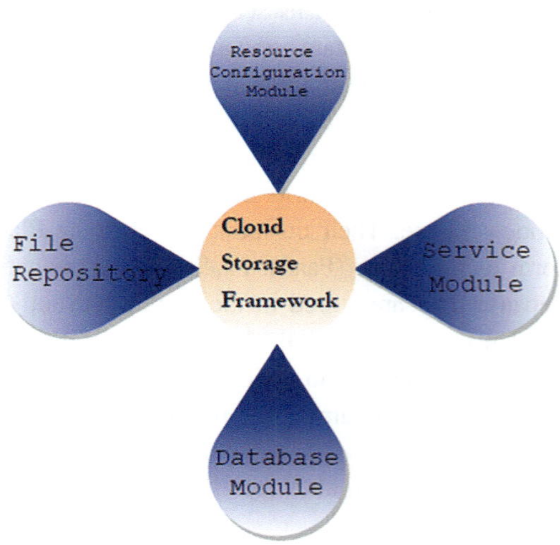

Fig. 5.8 Modules included in data storage framework

C. Service Module

The automatic operation of RESTful service can be constructed by the service module. The service module implements the extraction of metadata through configuration. Then, according to extracted metadata, it maps these metadata to the data entities and files stored in the databases and file repository. Finally, a related RESTful service is generated by this module.

D. Resource Configuration Module

This module supports dynamic and static data management regarding predefined meta-models. Therefore, the configuration of data resources and related services can be implemented in terms of tenant requirements. Moreover, load-balanced, isolated preferences and various disposing of data mechanisms can also be executed (Jiang et al. 2014).

5.3.3.2 Data-Processing Issues

The virtualized limitless cloud computing capabilities and its on-demand model enable IoT to work and analyse unprecedented complexity with more flexible mechanisms. Data-driven decision-making and prediction algorithms are utilized to handle complex data bringing about reduced risks and incremental incomes. Moreover, the real-time process, scalability, sensor-centric applications, task offloading focusing on energy savings are the other utilities that cloud-based IoT platforms provide more collaboratively (Pourqasem 2019). The stored data on the cloud is used to run applications with artificial intelligence algorithms. Analysing data sourced by sensors, learning from the user's general behaviour pattern, and adopting AI algorithms can help attain intelligent control. Centralized management can make electronic decisions, such as tracking, enhancing comfort, convenience, managing environmental conditions, and supplying the information required.

Data processing, decision-making, prediction, voice recognition, activity recognition, and image recognition are the six clusters utilized with AI in smart home systems. Intelligent interaction, security, energy management, and healthcare are the four functions deployed with multiple products. Figure 5.9 summarises these functions.

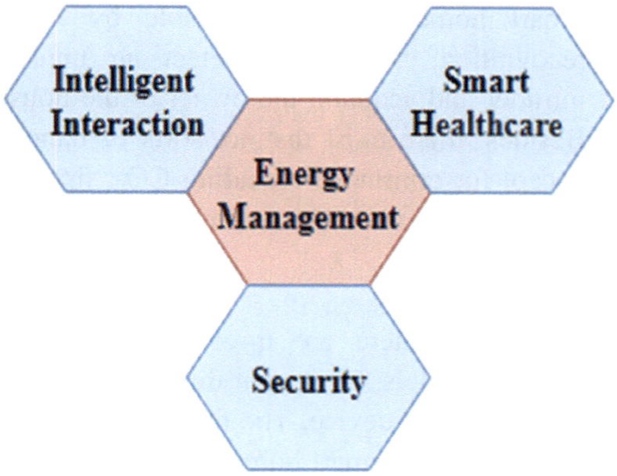

Fig. 5.9 Overall Functions of IoT

AI with decision making is the cluster utilized in smart home energy management. The energy consumption of smart appliances is coordinated to realize higher efficiency. Then AI predicts daily electricity demand by analysing energy consumption patterns and their associations with surrounding conditions. Relation of activities and available appliances are regularized by the aid of AI with activity recognition and then make suggestions to users if energy waste is detected.

- Smart-home healthcare utilizes four AI functions; voice recognition, decision-making, image recognition, and activity recognition. By applying an unsupervised clustering algorithm, recurrent neural network model, and genetic algorithm, AI systems can continuously monitor sensor data to monitor and detect shifts within the resident's behavioural pattern and emergency mode activation.
- Prediction-making and voice recognition can support applications of smart home intelligent interaction. Due to the vast number of smart home devices, intelligent interaction becomes more profound and preferable to implement. Artificial neural networks are mostly used to classify inputs from users to constitute a natural dialogue, thereby enabling users by text or voice commands.

- Smart home security is viable by image recognition, which can detect an unusual intruder and acquaint the owner of the house. Besides, there exist distinct sorts of dangers except for criminals, including CO_2, fire, etc. HMS can monitor the sensor data and detect alarm sounds.

Altogether, there are three basic types of interaction models. Users can directly command each smart home device. The first is where users directly give each smart home device commands where the specific device itself is utilized by AI embedded in every device. This pattern is mostly used in smart home energy management, healthcare, and security. The second is applied by giving instructions to the AI by the users, and then AI controls every device following the directives. Smart home device management and intelligent interaction are implemented by benefiting this pattern (Qi 2019).

5.3.3.3 Communication Issues

A dedicated gateway has a crucial role in forwarding the sensor data to the cloud. It is responsible for handling communication protocol issues between two services, including the cloud-based HMS software and the smart appliances. Widely used communication protocols in recent times perform hole-punching and maintain continuous connection with cloud platforms. IoT gateway constitutes an outbound request to the hosted cloud-based HMS software to implement hole-punching. The building router transfers the request and stores the routing history on a routing table. The request's entry on the routing table temporarily becomes an unprotected open port in the router's firewall rules. As the operations are based on the routing table entry, once the HMS software's response is ready, it is transferred back to the appliance. Then, cloud-based HMS acquires a bidirectional tunnel with the gateway provided that an entry is present on the routing table. The gateway provides periodical keep-alive messages to the cloud-based HMS system to ensure that it is available to any coming request. The protocol to be used in this structure then becomes incredibly profound to be investigated.

A number of platform protocols are popularly used, and Table 5.1 summarises these protocols with a number of features, namely transport, Quality of Service(QoS), architecture and security (Nugur et al. 2018).

Data transfer protocols can be divided into two groups, namely, file transfer protocols and messaging protocols.

File transfer protocols are ideal for web applications, while protocols for messaging are better suited for IoT platforms. File transfer protocols are not appropriate for IoT in the current format, as IoT nodes are superficial sensor nodes used to collect raw data from the system application scenario.

IoT network implements its application in many protocols, namely, COAP (Constrained Application Protocol), MQTT, AMQP, and XMPP.

The next generation of IoT applications will be supported by emerging messaging technologies, where each of them is capable of connecting various smart devices in a distributed network. Small sizes of messaging, message management, and lightweight message overhead qualities of MQTT and CoAP make them more adequate in addressing these needs. Figure 5.10 illustrates a network architecture with possible protocols.

A. CoAP

The CoAP (Constrained Application Protocol) is a synchronous application-layer request/response protocol. CoAP aims to utilize RESTful interactions by small devices with low power, computation, and communication capabilities. CoAP is capable of becoming the standard protocol to enable smart devices to interact and support IoT applications. The general characteristics of CoAP are listed in Fig. 5.11.

B. MQTT

Message Queue Telemetry Transport (MQTT) mainly aims for lightweight M2M communications via an asynchronous publish/subscribe pattern. It suits M2M, WSN and ultimately IoT ecosystem in which sensors and actuators can

Table 5.1 Comparison of various cloud protocols

Protocol	Transport	QoS	Architecture	Security
Message Queuing Telemetry Transport (MQTT)	TCP	Yes	Pub/Sub	TLS/SSL
Constrained Application Protocol (CoAP)	UDP	Yes	Req/Res	DTLS
Representational state transfer (RESTFUL)	HTTP	No	Req/Res	HTTPS
Advanced Message Queuing Protocol (AMQP)	TCP	Yes	Pub/Sub	TLS/SSL
Web socket	TCP	No	Req/Res Pub/Sub	TLS/SSL
Data Distribution Service (DDS)	TCP/UDP	Yes	Pub/Sub	TLS/SSL
Secure MQTT (SMQTT)	TCP	Yes	Pub/Sub	Own security
Extensible Messaging and Presence Protocol (XMPP)	TCP	No	Req/Res Pub/Sub	TLS/SSL

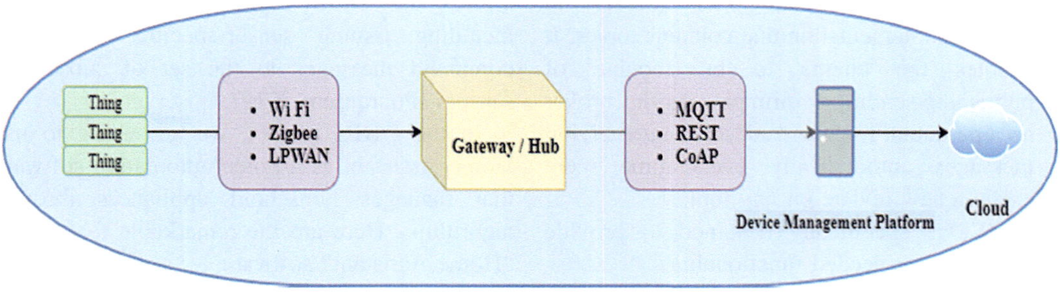

Fig. 5.10 IoT network architecture

Fig. 5.11 General characteristics of CoAP

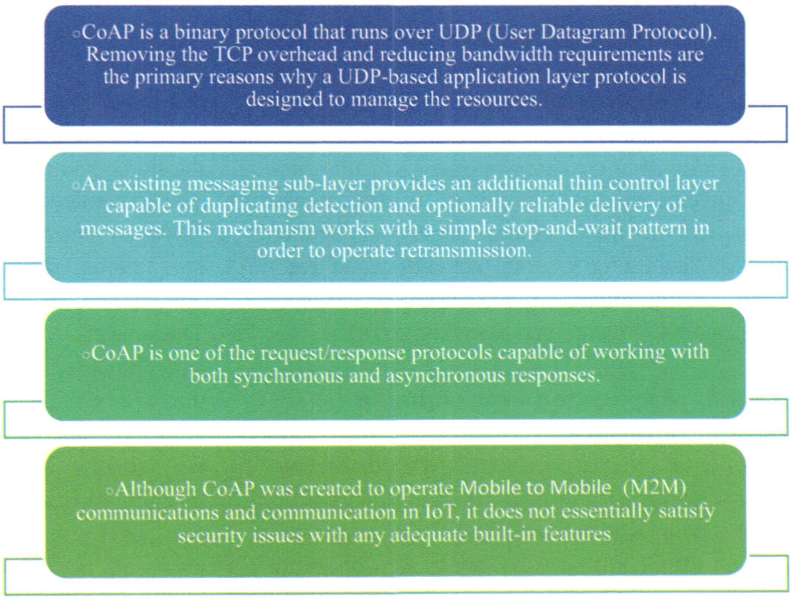

communicate with applications via the MQTT message server. It runs on top of the TCP stack targeting to connect embedded devices, thereby creating a network with various applications.

General characteristics of MQTT are listed below:

- It utilizes a publish/subscribe pattern to ensure flexible transition bringing about simplicity in implementation.
- Capable of routing for small, cheap, lower and low memory devices in low bandwidth and vulnerable networks, which makes it an ideal messaging protocol for IoT and M2M communications.
- There is a broker component (server) in MQTT implementation that contains topics. It enables the clients to be capable of publishing/sending information to the broker on a particular topic and subscribing/receiving messages automatically every time there exists a new update on any topic.
- MQTT is specifically designed to provide three distinct needed functionalities.

Once-and-once-only assured delivery mode to provide reliable transfer of a message through the path from a sensor node to a terminal application.

The availability of maximum lightweightness across any communication medium, minimizing every message's overhead; thereby, most of the remote telemetry is implemented over low bandwidth, high-cost networks.

Convenience and simplicity in the implementation of the protocol on the embedded devices, including sensors, actuators, and gateways, are the crucial aspects in the selection of the protocol (Tukade and Banakar 2018).

5.3.3.4 Application Programming Interface

Cloud-based IoT works with APIs to achieve interoperability. The sensor APIs and client APIs are the main APIs that are utilized. The sensor API is used for cloud-based IoT registration of sensor modules, publishing data, and receiving control messages sent by applications or clients. Thus, it enables the end-users to define a sensor-specific data format. After the sensors are connected and registered to cloud-based IoT, these sensor APIs can publish data that is collected from a physical sensor or other data sources. The functions provided by the sensor API enable the submission of data to the cloud-based IoT platform.

Differently, to register a client with the cloud-based IoT and subscribe to sensors of interest by URL, client API comprises some required classes. Subscribed clients can obtain information regarding the sensors connected to the system by utilizing the names or ID of the target sensor. Moreover, clients may have management capabilities on sensors by realizing these APIs, including issuing sensor-specific control or command messages to the set of subscribed sensors (Pourqasem 2019).

In the ODIH project, we are going to use Home Assistant as a home automation software that manages household appliances through algorithms. Here are the remarkable features of "Home Assistant" software;

- The home assistant provides a free and open-source hub.
- It enables the IoT network to have 100% local home automation.
- Management by using the web interface integrated into Home Assistant.
- Ease of creating and restoring full backups of the home's entire configuration.
- The convenience of control and observation of Z-wave connected devices is a popular wireless communication protocol for home automation.
- Thermostats, switches, locks, lights, covers, and climates are the basic house appliances that will be picked up automatically and easily after the platform's configuration.
- By utilizing MQTT, which is one of the machine-to-machine top connectivity protocols, light publish/subscribe messaging transport will be implemented (Home Assistant 2009).

5.3.4 Water, Energy and Food Security (WEF) Nexus and IoT

5.3.4.1 Energy Management, Consumption and Efficiency

Concepts for smart homes are built through the implementation of the IoT. IoT-based Home Energy Management systems provide several benefits, the first one being energy usage monitoring. In order to achieve that, the Demand Side Management system gathers data from smart plugs and IoT devices to calculate the total house demand as a function of House Energy Management, and all that information can be observed from the system interface. Secondly, finding energy waste resources can be achieved by comparing energy consumptions. Demand Side Management system can prioritize IoT devices and smart plugs by comparing energy consumption, and the highest one is the top priority. When the energy demand is exceeded the energy supply, the lowest priority should be turned "OFF". The third and last one is reducing energy costs. In order to reduce energy cost, several Demand Side Management techniques can be considered (Fig. 5.12).

Peak clipping and valley filling techniques are used for equalizing peak and valley load levels to alleviate peak demand load. The Peak Clipping technique reduces total energy demand and peak demand, but the valley filling technique increases total energy demand, and peak demand stays still.

The energy conservation technique is based on reducing overall energy demand. It decreases both total energy demand and peak demand. On the other hand, load building or load growth increases total energy demand, and peak demand may increase.

The flexible load technique is related to smart grid systems that detect flexible load demands and control the demand during critical periods. The flexible load technique decreases peak demand, and total energy demand may decrease.

The most reliable technique for reducing energy cost is the load shifting technique. Load shifting can be defined as shifting the total energy demand according to the energy usage chart. Energy price increases proportionally to total energy usage. During peak load, shifting the energy demand to the off-peak load period does not change total energy consumption, but this strategy redistributes the energy utilization time (Logenthiran et al. 2012). Several solutions can be considered in order to shift load, such as reducing energy consumption at peak time and battery usage to store energy. The load categorization of appliances is necessary to reduce

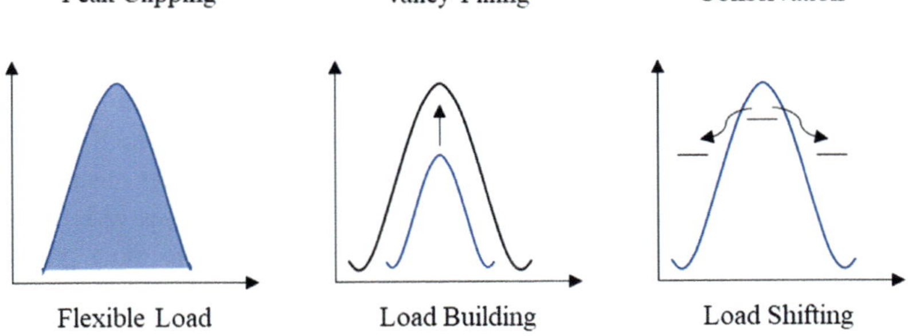

Fig. 5.12 Demand side management

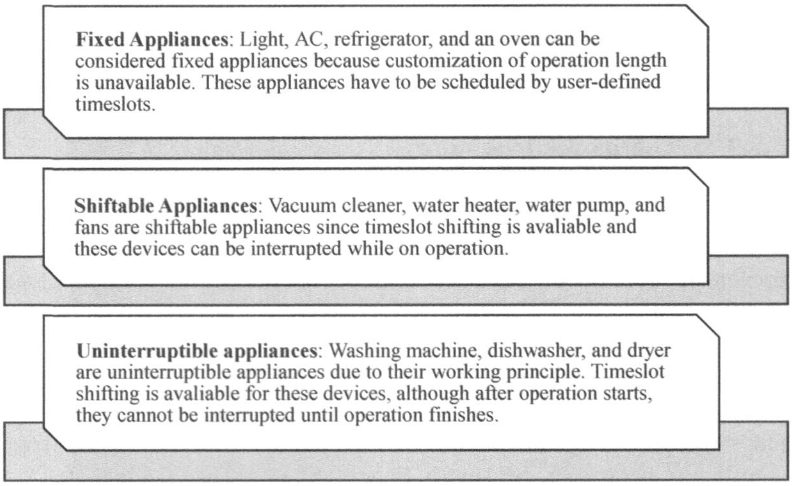

Fig. 5.13 Household appliances (Kazmi et al. 2017)

energy consumption. Appliances can be classified into three categories; fixed, shiftable, and uninterruptible appliances and details are given below (Fig. 5.13).

According to this categorization, reducing energy costs can be done by reducing shiftable and uninterruptible appliances during the peak load period. However, this strategy is not practical because energy saving is limited, and the user's life quality decreases. Moreover, battery usage for load shifting is considered in the ODIH project. To do that, we will use a battery to save energy during off-peak load and use that saved energy inside the battery during the peak load period.

5.3.4.2 IoT and Agriculture

Thanks to smart agricultural lighting, growing strawberries indoors in the middle of winter has become possible. Agricultural products with the desired characteristics are grown thanks to LEDs that can be modified at the desired wavelength. Plants grow five times faster and naturally, with systems produced with smart agricultural lighting and a water tank with an ideal climate. Also, plants can be grown cleanly on the bench without contamination because the system is closed. Intelligent agricultural lighting can generate different light wavelengths, and it is possible to change the desired colour and energy light to our increasing tastes. Many fruits and flowers are grown more effectively with daylight and red LED light, with the quick growth of white LEDs and high yield growth of blue LEDs. In vertical farming, IoT-supported technologies allow phones and tablets to track and study agricultural areas. IoT vertical farming plans for collecting, processing, storage, and disseminating data as required to carry out the plant's operations and functions. IoT vertical farming in agriculture aims to optimize productivity. The cost decreases as natural resources are used at the appropriate level.

A. Intelligent Lighting System

The Intelligent Plant Growth Lighting Device (ILSys) is a technical breakthrough that can be used in plant or animal biological studies. It can also be used to assess plant growth's precise effects on the quality and quantity of light. An experiment to investigate the effects of pulsed light at different frequencies in tomato plants was performed to determine device efficiency. The machine had exemplary behaviour, according to the experimental review. It showed that there was a substantial improvement in the photochemical efficiency of Photosystem II (uPSII) and the electron transport rate (ETR) relative to continuous light at specific frequencies (0.1, 1 and 50 Hz). In order to determine the more precise effects of pulsed light on plant growth, the development of intelligent light systems may be of great importance (Olvera-Gonzalez et al. 2013).

B. Irrigation System with Sensors

There were two parts of the whole monitoring system: a wireless network of sensors and a monitoring centre. In crop-growing regions, sensor nodes, the controller node, soil moisture sensors, irrigation tubing, spray irrigation, and irrigation control valves were deployed. In mesh network topology, the ZigBee network was accepted. To achieve the coverage of the network and reduce the energy usage and expense of the node at the same time (Gao et al. 2013).

C. Architecture

Nodes used modular architecture; three types of nodes used standard core modules and several nodes with separate extension modules. The temperature and humidity sensor gathered information on temperature and humidity; the routing nodes were responsible for routing contact and data transmission; the coordinator node collected and transmitted data from the routing node to the host computer monitoring centre.

An example of an automated agriculture system regulates ecological variables such as temperature, moisture, the concentration of carbon dioxide and lighting by crop growth. A-node executes the application program after setting up the network topology. By turning on its sensors and sensing the greenhouse atmosphere, the application program starts its active phase. The application program reads from sensors the temperature, humidity, and luminance of a greenhouse and reports the result through its parents to the sink (gateway device). If it receives any packets from its children during this active period, it relays the packets to the parent. A-node awaits its working schedule after transmitting its sensing data, such as the sensing time in the sleep order message from the sink. The C-node (actuator node) is programmed to monitor the light intensity of the rising melon greenhouse. It has an additional A-node relay board to power the greenhouse light switch. It is solely powered, and no power-saving system is used. The C-node application program is waiting for an order from the sink. Whenever the command is received from the sink via the application server, it controls the relay to turn the lights on or off (Yoo et al. 2007).

D. Shelf-life prediction

By setting up different types of sensors on our agricultural IoT platform, we are able to collect information about a farm product during its entire life cycle, such as planting, storage, manufacturing, transportation, and sales. It is challenging to estimate a farm product's shelf life because many factors can influence it during its life cycle, such as temperature, the humidity of the air/soil, etc. We adapt the back-propagation method to the integrated sensor data streams after incorporating raw sensor data streams into batch id-based data streams to predict the shelf life of a farm product (Wu et al. 2014).

5.3.4.3 IoT for Water Management

Water is essential, and it is necessary to control the supply. With the growth of the population in urban cities, the need for water is rising exponentially. To better manage the supply-demand ratio, it is essential to provide structures to avoid water shortage. We have, therefore, developed an IoT scheme to schedule the use of water according to supply. The sensor wires would detect the water level in the overhead tank. As a sensor, a one-stranded wire is used. Through running the transistor in transfer mode, sensing would be performed. If a certain water level is detected, and the corresponding level is sent to a microcontroller, the corresponding water level is uploaded to the cloud. The Android program can fetch this cloud information and show it to the end-user. Present and previous water levels will be seen in this Android application, along with the date and time. The implemented device also performs based on water's electrical properties and consists of a microcontroller, LCD, differential amplifier, power amplifier, thermistor, turbidity sensor, pH electrode, and other components (Wadekar et al. 2017).

Basically, many parameters are needed to be measured for water quality analysis. However, the proposed system measures the most critical water parameters, which are shown is Fig. 5.14.

Contributing to solutions that integrate the IoT paradigm into water management processes can help address expected solutions regarding some considerations from the business, social, and technical points of view. There are some main benefits of providing IoT in water management scenarios:

A. Efficiency Increase

Water supply firms and organizations can make better strategic decisions and minimize maintenance costs through real-time operational monitoring. To track and develop water treatment infrastructures, they use real-time data from sensors and actuators, making them more efficient, reducing energy costs, and eliminating human interference. Cost savings: by increased capacity efficiency, operation efficiencies, and quality, water treatment costs can be decreased. Enhanced resource utilization (e.g., smart water irrigation units that reduce manual operation) and infrastructure enhancements (e.g., automated control of irrigation conditions) will help customers and organizations.

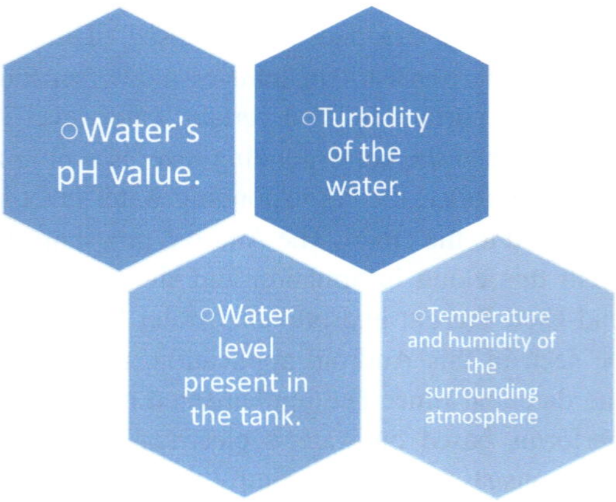

Fig. 5.14 Critical water parameters

B. Asset Utilization

Companies can benefit from their assets and supply chains' transparency and visibility through improved monitoring of assets (machinery, equipment, tools, etc.) using sensors and connectivity. On critical pieces of infrastructure and machinery, they can quickly locate assets and run preventive maintenance.

C. Productivity

Productivity is a fundamental parameter that impacts any organization's profitability. IoT enables real-time control, new business models, process optimization, conservation of resources, reduction of service time, and the ability to do all of these globally, reducing the mismatch of necessary versus available skills and improving labour efficiency (Robles et al. 2015).

5.3.5 Materials

While choosing the materials to be used in the project, attention was paid to the devices' IoT support and their compatibility with HMS. Besides, care was taken to ensure that the devices are energy-efficient and user-friendly.

5.3.5.1 Home Communication Network

A. Modem/Router

A modem is a device that receives and sends analogue signals from the world wide web and digital signals from the home network in both ways and conducts modulation/demodulation processes (Borth 2018). However, these signals must be directed and guided in proper order. As the name implies, routers manage the data traffic between local networks and the Internet (Cisco 2020). In ODIH project, an internet service provider supplied a 4G wireless router that connects to the cellular 4G Internet wirelessly with both modem and router capabilities. Figure 5.15 summarizes ODIH's connection to the Internet.

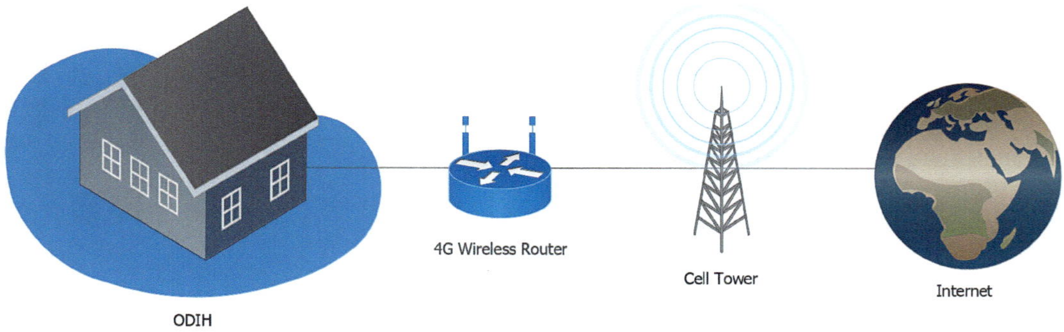

Fig. 5.15 ODIH's internet connection

B. Switch/Access Point

Switches allow users to use one ethernet port of the router for several devices by acting as a multiplexer. Besides, PoE (Power over Ethernet) enabled switches can supply power to peripheral devices through ethernet ports, so the necessity for external power adapters is ruled out. Access points extend the wireless connection coverage area of the router, so remote Wi-Fi devices could connect effortlessly even though they are far from the 4G router.

C. Samsung SmartThings Hub

To support a wide range of Samsung appliances and several Zigbee and Z-wave devices, SmartThings Hub comes in handy. Hub connects to the router with an ethernet cable and creates a sub-network comprised of mostly Samsung home appliances. Thus, communication between Samsung appliances and SmartThings Cloud is easily governed from one centre.

5.3.5.2 Home Appliances

A. Refrigerator

Many existing IoT refrigerators employ cameras integrated inside, which scan and list the foods placed in. In some refrigeration units (commercial etc.) RFID scanners (Moin 2015) or weight sensors (Edward et al. 2017) can be used to identify the foods. The refrigerator keeps track of the food's expiry dates, although these dates need to be entered manually. Based on both listed food and expiry date data, the refrigerator recommends recipes, which is an added advantage for preventing food waste. IoT refrigerators will allow scheduling defrost cycles easily, hence presenting economic impacts for the energy-management. Figure 5.16 illustrates refrigerator's connection to the home network.

The refrigerator that suits our requirements most is Samsung's Family Hub RS22T5561SG/AA. To go through the details:

- Has an 8-inch LCD screen to access HMS.
- Has a high price. However, we hope that this technology will be standardized for every house in the future.
- Has high energy use. The average working time of a refrigerator is assumed to be 8 h a day at their rated power, with three defrost cycles. This 623 L Family Hub refrigerator's estimated yearly energy use is 636 kWh, while an average refrigerator's energy usage is around 438 kWh. In comparison, LG's 623 L IoT refrigerator LFXC22596D consumes 664 kWh/year, and Bosch's 589 L B36CT81SNS consumes 568 kWh/year. While it seems Samsung has a high-power demand; it also has the most diverse functions. This high energy consumption disadvantage will be overcome by load shifting.

This refrigerator is managed by SmartThings Hub, which is also supplied by Samsung. It uses

Fig. 5.16 Refrigerator's connection to the home network

Wi-Fi 2.4 GHz with IEEE 802.11 b/g/n and Soft-AP protocols, with WPA/TKIP and WPA2/AES encryption. This is correct for almost all smart Samsung home appliances.

We are unable to verify which motherboard and controllers Samsung used in the design; these studies are confidential due to the nature of them. Nevertheless, it utilizes a Linux-based operating system, Tizen OS, and also, Samsung SmartThings is an open-source project.

B. *Stove*

A fundamental concept behind the IoT stove is eliminating the concern of whether any burner is turned off completely. Nothing more is expected from a smart stove. An induction stove only heats the bottom of the pot, not the cooktop itself. Up to 90% of the heat produced is transferred to the food, while general electric stoves can transfer 74% of the energy and gas stoves can only 40% (Sweeney et al. 2014). These stoves have cooking temperature patterns for different kinds of dishes and directly control the power it demands. Energy efficiency is a key issue in the ODIH project, and an induction stove would be the ideal choice.

- We will hold on to Samsung as our choice, NZ30K7880US/AA Induction Cooktop. The range of actions that can be taken with the SmartThings app are as follows:
- Monitor the operating status and power level settings of the stove components.
- Set timers, check or change the settings, adjust the temperature.
- Once started cooking, the wirelessly connected Samsung hood turns on automatically.
- As mentioned above, this stove connects to SmartThings Hub with a built-in Wi-Fi module. Addedly, it can connect with the hood over Bluetooth too.

Figure 5.17 depicts stove's connection to the home network.

C. Washer Dryer Machine

Models fitted with sensors allows different temperature settings for different materials, measuring the garments' moisture and signalling when the drying work is completed. Machines also facilitate the ironing process by shrink-free straightening the garments.

Temperature Sensor: The sensor communicates through a single bus cable. It is used to read the water temperature coming in. If this temperature is low, the detecting filament is heated, and this sensor determines the temperature of this hot water.

Door Sensor: Whether it is opened or closed, it is used to understand the door's status. The magnetic sensor makes it possible to incorporate this. Whenever the magnetic flux lines between the two magnets are severed, the door is considered open, and the status is changed.

Water Flow Sensor: The amount of water used for washing can be estimated using this sensor. A plastic valve body, a water rotor, and a hall-effect sensor are components of a water flow meter system. The rotor rolls as water pass into the rotor. Its velocity varies at various flow speeds. The corresponding pulse signal is sent out by the hall-effect sensor.

The washing machine comprises a sensor indicating information about the load state.

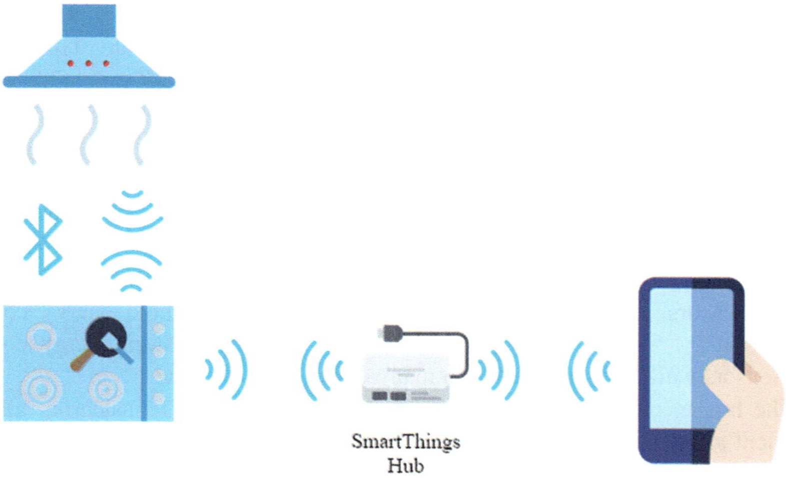

Fig. 5.17 Stove's connection to the home network

Load Sensor: This sensor icon lights up when a cycle that supports load sensing is selected. However, loading sensing could not be assisted by some of these cycles. Once the cycle is started, the icon remains solid or blinks throughout the load sensing process, and once the process is complete, the icon turns off.

The smart washing machine must detect the type of clothing (white, coloured, specialized), the exact amount of detergent required depending on the load, and even the best cycle to use depending on how dirty the load is. If a problem such as power loss or load imbalance occurs, or if dirt is detected (common for front-loading washers), or any other predefined issues, the system must also submit alerts. If hot/warm water is required for other uses, such as a shower bath, it may stop and restart itself during a loop and take other corrective action if the power is cut off or a mechanical problem occurs. Also, through the HMS, the washer and dryer will communicate with other home appliances to decide operating priorities. For example, if the combined washing/drying process uses the most energy, as determined by the utility, it will operate at the most energy-efficient time, allowing less energy-intensive devices to bid for the remaining periods (Kiruthika and Arulanantham 2016).

Samsung WD6800 Quick-Drive Washer Dryer is going to be run by SmartThings hub. The integration of the washing machine (washer and dryer) is implemented as follows:

- To connect the SmartThings hub and washer dryer, WPA/TKIP and WPA2/AES are recommended as encryption methods. Nevertheless, unapproved Wi-Fi protocols and newly developed Wi-Fi protocols are not supported.
- The Wi-Fi 2.4 GHz protocols are supported by the Samsung washer-dryer machine.
- Surrounding wireless environments can influence the reception sensitivity of connection.
- The IEEE802.11 b/g/n (2.4 GHz), Soft-AP protocols are supported by the Samsung washing machine.
- Particularly IEEE802.11n is recommended.
- To provide a successful connection to the Samsung washing machine, an approved wired or wireless router should be utilized (Samsung, 2018).

D. Television

Smart television allows the user to receive notifications on the television screen. It can access many websites on the Internet through applications. Alert system streams directly from smart TV, from controlling the lighting to checking the thermostat, enabling you to get a full view of the house at a time. By using the internet connection, videos on many devices such as security cameras or phones can be viewed on the screen. It can access Netflix, YouTube, Twitch, Twitter via ethernet. By using the product "Samsung—75 Class TU7000 Crystal UHD 4 K Smart TV (2020)" in ODIH, smart functions above can be acquired.

E. Sensors

Sensors work synchronized with many devices in the house, making the devices work more efficient and useful. With gas sensors, the air inside the house is healthier and prevents dangerous situations in the release of dangerous gases. With the curtains' control, the amount of light in the house is adjusted according to the sleep and temperature pattern. A safer home environment can be created with motion sensors that can be placed in many parts of the house. By using sensors on the input and output lines of the sources used at home, it is possible to receive notifications of any interruption or leakage. Figure 5.18 visualizes sensor network of ODIH.

F. Thermostat & Air Conditioner

The smart temperature control system offers the user the ability to decide the house's temperature without the need for being at home from the phone. The system, by using the local weather data, automatically adapts to the user's daily life and seasonal changes. Energy consumption can be monitored instantly on the phone. It adjusts the temperature of the rooms according to the usage habits of the user. Unused rooms of the house are detected, and energy usage is limited in these rooms. The system can use sensors and residents' phone location to check if he/she has left home and then set itself to an *eco-mode* to save energy. A smart thermostat will command the refrigerator to hold off on the defrost cycles until off-peak hours. It can save up to 23% annually on heating and cooling (compared to a hold of 22 °C). It can also be automated with hubs like SmartThings and set it to turn on or off when doors are opened, as people come and go, and much more. The thermostat should have Temperature Humidity Occupancy Proximity Sensors Compatible with Daikin Heat Pump used in ODIH.

Some rooms may naturally be airier than the others. These situations are detected with the help of sensors, and the air is directed to the less airy rooms through smart vents. Heating and cooling often account for around 50% of household energy consumption. Smart systems can lower energy consumption and save users money.

By considering all IoT enabled applications, smart thermostats are the essential ones for energy saving because HVAC systems consume most of the energy in households, and %65 of the energy usage of a building is taken by temperature regulation (Chaloulákou, 2019).

G. Kettle

Modern life and business conditions made humanity care and save every minute in casual life; morning routines are even more critical in the meaning of time efficiency. To reduce breakfast preparation times, several electric appliances must be controlled remotely and prepare the hot water, toast, or coffee until the time resident arrives at the kitchen for breakfast. Home Assistant can deal with this precise timing automation by obtaining data of wake-up calls, presence sensors, and resident activity by processing artificial intelligence. Nonetheless, residents may experience an unusual and unexpected routine for Home Assistant, and kettle or coffee machine may have already been turned on and completed their processes. These kinds of situations naturally provide different data for the morning routine to train artificial intelligence, but energy-saving and resident comfort must be preserved. Keeping the water at a constant temperature requires a bit more energy, but heat insulation can significantly compensate for unnecessary energy using for water temperature immobilisation.

Water heating is one of the most energy-consuming processes in a house. Kettles draw high amounts of current to satisfy the thermal

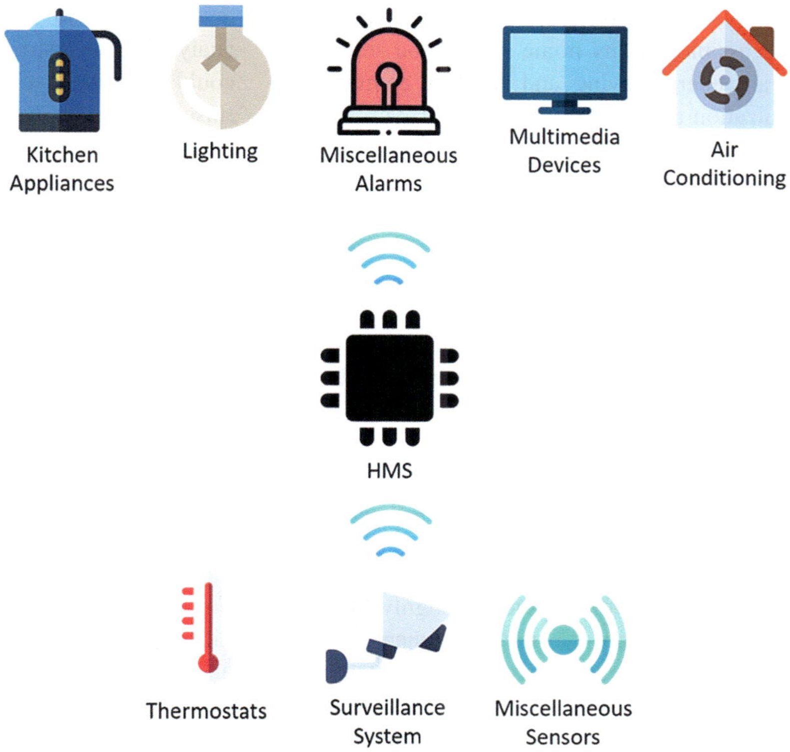

Fig. 5.18 Sensor Network and HMS

energy needs of water to boil or heat. Most residents heat their water to the boiling temperature even if the usage area does not need boiled water. For instance, baby foods must be prepared below 37 °C, and yeast solutions must not exceed 48 ° C; also, various hot drinks have their unique preparation temperatures. Furthermore, users usually are not aware of the amount of water needed and going to be heated. Water can hold lots of energy due to its specific heat value and wasted energy when excess amounts of it are boiled. To boil one glass of water (200 ml) at 27 °C, 61 kJ of energy is needed. If 1 litre of excess water is boiled per day in a 2-resident house, one month of energy wastage equals 5 kWh, which means energy consumption of 5 days of lighting or weekly dishwashing (Specific Heat Capacity 2020).

H. Range Hood

The home environment's air quality could be considered an essential constituent of resident comfort and mostly regulated by air conditioning systems. However, the cooking process may emit high amounts of smoke and steam that the air conditioning system cannot tolerate instantly, so range hoods take place at this point. Conventional range hoods include several air suction power levels and lighting bulbs, but controls are entirely left to the user, and the appliance is vulnerable to misuse. A smart range hood should suggest the most appropriate level of suction or lighting to the user to avoid abusage or to provide maximum comfort. Traversing air must be analysed in terms of moisture, smoke, oil, and odour density levels by the sensors inside the airway of the range hood and the sensors just outside of the filters. Through sensor data, the user must be notified about filter condition and utilization analysis. The user must be suggested for more comfortable or economic air suction or lighting levels. Furthermore, a work cycle containing stages such as powering on, suction, and lighting level adjusting, turning off must be automated to discard errors caused by human intervention. Also, usage behaviour data must be accessible for any home automation software to train energy-saving and resident comfort optimization algorithms ('5 Ways IoT Makes Better Range Hoods' 2017).

I. Microwave Oven

Such as small home appliances like kettle and coffee maker, microwave ovens are used frequently for short time durations like 1–3 min. Numerous packaged foods have standardised heating instructions, but handmade foods such as leftover meals or bakery goods do not have optimized heating adjustments. Latter kinds of foods are more susceptible to heat calibration errors when a human makes an adjustment decision. Because of excess heating, energy waste and user discomfort are highly likely due to extremely hot or dried out meals. Sensors should check several aspects inside of the oven and the condition of the meal going to be heated. Firstly, a couple of temperature sensors should check the meal's temperature at different angles in case of unbalanced heating. Secondly, moisture in the air and the meal should be checked by moisture sensors to prevent overheating or drying. Finally, at least two cameras should monitor inside the oven to determine the mess condition of the meals' oven and package material through image processing. All sensor data should be accessible for home management systems and learning algorithms. Notifications, warnings, and usage behaviour data must be completely clear to the user (Mangla 2017).

J. Dishwasher

Dishwashers use hot water and environmentally toxic detergents in order to clean kitchenware without rubbing or brushing. Thus, dishwashers must be manipulated to give maximum comfort with minimum electric energy, water, and detergent usage for the environment's sake. Several cameras and image processing cooperated with weight sensors would be sufficient to determine the dishes' type and variety. Unique calibration of hot water and the amount of detergent should be adjusted according to the analysis of the dishes. A recycled water reservoir could allow the machine to reuse useable residual rinse water in the washing cycle. Remote control and compliance with several home assistance software are essential to load shifting and energy saving.

K. Surveillance System

A smart security system works more effectively than a conventional one with sensors located all around the house. The system can recognize the household through security cameras or their authorized smartphones. Image processing and face recognition algorithms should accurately interpret camera data and notify the resident immediately through home assistance software. For instance, the system should recognize the difference between a ball that bumps the door and the door being forced open. Smoke and carbon monoxide sensors should also notify or awaken the resident in an emergent manner, or the system should automatically call for emergency services. Furthermore, the system should send an instant notification if a potential accident does not wait for the accident to occur.

L. Smart Lighting

As for lighting, LED technology is the best decision because of its energy efficiency. Manufacturers develop their products with LED, not just for their low energy consumption or long-life robustness, LED bulbs are also easy to implement IoT applications. Usually, smart LED bulbs communicate via Bluetooth or Zigbee. Bluetooth connection is direct to mobile devices, but its range is very limited, and also Bluetooth technology uses more energy compared to Zigbee. On the other hand, Zigbee communication protocol has a limited range too, but a smart LED bulb connects to a hub or gateway via Zigbee, and it can be controlled with smartphones over a cellular network.

Philips Hue LED bulbs support both Zigbee and Bluetooth protocols. Bluetooth communication protocol does not require any additional item; bulbs connect to a smartphone directly. However, the Zigbee protocol requires an additional bridge to connect SmartThings Hub. Hue Bridge allows using a maximum number of 50 bulbs, while Bluetooth allows a maximum of 10 bulbs. Hue Bridge connects to the local network via Ethernet cable. In this way, control of bulbs from anywhere with a SmartThings app installed smartphone as long as it is connected to the Internet is possible.

The estimated monthly energy use of 20 Philips hue bulbs (9 W) is 32.4 kWh, despite that incandescent 20 bulbs (100 W) consume 60 kWh of energy in a month.

Figure 5.19 illustrates the components of the IoT network in ODIH and the protocols which will be utilised in their communication.

5.4 Results

In 2019, the average annual electricity consumption for a U.S. residential utility customer was 10,649 kilowatt-hours (kWh), an average of about 877 kWh per month. The annual average energy consumption of household appliances in ODIH is 4548 kWh, and the average monthly consumption is 385 kWh (except heating).

Water heaters, dishwashers, and washing machines are the loads with shiftable consumption-about 54 per cent of the overall consumption. Refrigerators, boiling kettles, coffee machines, floor heating, iron, and vacuum cleaners are virtually shiftable loads, with about 10 per cent of total consumption. TV sets, PCs with a modem, home cinema, music centres, cooking stoves, kitchen ventilation, bathroom lighting, and ventilation are non-shiftable loads-roughly 36 per cent of total usage (Zimmermann et al. 2012).

5.4.1 A Day with IoT

In the early hours of the morning, the resident wakes up with the sunrise simulation made by the lighting system and the audible alarm. Thanks to the motion sensors, HMS perceived that the resident woke up, and it started the morning routine automation. At the beginning of the routine, the television automatically turned on as the resident walked in and gave the favourite podcast to the speakers. Meanwhile, in the kitchen, the coffee machine started to prepare the morning coffee according to the HMS estimation of the arrival time of the resident at the kitchen. Knowing the resident's sleeping habits, HMS commanded the ventilation system to bring the temperature of the rooms to the optimum level for the start of the day. While the resident was preparing breakfast, he/she followed the podcast from smart fridge and checked the traffic and weather conditions thanks to the audible home assistant. While the resident was sleeping at night, clothes were automatically washed and dried. In the morning, the resident ironed them at the appropriate temperature and time the iron advised.

After leaving, the smart lock system locked the main door of the house. HMS noticed the resident's absence and ordered an air conditioning system to adjust the house's temperature to an economically ideal mode. When the resident was not at home, the smart camera system is watching the house, and the resident can monitor it instantly from a phone application. Realizing that the house is empty, HMS ordered a robot vacuum and robot mop to start cleaning the house. User can remotely check the status of each smart device, analyse instant data and give commands with the phone application. Status of home appliances that may cause problems could be tracked, and if resident mistakenly left any of them open or turned on, he/she can act over the phone application according to the situation.

While shopping, the resident connected to the HMS phone application and got missing foods by checking the refrigerator's indoor camera. When the resident returns home, the house was cleaned. The refrigerator recommended a recipe for dinner considering food expiration dates. The resident can prepare dinner by following the steps on the refrigerator's screen. On the days when the resident comes home late, the house turns on some of the lights to mislead burglars.

Fig. 5.19 Home network. Automation administrators (orange), communication devices (blue), home appliances (yellow)

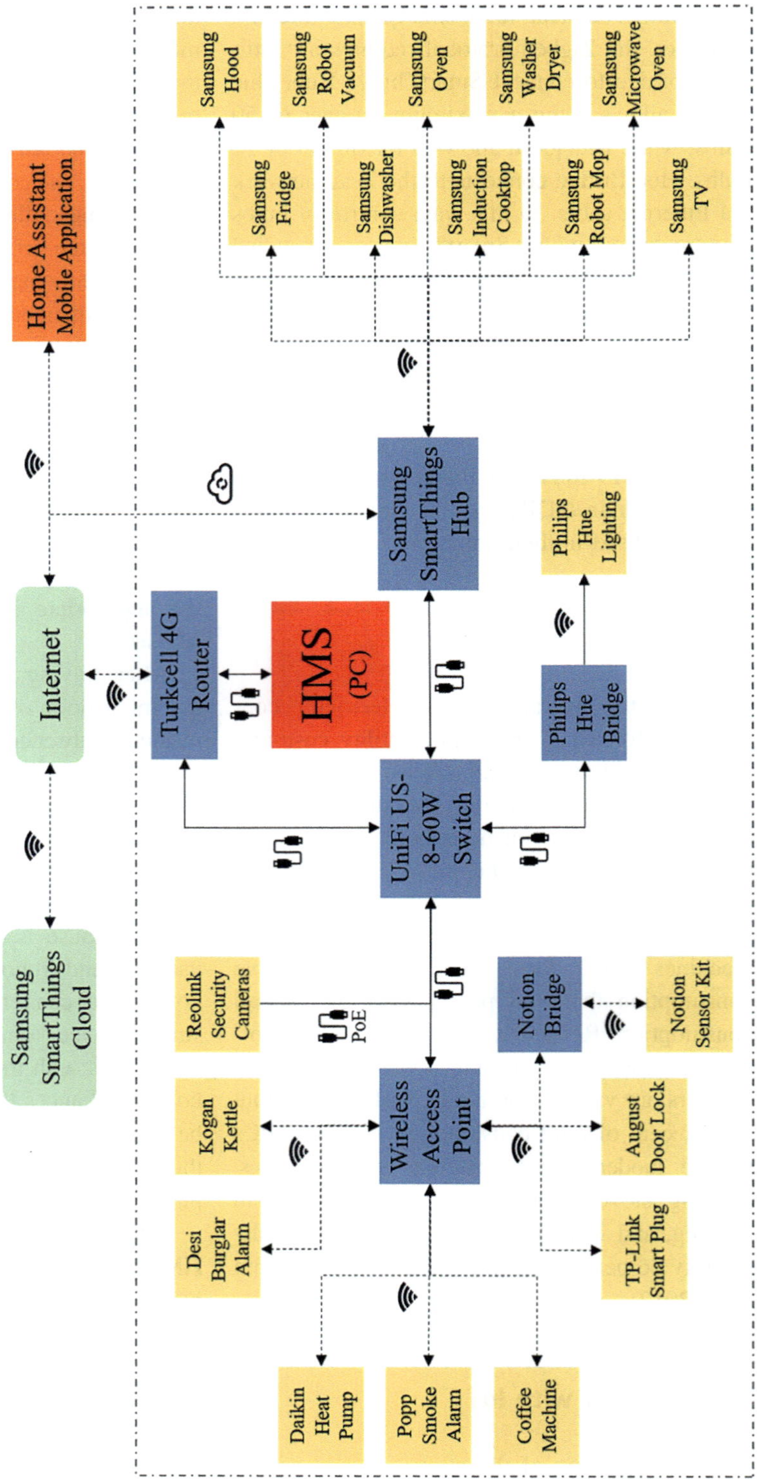

The house remains alive at night-time routine even though the resident asleep. HMS adjusts itself and the devices according to the appropriate electricity tariffs, based on the method called load shifting, as shifting the use time of the most energy-consuming appliances to the off-peak times, which is the late night to morning hours in Turkey.

The washer-dryer machine can wash and dry the laundries in the most efficient and low-cost way. Moreover, the refrigerator fulfils deep-freezing tasks in the later hours of the night due to the pricing tariff. The refrigerator provides an information level of the emptiness of shelves and freshness of content; thereby, efficiency in terms of food management can be achieved via combinations of the given information by the fridge. In aspects of water and food consumption, efficiency can be succeeded by resident's smart home. By the utilization of smart meters, the water that is made use of can be smartly controlled anytime by the home management system and resident. All these processes consisting of consumption of energy, a load-shifting activity that leads to efficiency, can be monitored by the computer. Besides, the security system and fire detection system can work together to enable the total security and safety of the ODIH.

When an intruder haunts the ODIH, devices, which are normally smoke detectors, can act as a burglar alarm. Even if there is nobody at home, the lighting system can turn the lights on and off at regular intervals, giving the impression that someone is at home. If there is a water leak on any part of the house, the resident can know immediately; the same is true if a cat walks through the window. What makes them possible is sensors measuring various values such as motion, gas, temperature, humidity, and light scattered throughout the house.

By doing that, the energy efficiency of home appliances and user comfort significantly increased, and energy consumption decreased considerably compared to a conventional house.

In white appliances, the fundamental energy consumption is mostly based on user experience, and it would be a subjective judgment to say that IoT devices provide a decisive advantage over normal appliances. However, the energy consumption of small household appliances with a single function and high energy consumption such as a kettle can be used more purposefully, and energy can be used more efficiently thanks to IoT. In more complex home appliances and device ecosystems, efficiency is provided by changing the user experience itself, thanks to the HMS and IoT devices, rather than fundamental energy consumption (e.g. load-shift).

5.5 Discussion

IoT integrated devices have disadvantages besides advantages. IoT enabled devices are more expensive compared to non-IoT devices. Also, because devices are IoT enabled, they may use more energy compared to non-IoT ones. The most significant example of energy comparison is the fridge. It consumes around 650 kWh of energy per year, and this consumption figure is twice of a non-IoT fridge. Another difficulty is that some of the IoT enabled devices are not easily available for purchase in every country.

In the ODIH project, the Home Energy Management System (HMS) is going to be used for a residential case. However, similar Energy Management Systems for the industry sector is more advanced compared to domestic systems, and most of the manufacturers focus on the industry sector rather than households.

The development of IoT technology is mostly aimed towards user comfort; however, the industry designing and manufacturing IoT technology should also consider and focus on sustainability. This will provide broader advantages to everyone whether or not one owns IoT compatible services.

5.5.1 Device Compatibility & Communication Protocols

A home environment includes a wide variety of appliances such as washing machine, security systems, lighting system or air conditioning. Various suppliers might have different products

Table 5.2 Comparison of communication protocols

ZigBee	Z-Wave
Uses 2.4 GHz	Uses 908 MHz
Complicated integration	Easy integration
Cheap	Expensive
Short-range up to 20 m	Long-range up to 65 m
Hardware is available from several manufacturers	Hardware is manufactured only by Silicon Labs
High data rate (250 kb/s)	Low data rate (40 kb/s)

for residential customers. In this context, the communication procedures of various appliances bought from different suppliers should be open source and articulable by end-user. In current systems, communication protocols and integration dynamics usually remain limited within the boundaries of a dominant brand due to non-compatibility with open-source systems. Furthermore, communication protocols that appliances utilise have certain advantages and disadvantages (Table 5.2).

5.5.2 Open Source Problem

We needed open-source software for our IoT devices for reasons such as integrating the device to HMS, customisation, and compatibility. Not every device on the market meets this need, so we chose our equipment carefully. However, finding open-source hardware was a challenge. There are many open-source IoT hardware platforms (e.g. Arduino, Intel, Raspberry Pi). Nonetheless, even though companies share part of the software of their devices, they do not share the hardware they use in their designs at all, which limits our capabilities on the device. On the other hand, Google Nest cut support for third-party apps for a while. It was possible to integrate Nest products into their home management systems through reverse engineering. However, this is far from standardizing IoT technologies for everyone.

5.5.3 Cloud Connection or Local Network

Even though companies allow their products to be accessible for open-source home automation systems, the data communication method of appliances depends on the manufacturer's choice. Cloud-based integrations use companies' cloud services through an uninterrupted internet connection, and appliances cannot use IoT enabled services or stop supplying software upgrades. Furthermore, a company may disable some features of the appliance or restrict particular aspects of the device by withdrawing IoT support. Locally based integrations only work within a local network, and internet connection problems do not cause the whole system's shutting down. The decision of cloud or local integration of an appliance should be left to the end-user. Both types have their advantages and disadvantages, and residents should adjust the balance of benefits by choosing an integration type.

5.5.4 Discussion and Policy Recommendations

According to the United Nations (UN), there is a lack of access to modern energy for 13 per cent of the world population, and the goal is to establish worldwide access to modern and reliable energy by 2030 (United Nations 2016). Today, renewable energy usage reached almost

28 per cent worldwide, with a 2 per cent increase over the previous year (Nicholas et al. 2019). On the other hand, there is growing attention on energy efficiency measures as well. IoT is the major enabling platform for technological solutions for energy efficiency programmes. Data collection is crucial in terms of tracking consumption and generation of energy.

In 2018, 96 per cent of the world's population lived within the scope of a mobile cellular signal, and with a third-generation (3G) or higher-quality network, 90 per cent of users could access the Internet (UNDESA 2020). Authorities should pave the way for investments in network infrastructure and encourage development and innovation by structuring taxes - tariffs in this area. In this way, sustainability can be achieved in accordance with the UN's "Industry, Innovation and Infrastructure" sustainable development goal.

According to the UN, today, half of the world's population lives in cities, and it is estimated that 5 billion people will live in cities by 2030 (UNDESA 2018). Policymakers can facilitate the transformation of homes, buildings, and cities to be connected to each other through new technologies such as IoE to create safer, more inclusive and environmentally friendly ecosystems. By doing this, we can achieve the "Sustainable Cities and Communities" target faster, which is one of the goals foreseen by the UN for sustainability.

IoT technologies for energy consumption and production efficiency mainly focus on industrial applications due to gigantic amounts of energy consumption compared to minor household appliances. To foster responsible energy consumption objectives, besides industrial facilities, policymakers should encourage public institutions and municipalities to save considerable amounts of electricity in public buildings with IoT technologies. According to Martirano, using presence and light sensors in a traditional education building results in up to 54% energy saving by utilising power management and switching control algorithms (2011). In addition to buildings, highways and street lighting systems have the potential to be automated with IoT systems according to the particular day, weather and time conditions (Nagamani et al. 2019). The ODIH project proposes an efficient pilot for homes of the future. Even though a single home does not yield a big difference to the whole city's energy toll, multiple houses like the ODIH, on a large scale, could lead to responsible consumption amounts comparable to the large industrial facilities. Residential area efficiency goals should be set by the cooperation of the state and the municipalities, and appropriate regulations must be put into effect by the state.

By 2050, it is expected for urban areas to host two-thirds of the population of the world. Transport systems, water, sanitation, waste management, urban planning, disaster risk reduction, access to information, education and capacity-building, which constitute the relevant issues to sustainable urban development, should be enhanced by the integration of IoT to homes and many more fragments of the sustainable cities (The United Nations 2019).

5.6 Conclusion

In this study, we aimed to establish an automated network of smart devices for ODIH, which focuses on high efficiency in energy. When we examined the working principles of the IoT, we first looked into the compatibility between sensors, such as temperature and motion sensors and home appliances. Secondly, we inspected the connection of sensors to a gateway device, and finally, the processes that occur during device management. Then, we have dealt with efficient and innovative systems for the connection used in the IoT ecosystem, such as ZigBee and WIFI. We continued with architectures such as Home Assistant, together with the user interface preferences of the modules where the latest information is transferred and processed with the middleware such as hub or gateway used during communication in IoT. In addition, in our research regarding efficiency in IoT ecosystems, we discussed the field experiments previously conducted and highlighted the benefits of adopting these experiments to ODIH. We have designed our own model for this study by examining the place of IoT in energy, food and water production as well as the transfer and consumption of these sources at ODIH.

Moreover, three major issues related to cloud-based IoT systems, storage, data processing and communication problems, were discussed. By doing so, we clarified the infrastructure of storage of data in the cloud system and various algorithms to be utilised addressing developed functions to exist in a smart home environment which is included in data-processing issues. Furthermore, the protocols to run the communication between devices and systems are classified into two groups: file transfer protocols and messaging protocols. Thus, these issues regarding cloud-based IoT systems were addressed and explained with essential technical information.

The Demand Response draws more attention as it promises greater flexibility during the management of energy consumption and supply, especially when assuming the large consumer base and a growing number of smart appliances in residential customers. With the increasing number of smart home appliances, the data flowing from the ecosystem to the management system significantly rises. Thus, we emphasised the utilisation of algorithms of artificial intelligence. What is more, the contribution of these expanding raw data to data science is mentioned.

We discussed that the project ODIH offers an in-home solution to manage energy, water, food, and air quality. To collect and transfer data to the HMS, all electric devices must be IoT compatible. The data collected and processed will be delivered to the user with an appropriate interface, such as in-home displays, websites, or smartphone application.

Smart homes pose many benefits with the integration of IoT-based systems, such as energy, water and food consumption control and monitoring. Besides, it can recognise the waste and provide an opportunity to take customised measures according to the user's behaviour. Also, we can provide many benefits by using IoT in agriculture. By calculating the appropriate light, temperature, humidity, and the most suitable architectural structure for those, IoT can assure that the food will grow more efficiently. It will also be beneficial to use IoT to manage the food storage most properly during the consumption phase. Besides, IoT is needed for calculating the necessary amount of water production and consumption by using effective measurement methods in water management and energy control and helps to create a management plan according to users.

With IoT, in cooperation with developing AI and ML, our capabilities will keep expanding. More tasks can be integrated and automated, human decision-making will be minimised, and quality of life will be improved. Thinking of a broader perspective, IoT will be an essential building block of the Smart Cities of the future. Smart grids, driverless cars, and smart farming are just a few examples. ODIH will constitute the core of Smart Cities. All of these application fields will evolve further with 5G technology. Further research should continue to explore growing IoT technologies and implement.

References

'5 Ways IoT Makes Better Range Hoods' (2017) Available at: https://technofaq.org/posts/2017/07/5-ways-iot-makes-better-range-hoods-infographic/

Borth DE (2018) Modem, Britannica. Available at: https://www.britannica.com/technology/modem. Accessed 20 Oct 2020

Chaloulákou A (2019) Smart thermostat: developing an IoT device using artificial neural networks. Available at: http://ikee.lib.auth.gr/record/308954. Accessed 3 Oct 2020

Chen X et al (2018) An Internet of Things based energy efficiency monitoring and management system for machining workshop. J Cleaner Prod 199:957–968. https://doi.org/10.1016/j.jclepro.2018.07.211

Chopra K, Gupta K, Lambora A (2019a) Future internet: the internet of things-a literature review. In: Proceedings of the international conference on machine learning, big data, cloud and parallel computing: trends, prespectives and prospects, COMITCon 2019. Institute of Electrical and Electronics Engineers Inc., pp 135–139. https://doi.org/10.1109/comitcon.2019.8862269

Chopra K, Gupta K, Lambora A (2019b) Proceedings of the international conference on machine learning, big data, cloud and parallel computing: trends, prespectives and prospects, COMITCon 2019. Institute of Electrical and Electronics Engineers Inc., pp 136, 137

Cisco (2020) What is a Router? Available at: https://www.cisco.com/c/en/us/solutions/small-business/resource-center/networking/what-is-a-router.html#~types-of-routers. Accessed 17 Sept 2020
Da Cruz MAA et al (2018) A reference model for internet of things middleware. IEEE Internet Things J 5:871–883. https://doi.org/10.1109/JIOT.2018.2796561
DeNisco-Rayome A (2017) The five industries leading the IoT revolution, ZDNet. Available at: https://www.zdnet.com/article/the-five-industries-leading-the-iot-revolution/. Accessed 1 Aug 2020
Din IU et al (2019) Machine learning in the internet of things: designed techniques for smart cities. Future Gener Comput Syst. https://doi.org/10.1016/j.future.2019.04.017
Edward M et al (2017) Home server based on Raspberry Pi 3, pp 148–151
Ejaz W et al (2017) Efficient energy management for the internet of things in smart cities. IEEE Commun Mag 55(1):84–91. https://doi.org/10.1109/MCOM.2017.1600218CM
Gao L, Zhang M, Chen G (2013) An intelligent irrigation system based on wireless sensor network and fuzzy control. J Netw 8(5):1080. https://doi.org/10.4304/jnw.8.5.1080-1087
Geetha S, Gouthami S (2016) Internet of things enabled real time water quality monitoring system. Smart Water. https://doi.org/10.1186/s40713-017-0005-y
Greengard S (2015) The internet of things. The MIT Press, London
Home Assistant (2009) Documentation—home assistant. Available at: https://www.home-assistant.io/docs/
Jiang L et al (2014) An IoT-oriented data storage framework in cloud computing platform. IEEE Trans Industr Inf 10(2):1443–1451. https://doi.org/10.1109/TII.2014.2306384
Kazmi S et al (2017) Towards optimization of meta-heuristic algorithms for IoT enabled smart homes targeting balanced demand and supply of energy. IEEE Access 7:24267–24281. https://doi.org/10.1109/ACCESS.2017.2763624
Kiruthika J, Arulanantham D (2016) Making washing machines smart through IoT. Int J Modern Trends Eng Sci 3(6):39–41
Knud A, Lueth L (2015) IoT basics: getting started with the internet of things, IoT analytics
Latre S et al (2016) City of things: an integrated and multi-technology testbed for IoT smart city experiments. In: 2016 IEEE international smart cities conference (ISC2). IEEE, Trento, pp 1–8. https://doi.org/10.1109/isc2.2016.7580875
Logenthiran T, Srinivasan D, Shun TZ (2012) Demand side management in smart grid using heuristic optimization. IEEE Trans Smart Grid 3(3):1244–1252. https://doi.org/10.1109/TSG.2012.2195686
Lutui PR, Cusack B, Maeakafa G (2018) Energy efficiency for IoT devices in home environments. In: 2018 IEEE International Conference on Environmental Engineering, EE 2018—Proceedings. Institute of Electrical and Electronics Engineers Inc., p 1. https://doi.org/10.1109/ee1.2018.8385277
Madakam S, Ramaswamy R, Tripathi S (2015) Internet of Things (IoT): a literature review. J Comput Commun. https://doi.org/10.4236/jcc.2015.35021
Mahdavinejad MS et al (2018) Machine learning for internet of things data analysis: a survey. Digit Commun Netw. https://doi.org/10.1016/j.dcan.2017.10.002
Mangla S (2017) Top 5 microwave usage mistakes people make. Available at: https://www.mrright.in/ideas/appliances/microwave/top-5-mistakes-people-make-when-using-microwaves/. Accessed 11 Oct 2020
Martirano L (2011) Lighting systems to save energy in educational classrooms. In: 2011 10th International conference on environment and electrical engineering, EEEIC. EU 2011–conference proceedings, pp 1–5. https://doi.org/10.1109/eeeic.2011.5874691
Mashima D, Chen WP (2016) Residential demand response system framework leveraging IoT devices. In: 2016 IEEE international conference on smart grid communications, SmartGridComm 2016. https://doi.org/10.1109/smartgridcomm.2016.7778813
Mattern F, Floerkemeier C (2010) From the internet of computers to the internet of things. In: Lecture notes in computer science (including subseries Lecture notes in artificial intelligence and lecture notes in bioinformatics), p 242. https://doi.org/10.1007/978-3-642-17226-7_15
Miragliotta G, Perego A, Tumino A (2012) Internet of things: smart present or smart future?, pp 1–6
Moin E (2015) Smart refrigerator for grocery management. Technical disclosure commons. Available at: https://www.tdcommons.org/dpubs_series/75/
Nagamani S et al (2019) Smart street light management system using internet of things. In: 2019 International conference on intelligent computing and control systems, ICCS 2019, (Iciccs), pp 103–107. https://doi.org/10.1109/iccs45141.2019.9065477
Namani S, Gonen B (2020) Smart agriculture based on IoT and cloud computing. In: 2020 3rd International conference on information and computer technologies (ICICT). IEEE, San Jose, CA, USA, pp 53–556. https://doi.org/10.1109/icict50521.2020.00094
Nicholas K et al (2019) Tracking SDG 7: the energy progress report. Washington DC, pp 1–176
Nugur A et al (2018) Design and development of an IoT gateway for smart building applications. IEEE Internet of Things J. https://doi.org/10.1109/JIOT.2018.2885652
Olvera-Gonzalez E et al (2013) Intelligent lighting system for plant growth and development. Comput Electron Agric 92:48–53. https://doi.org/10.1016/j.compag.2012.11.012
Patel KK, Patel SM, Scholar PG (2016) Internet of things-IOT: definition, characteristics, architecture, enabling technologies, application & future challenges. Int J Eng Sci Comput 6(5):1–10. https://doi.org/10.4010/2016.1482

Pourqasem J (2019) Cloud-based IoT: integration cloud computing with internet of things cloud-based IoT: integration cloud computing with internet of things, Dec 2018. https://doi.org/10.22105/riej.2018.88380

Qi BW (2019) Analysis on the application of artificial intelligence in classroom. In: Journal of Physics: Conference Series. Institute of Physics Publishing. 10.1088/1742-6596/1345/4/042075

Raj P, Raman AC (2017) The internet of things: enabling technologies, platforms, and use cases. CRC Press, Boca Raton, FL

Ramakala R et al (2018) Impact of ICT and IOT strategies for water sustainability: a case study in Rajapalayam-India. In: 2017 IEEE international conference on computational intelligence and computing research, ICCIC 2017. https://doi.org/10.1109/iccic.2017.8524399

Robles T et al (2015) An iot based reference architecture for smart water management processes. J Wirel Mob Netw Ubiquit Comput Dependable Appl 6(1):4–23

Sagahyroon A, Aburukba R, Aloul F (2018) The Internet of Things and e-Health: remote patients monitoring. In: Hassan QF, Khan AR, Madani SA (eds) Internet of Things: challenges, advances, and applications. CRC Press, Boca Raton, FL, pp 303–319

Samsung (2018) Washing machine, pp 1–76. Available at: https://downloadcenter.samsung.com/content/UM/201802/20180226112001760/WD90N645OOX_03924A-01_EN.pdf

Satish T, Bhavani T, Begum S (2017) Agriculture productivity enhancement system using IOT. Int J Theoret Appl Mech

Shrouf F, Miragliotta G (2015) Energy management based on Internet of Things: practices and framework for adoption in production management. J Clean Prod. https://doi.org/10.1016/j.jclepro.2015.03.055

Specific Heat Capacity (2020). Available at: http://www.dynamicscience.com.au/tester/solutions1/chemistry/energy/specificheatcapacity.htm#:~:text=So10mlofwaterhas,73°C%3D30.514KJ. Accessed 17 Sept 2020)

Sweeney M et al (2014) Induction cooking technology design and assessment. In: 2014 ACEEE summer study on energy efficiency in buildings, pp 370–379. Available at: https://aceee.org/files/proceedings/2014/data/papers/9-702.pdf

The United Nations (2019) Sustainable cities and human settlements, Sustainable Development Goals. Knowledge Platform

Tukade TM, Banakar RM (2018) Data transfer protocols in IoT-an overview. Int J Pure Appl Math 118 (16):121–138

UNDESA (2018) World urbanization prospects, demographic research. Available at: https://population.un.org/wup/Publications/Files/WUP2018-Report.pdf

UNDESA (2020) Take action for the sustainable development goals – united nations sustainable development. https://www.un.org/sustainabledevelopment/sustainable-development-goals/. Available at: https://www.un.org/sustainabledevelopment/sustainable-development-goals/. Accessed 17 Nov 2020

United Nations (2016) United Nations sustainable development agenda—Goal 12: Ensure sustainable consumption and production patterns. Available at: http://www.un.org/sustainabledevelopment/sustainable-consumption-production/. Accessed 9 Dec 2016

Vermesan O, Friess P (2013) Internet of Things: converging technologies for smart environments and integrated ecosystems (River Publishers Series in Communications)

Wadekar S et al (2017) Smart water management using IOT. In: 2016 5th International conference on wireless networks and embedded systems, WECON 2016, pp 3–6. https://doi.org/10.1109/wecon.2016.7993425

Wu MZ, Wang YT, Liao ZC (2014) A new shelf life prediction method for farm products based on an agricultural IOT. In: Advanced materials research. https://doi.org/10.4028/www.scientific.net/AMR.846-847.1830

Yang J et al (2018) Device-free occupant activity sensing using wifi-enabled IoT devices for smart homes. IEEE Internet Things J 5(5):3991–4002. https://doi.org/10.1109/JIOT.2018.2849655

Yoo SE et al (2007) A2S: automated agriculture system based on WSN. In: Proceedings of the international symposium on consumer electronics, ISCE. https://doi.org/10.1109/isce.2007.4382216

Zhao X (2010) The strategy of smart home control system design based on wireless network. In: ICCET 2010–2010 international conference on computer engineering and technology, proceedings. https://doi.org/10.1109/iccet.2010.5486365

Zimmermann J-P et al (2012) Household electricity survey: a study of domestic electrical product usage. Intertek, p 600. Available at: https://www.gov.uk/government/uploads/system/uploads/attachment_data/file/208097/10043_R66141HouseholdElectricitySurveyFinalReportissue4.pdf

6 Home Management System: Artificial Intelligence

Abstract

Artificial Intelligence is changing the way we work and the way we live. Today, we are witnessing the first moments of a long and profound revolution that will affect the future of industries, jobs and lives. Even though electronic devices have become smarter thanks to IoT, they cannot provide as much service as desired. Therefore, to experience smart homes in theory and daily life, there is a need to develop an automatic system capable of being self-sustaining and responding to modern life needs. This research aims to study the possibilities of Artificial Intelligence applications for domestic use, particularly in establishing a Home Management System (HMS) that reduces consumption, wastes and optimises resources to impact on a small and large scale positively. For this purpose, this study provides a comprehensive analysis of machine learning techniques such as deep learning and reinforcement learning on sustainable smart homes considering several application fields. We present how this system can be used at homes, such as energy management, food & agriculture, water consumption & generation, waste management, healthcare, customisation & entertainment and security.

6.1 Introduction

As the number of people living on earth increases steadily, their need for more resources grows as well. In order to use the living spaces better, water, food, and energy supplies need to be used more efficiently by the communities. Indeed, this "need perception" has enabled people to create new technologies and useful inventions, starting from the Industrial Revolution to World Wars. Production alone is not sufficient to sustain daily life. Communication of information and transportation of goods are also essential. The inventions that can be examined in Fig. 6.1 are the basic needs that emerged due to the communication and transportation needs of people living in these times.

Over time, significant inventions shown in Fig. 6.1 were deemed inadequate for daily use, and product development has become mandatory. Thus, other devices shown in Fig. 6.2 are invented to increase comfort and usability under the requirements of the time.

At first, these inventions were regarded as useful for daily life and helped to do housework easier rather than doing it manually. However, with the rapid development of industrialisation, it has been observed that the number of people participating in the workforce is increasing. Consequently, this has caused those who work

The author would like to acknowledge the help and contributions of Mehmet Mücahit Kaya, Ada Kanoğlu, Asiye Demirtaş, Emirhan Zor, Fatma Tuğçe Akgül, İlke Burçak, Mustafa Can Nacak, and Yusuf Taşkıran in completing of this chapter.

S. Küfeoğlu, *The Home of the Future*, Sustainable Development Goals Series,
https://doi.org/10.1007/978-3-030-75093-0_6

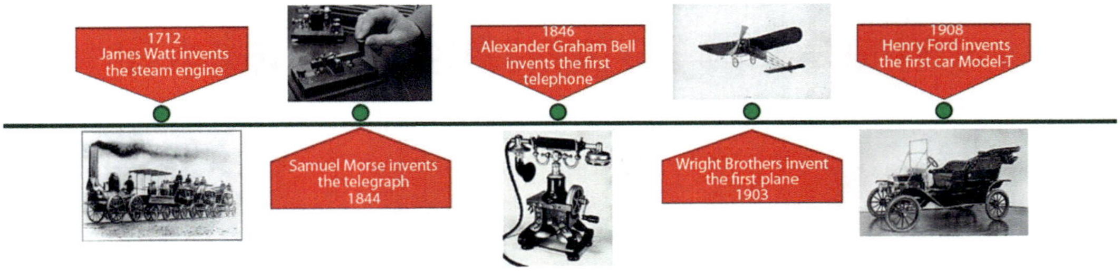

Fig. 6.1 The most important inventions from the industrial revolution to World War I

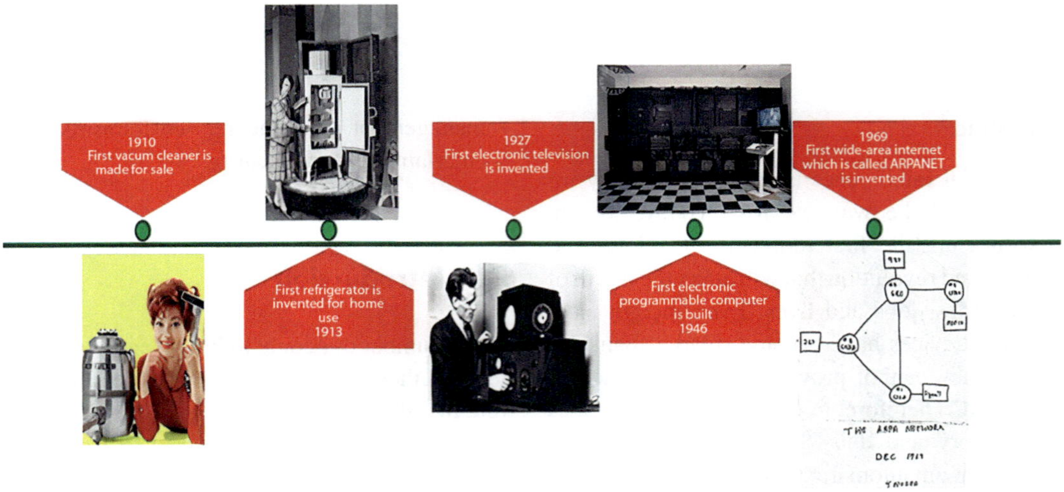

Fig. 6.2 Various significant inventions of the twentieth century

long hours not to find enough time for household chores in their daily lives. According to the research conducted in 1974, employed and non-employed women were compared. While a working woman spent only 26 h on household chores per week, a non-working woman spent 55 h per week. It is almost twice of it (Vanek 1974). This result simply shows that inventions that can be seen in Fig. 6.2 were not at the scale that could meet the needs of billions of people due to resource constraints. As a result, people dream of developing "smarter" systems in their homes, offices, or hospitals to be done all the seemingly complicated tasks without spending any effort for themselves.

If today's world is examined, it can be observed that human beings are surrounded by intelligent systems. By simulating human actions, these systems provide a lot of assistance in daily life. Without being aware, intelligent systems can communicate with people by recommending a movie, providing the best stock advice, showing a proper product with a relevant search, or giving the shortest direction for a faster drive. Using Machine Learning, an intelligent system is defined as software that finds a solution to a specific task or a problem. Consequently, usage of Machine Learning along with Artificial Intelligence technology in such systems makes everyone's life easier.

As a term, Artificial Intelligence (AI) is a programmable system that makes machines "smarter." By using the appropriate methods such as machine learning, deep learning, and reinforcement learning with an extensive data set, the system grasps, learns, makes decisions, and takes rational actions to reach a specific objective. Important methods mentioned above will be explained later in different sections. Nevertheless, understanding the importance of the data used in systems is groundbreaking. To explain with an example, this can be likened to the learning stage of a newborn child. When people are born, they observe, perceive, and make decisions thanks to the information that comes from the outside world. Similar to this, the development of artificial intelligence was created based on this question: "How does a person think?". As a result, the lack of data for such systems that make inferences using the same thinking system will cause the system to stop working. Figure 6.3 shows the data evaluation stream of a typical AI system.

According to the steps that can be seen in Fig. 6.3, for a system that learns with concepts such as experience and behaviour in the first step, we should point out that the datum which is incorrectly evaluated at this step will naturally affect the eventual "decision making".

A vast generalisation has been made when defining artificial intelligence before. However, because of the "artificial intelligence delusion" created by many Hollywood movies, types of artificial intelligence should be mentioned to understand the concept better. Figure 6.4 shows the details of these types in detail.

As it can be observed from Fig. 6.4, constantly human beings have been able to obtain only the type of intelligence called "artificial narrow intelligence". For this type of artificial intelligence, which is programmed to perform specific tasks and used for a wide range of activities, from medical diagnostics to electronic commerce platforms or agriculture to education, numerous applications exist. However, in this study, only artificial intelligence applications that can be used in smart homes will be analysed.

The first application we should mention is energy. The energy industry across the world is facing some challenges driven by increasing demand for mechanisation, a relatively low increase in efficiency, and radical changes in people's energy consumptions. Previously, one of the most frequently used energy sources was fossil fuels. Failure to use these non-renewable fuels such as coal, oil, and natural gas for the correct purposes is questioned in terms of efficiency and environmental pollution. Therefore, today's people have turned to "cleaner" and highly efficient energy sources that can compete with fossil fuels. However, this "energy transformation" intended to be made for every existing system and device has become impossible due to the characteristics of the machines. As a result, this has led human beings to save energy. Although today's people spend most of their time

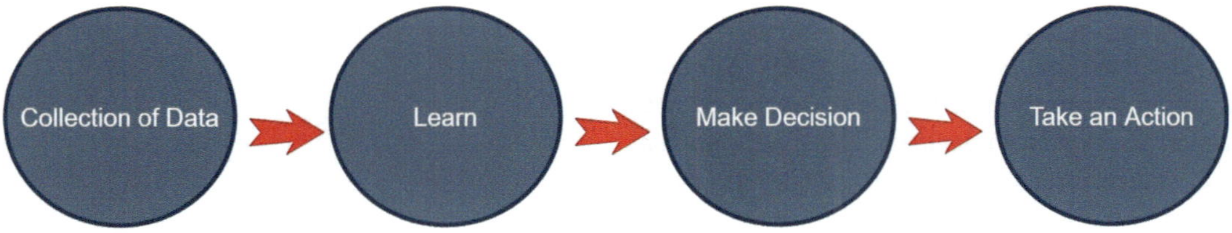

Fig. 6.3 Data evaluation progress of artificial intelligence

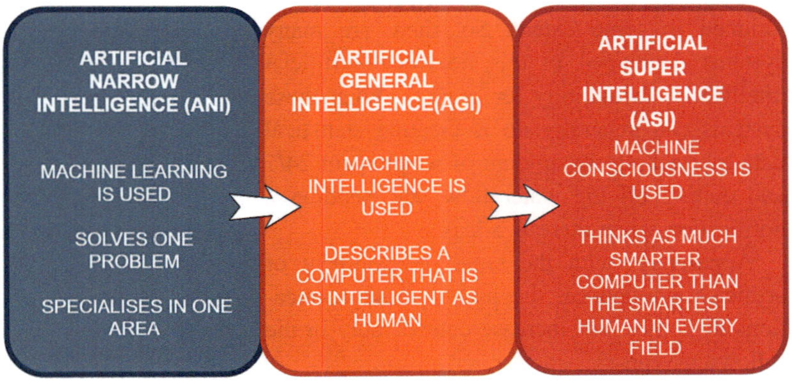

Fig. 6.4 Types of artificial intelligence

outside, the time spent at home is also quite high. Hence, this raises questions about energy consumption in homes leading the usage of artificial intelligence for increasing energy savings. In terms of AI that is used in Home Management Systems (HMS), with the help of machine learning, user's movements and behaviours are recognised, and unexpected behaviours are detected from the data obtained. In this way, the ideal energy to be spent on a device is arranged by the smart system that uses artificial intelligence technology. Currently, there are various numbers of smart devices and systems such as smart plugs, learning thermostats, autonomous lighting, monitors, or photovoltaic solar panels.

The second area is agriculture. Nowadays, obtaining "quality" food in the world has become difficult due to reasons such as the high need for food, ever-increasing waste, destruction of agricultural lands, and unconscious construction in the fertile lands. Consequently, it has become necessary to use "smart farming practices" for the good of human beings. Thanks to new artificial intelligence applications in this area, controls such as time and nutrients required for a plant to grow can be provided. Thus, a more efficient and faster harvest is created. To create such advancement in this field, different artificial intelligence techniques such as greenhouse automation, simulation, modelling, and optimisation are used.

The third one is the water supplies. Although a significant portion of the earth is covered with water, using water resources wisely will prevent the water shortage problems that will be experienced in the near future. Considering that people spend a significant amount of their time at homes, it can be assumed that household water consumption is also relatively high and can be rationed to some extent. So, there are many systems developed in this field with artificial intelligence technology for savings. What these systems can generally do is creating statistical data by tracking the amount of water spent 24/7 and cutting off the water flow, and sending a notification when there is a leak.

Apparently, with the rapid development of artificial intelligence, the use of these applications in homes will increase considerably in the following years.

6.1.1 Machine Learning

Machine learning is simply defined as estimating outputs based on a statistical analysis method without simple "if-else" programming. As can be seen in Fig. 6.5, Machine Learning (ML) includes Deep Learning (DL). So AI covers both ML and DL. With the algorithms developed in this learning type, input data are taken, and the output is estimated. This output value is updated based on statistical analysis as new data are added. An artificial intelligence application improves itself with increasing experience. Machine Learning allows optimisation of processes and better allocation of resources as the algorithms learn from their experiences. As the number and size of the training set increases, the accuracy of the algorithm also goes up. Bigger data set of known variables allow the software to make more accurate "operations" when the software is presented with a set of unknown variables. Training of the software through feedback loops make it possible for the algorithm to learn right and wrong decisions. The combination of these factors makes the software to become more accurate and more capable than other algorithms that are manually programmed. When algorithms are fed with large amounts of training data, Machine Learning makes it possible for the algorithm to be more proficient than any other human at the same given

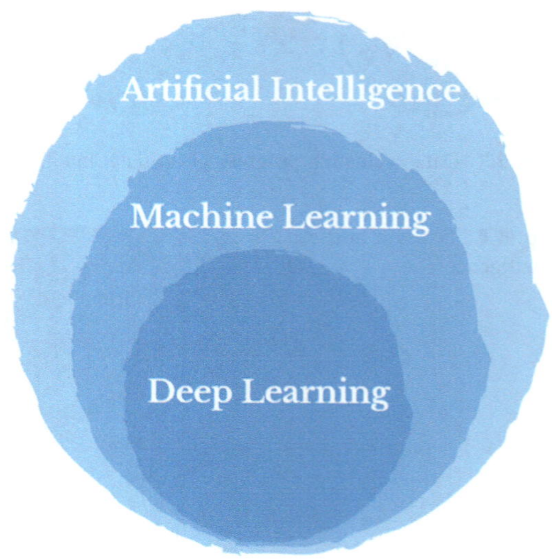

Fig. 6.5 Relationship between AI, ML and DL

task (Bini 2018). Figure 6.5 illustrates interrelation among AI, ML, and DL.

To clarify the subject with an example, IBM's Watson Health is an outstanding example of Machine Learning software in healthcare (Bini 2018). Watson Health was fed with all available information related to cancer diagnosis and its treatment in any language. Compared to human doctors, computer programs have had higher accuracies at the diagnosis of skin cancer from dermatological images (Brynjolfsson and Mitchell 2017). These programs were trained with results of subsequent biopsies. Therefore, Machine Learning shows promise in Decision Making, Prediction and Classification, and Diagnosis.

As seen in Fig. 6.6, Machine learning has evolved from past to present as Supervised Learning, Unsupervised Learning, Reinforcement Learning, Deep Learning, Deep Reinforcement Learning.

In supervised learning, inputs and outputs are given to the system. With the help of methods such as regression and classification, a function that gives the relationship between inputs and outputs is generated. Thus, the learning phase is completed. For example, we want to design a system that decides whether there is a cat in the images we give. If a cluster is created with different cat images first and then images without cats, this is supervised learning. As a result, training data for supervised learning is prepared. The information given in the training data is learned. New information is evaluated in line with these, and a decision is made.

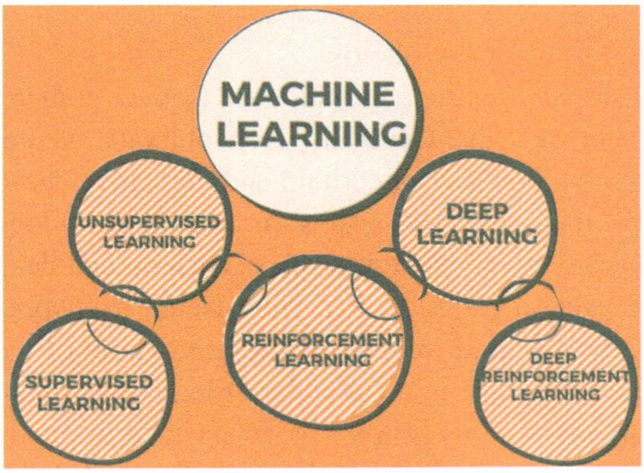

Fig. 6.6 Types of machine learning

The information in the input data applied in the unsupervised learning method is processed, and the output is then determined (Tutorialspoint 2020). Unlike supervised learning, no relation is established between input and output. Therefore, it needs more datasets than supervised learning. Unsupervised learning tries to find a pattern that can be useful for categorisation. Clustering algorithms used in unsupervised learning find natural clusters formed in the data it processes. If we elaborate, it is an example of the Hierarchical cluster used in unsupervised learning to cluster customers as women and men and then group them according to their ages.

Reinforcement Learning is a type of machine learning that is based on guiding to the desired action by giving a penalty or reward during each movement of what we want to train. Words like Agent, Environment, Reward are used in Reinforcement Learning terminology. The agent acts according to the environment and waits for a reaction by the environment (Rusen 2020). According to the reaction in the environment, it is subjected to a predefined reward system. As the reward wins, the agent is trained, and the situation of being right or wrong becomes clear in his actions. Various actions should be tried and applied many times to increase the accuracy in training. For example, we want our cat to take an object and bring it to us at home. In this example, the environment is home, and the agent is the cat. As soon as the cat starts moving, it gets a right or wrong response. Thus, the cat begins to make sense of its movements. After many attempts, the cat quickly takes the object we want and starts to bring it to us. Therefore, this method is called Reinforcement Learning. The working principle of this learning type is summarised in Fig. 6.7.

Deep learning is a method of machine learning which trains a machine to perform human tasks, such as speech recognition, face recognition, image enhancement, image identification, or prediction. An in-depth learning framework defines basic data criteria and teaches the machine to learn on its own through correlations over several computing layers instead of organizing data to operate across predefined balances (Sas 2018).

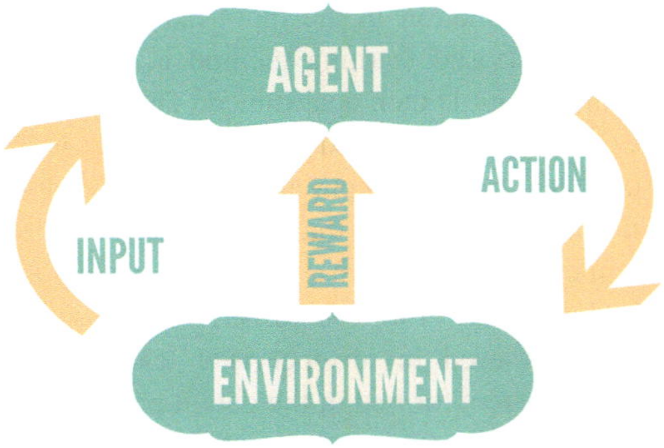

Fig. 6.7 Working principle of the reinforcement learning

Deep Reinforcement Learning (DRL) combines deep and reinforcement learning methods. DRL, one of the most advanced and active research fields in Machine Learning, has solved many problems that seemed unsolvable such as achieving super-human performances at complex games (Silver et al. 2016). Moreover, it promises to solve more difficult problems in robotics, resource management, and almost every field that requires decision-making capabilities in the future. To sum up, machine learning has provided many benefits in many areas such as health, finance, and robotics. As seen in Fig. 6.8, machine learning is used in many areas besides smart homes.

The key to machine learning applications in smart homes is sensor data. According to real-time and historical data, the decision-making mechanism of smart homes is taking shape. With a simple example, every day, the inhabitant comes home at 7 p.m., take a shower, makes his coffee, and then turns the TV on and continues from where he left off. Thanks to the smart home sensor data, the user can learn the daily routine and predict his next move. When the user opens the door of the house, the bathroom lights will be turned on, then the coffee machine will work, and when he takes his coffee, the television will turn on. In this way, the lighting of the house is controlled according to the user's routine, and energy management is provided. Figure 6.9 compiles numerous ML applications in smart homes.

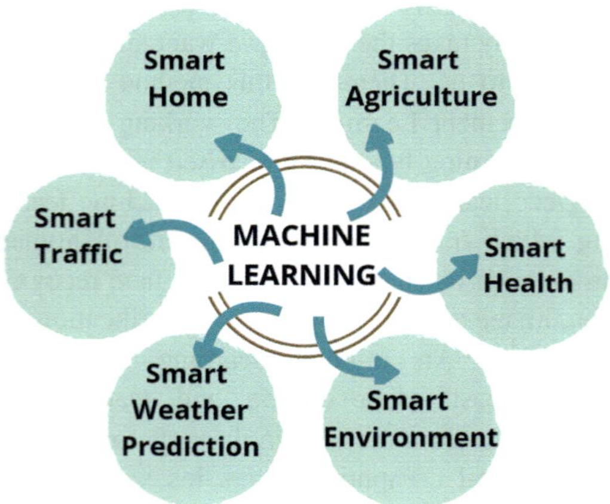

Fig. 6.8 Sectors in which ML is used

6.1.2 Deep Learning

Deep Learning (DL) is a machine learning subfield that relates to algorithms and simulates data processing of the human brain to understand human speech and visually recognise objects. The "deep" in DL refers to having more than one hidden layer. It can be seen in Fig. 6.10 that while neural networks have one hidden layer, deep learning neural networks have more than one hidden layer. Deep Structures consist of three different layers. These are the input layer, hidden layers, output layer. The input layer receives the data, the hidden layers perform mathematical computations on the data, and the output layer returns the output data. Each connection between neurons is associated with the weights. The initial values of these weights are assigned randomly. These weights dictate the importance of the input value. After that, these weights are trained by using labelled data such as cat photos or car photos. Finally, our deep neural network, which is just trained, learns how to know cat photos or car photos.

Deep learning algorithms are applied in many different fields. Generally, Convolutional Neural Networks (CNN) algorithm and derivatives have obtained the best results in image and video processing compared to other machine and deep learning algorithms. Recurrent Neural Networks (RNN) is used to understand the structure of the incoming data in a particular order, such as text, speech, various sensors based on time or statistical data. Long Short Term Memory (LSTM) architectures give quite good results in speech

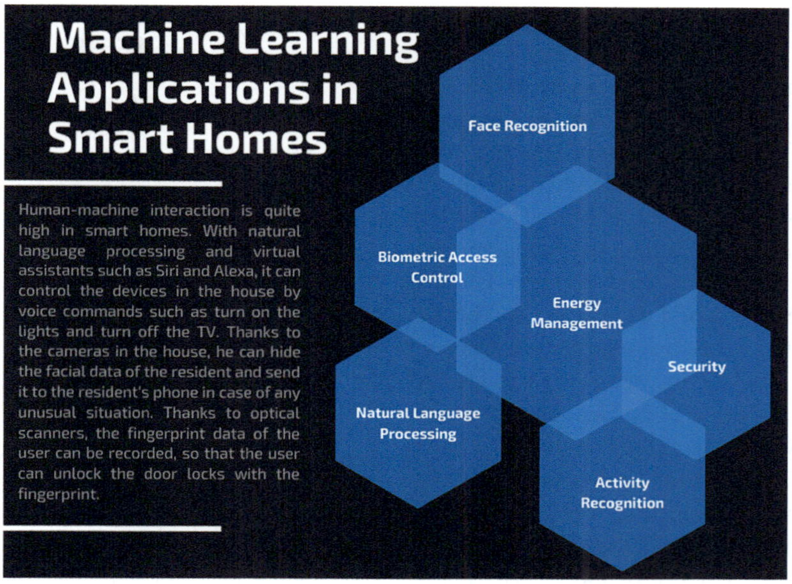

Fig. 6.9 Machine learning applications in smart homes

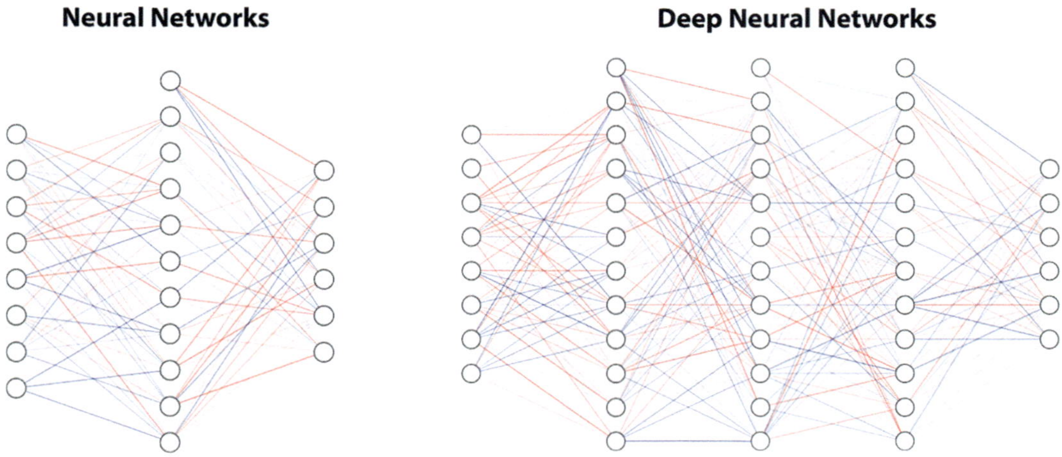

Fig. 6.10 Difference between artificial neural networks and deep neural networks

and text processing. In smart cities, these algorithms are used for object detection (Cai et al. 2016, October), facial recognition (Matsugu et al. 2003), landmark detection (Wu et al. 2017), real-time motion analysis problems (Mutis et al. 2020), emotion detection (Kanjo et al. 2019), face detection (Yang 2018), face verification (Sun et al. 2015), and iris recognition (Nguyen et al. 2017), chatbots (Xu et al. 2017), sign language translation (Huang et al. 2018), anomaly detection (Du et al. 2017). The self-sustaining concept is improved by deep learning algorithms such as enhancing suggestions and decision-making ability of automatic systems under less human control (Peng et al. 2020), human activity recognition (Fang and Hu 2014), pain detection (Nugroho et al. 2018) and controlling energy consumption (Dey et al. 2017).

6.1.3 Reinforcement Learning

Reinforcement Learning (RL) is one of the most prevalent and exciting topics for the last years. Especially after the integration of RL algorithms with deep neural networks, the effect of the RL techniques is considerably improved. In 2016, AlphaGo, DeepMind's computer program that was trained on a deep reinforcement learning algorithm, beat the world champion Lee Sedol at the game of Go which has long been considered as the most challenging classic game for AI (Silver et al. 2016). This incident caused a huge impact all over the world and proved the capabilities of RL. Since then, the development of RL algorithms has accelerated further.

In supervised learning, the learning process occurs with the existence of lots of data that is labelled by a competent external expert. On the opposite side, there is no external instructor in unsupervised learning, and the aim is to find hidden patterns in a stack of unlabelled data. Unlike these two categories of machine learning, reinforcement learning approaches the problem from a different perspective. Reinforcement Learning is composed of an agent that takes actions in an environment. The agent is not told what actions to take, so it must discover which actions offer the highest reward based on the environment's state. The most significant distinctive characteristics of reinforcement learning are trial and error search and delayed reward (Sutton and Barto 2018). Four key components of RL are a policy, a reward signal, a value function, and optionally a model of the environment (Sutton and Barto 2018). Often, the RL problems are formalised as a Markov Decision Process (MDP). The MDP framework consists of a finite set of states S, a finite set of possible actions A, the transition function T, and the reward function R (Otterlo and Wiering 2012). Figure 6.11 illustrates the agent environment interaction in Markov Decision Process framework. Today, there are many RL methods applied in a variety of areas.

Studies in the field of Reinforcement Learning are generally concentrated on energy management. Despite this, some works continue in several different areas. An RL algorithm for human motion prediction was created in one of these studies. Thanks to the prediction of human movements, residents with memory disorders can be helped. In addition, devices can be prepared in advance for some activities, such as cooking (WeiWei and Wei 2019). Cybersecurity is one of the biggest concerns in smart home applications. The security system for smart home units was developed with multi-criteria reinforcement learning in research (Chen and Luo 2012). Khalili and Aghajan developed a reinforcement learning algorithm that predicts the music and light preferences of users (Khalili et al. 2009). Anomaly detection algorithms detect unexpected abnormalities in the system for security. Finally, in a study, an algorithm was developed using deep reinforcement learning for anomaly detection (Zhong et al. 2019).

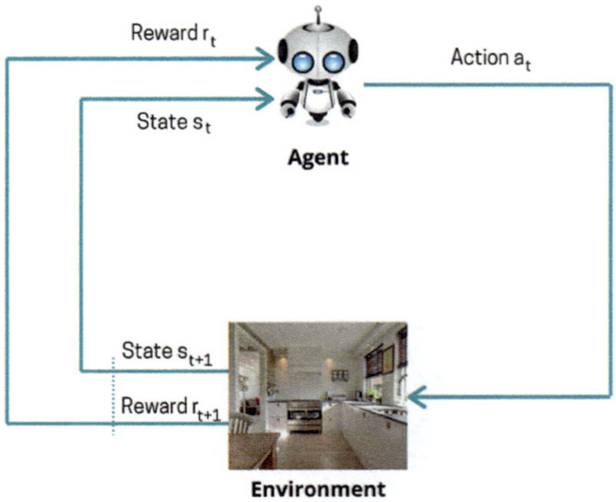

Fig. 6.11 Agent environment interaction in RL

In smart self-sustainable environments that can accomplish the Sustainable Development Goals, there is a need to build a Home Energy Management System (HEMS) that masterminds the energy resources and IoT devices (United Nations 2020). Such a system should control HVAC (Heating, ventilation, and air conditioning), water heater, lighting, electric vehicle charging, and carry out appliance scheduling. Reinforcement learning methods have been applied to this sort of task to reduce energy costs without sacrificing comfort.

There are many RL approaches for energy saving in a smart home that tackle the tasks separately. For example, lots of studies have discussed HVAC control for reducing costs because of its dominant power consumption. The proposed solutions to this problem include the algorithms: Q Learning, Monte Carlo Actor-Critic with neural networks, Neural Fitted Q-Iteration, DQN, A3C etcetera (Ruelens et al. 2015; Wang et al. 2017; Marantos et al. 2018; Wei et al. 2017; Zhang et al. 2018). Considering these studies, we can claim that the RL algorithms outperformed the other approaches. However, there is a vast potential to reduce costs by applying RL to build an energy management system that conducts all devices, energy storage batteries, and solar power rather than simply performing the algorithms to appliances individually. With this approach, energy costs can be reduced by 30–50% (Mason and Grijalva 2019). Yet, the number of studies applying reinforcement learning for energy management keeps increasing. With the obvious trend towards DRL methods, the future seems promising.

In general, AI technology in smart homes can be categorised into six main functions: activity recognition, data processing, voice recognition, image recognition, decision-making, and prediction-making. We can classify these functions with related smart home application fields, as can be seen in Fig. 6.12. Smart home energy management utilises AI for decision-making, while entertainment systems make use of voice recognition. On the other hand, healthcare implementations in smart homes are supported by not just decision-making and voice recognition, also activity recognition and image recognition. Besides, intelligent interaction systems exploit prediction-making and voice recognition functions. Furthermore, the security of smart homes is assisted by three AI functions, which can be sorted as image recognition, data processing, decision-making. And finally, smart home personal robots utilize prediction-making, image recognition, and voice recognition (Guo et al. 2019).

6.2 Aim of the Study

Achieving sustainability in living spaces is one of the most critical goals for the future. Our purpose in this research is to build a self-sustaining house that can manage all appliances to optimise water consumption, reduce waste and energy costs while maintaining the households' comfort. There is a need to build intelligent systems that can make decisions in complex situations to carry out full sustainability. Hence, we aim to develop an HMS to mastermind the Open Digital Innovation Hub (ODIH) with minimum human intervention.

Our main goal in the project is to build a digitalised and self-sustaining ODIH by using AI

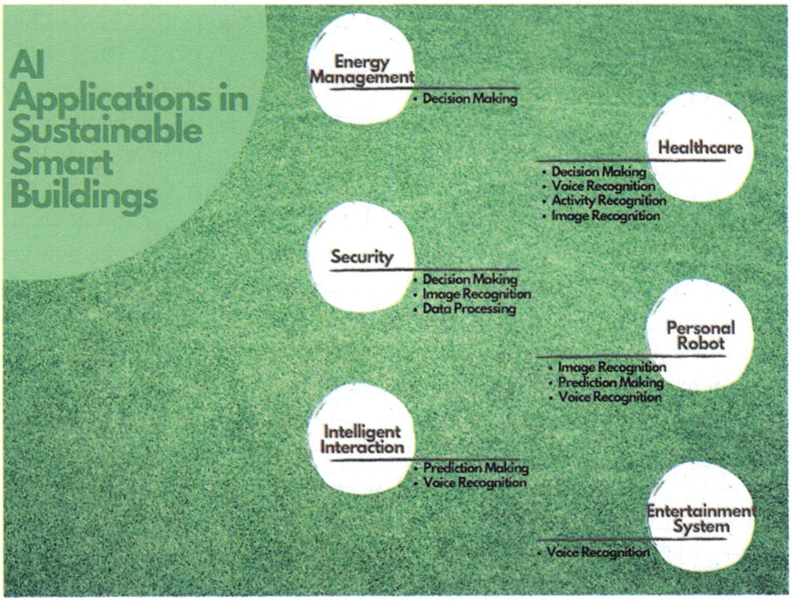

Fig. 6.12 AI applications in sustainable smart buildings

technologies. To expand this goal further, we can describe the primary targets of this study as follows:

- Defining standards for HMS to comprehend its abilities and make a clear pathway for further development
- Examining the AI techniques to enhance the capabilities of home
- Analysing the ways to optimise energy, food, and water consumption/production
- Exploring a design of the HMS software to ensure reliable control of IoT devices.

To achieve these goals, it is necessary to conduct research in subareas such as energy management, food & agriculture, water, healthcare, customisation/entertainment, and security. We aim to create the best environment for a sustainable smart home by using these areas. In subsequent sections, we will review the AI applications and discuss the methods for HMS.

6.3 Methodology

6.3.1 The Home Management System

The number of smart homes is expected to rise to 481 million by 2025 (Statista 2020). The future homes will get smarter year by year. Hence, the capabilities of HMS will be more advanced in the future. HMS is the brain of a smart home. It can give householders the essential tools to control their homes or can even control homes autonomously. Today, there are lots of HMSs with different sets of skills. There is a need to classify the HMSs to be able to examine them more clearly. Therefore, we will analyse with a five levelled analogy.

First, HMS is in its simplest form in Level 0. It is connected with only a limited number of devices, and its management structure is restricted. There is not any kind of automated system. It just provides an interface for adjustments of devices like lighting, thermostat, or lock systems. The users are not allowed to add or remove appliances. Usually, these systems do not have mobile apps. They provide the control mechanism with just their related hardware. Brilliant Smart Home Control can be given as an example to this level (Brilliant 2020).

In Level 1, HMS could connect with a wide range of devices all-around home. Manual management can be executed with a touch panel, mobile app, or voice-enabled system. Generally, the users are allowed to add their devices to the HMS. Still, we do not encounter any automation at this level. Amazon Echo Show is an example of this level (Amazon 2020).

In Level 2, HMS starts to enable the user to carry out some basic automation. However, at this level, automation is entirely manual. It is restricted to the ability of the product and the preferences of the user. At this level, there is not any intelligent system that could take action autonomously. Nowadays, most of commercial products could only reach Level 1 or Level 2 management system. Homey could be given as an example (Homey 2020).

In Level 3, from now on, HMS has some sort of intelligent system powered by AI. The system could recognise the patterns and give suggestions to the users. It can manage some significant tasks all by itself. Generally, the system behaves as an assistant and supports the residents of the household. Today, we have the AI technology and computational power to build such systems. With the knowledge we have, we estimate that Kirio and LG ThinQ powered smart home systems of LG could lie somewhere around Level 3 (WIRED 2019; Kirio 2020). However, we can say that no system has proved its capabilities yet across different circumstances.

Ultimately in Level 4, HMS reaches a level such that it could manage all tasks and details. The system knows what to do and which action to take for the household better than the residents themselves. It behaves as a master and sage for users rather than just an assistant. Today no such system could reach this level, and some researchers are quite doubtful that AI could ever get to such kind of intelligence (Fjelland 2020).

The functions of HMS can be considered in seven subgroups besides five levelled analogy: energy management, food, water, waste management, healthcare, customisation/

entertainment, and security. In Fig. 6.13, the functions can be seen in the order of our analogy. In the subsequent sections, we will examine the functions of subgroups separately.

6.3.1.1 Energy Management

The world's energy consumption is estimated to rise by nearly 50% between 2018 and 2050 (U.S. Energy Information Administration 2019). The future of our civilisation depends highly on managing energy needs efficiently and sustainably. Therefore, energy management systems have a significant role in achieving full sustainability. With advanced digital technologies like sensors, actuators, and communication systems, home energy management systems have become more accessible. Especially the vast advances in the field of machine learning enable researchers to design state-of-the-art energy management systems.

Energy management systems have two major components: monitoring and controlling (Lee and Choi 2019). Basic techniques only have a monitoring component. These systems track energy consumption and generation, providing a comprehensive overview using data with charts and graphs by its dashboard for situational awareness. They may procure a manual control for the household appliances, but there is no automated system, and the system could not arrange the flow to use energy optimally. Such systems correspond to the Level 1 management system in our analogy. Today, the majority of commercial applications provide such kind of Level 1 management system. For example, Powerley and Neurio provide energy management systems as a product (Powerley 2020; Neurio 2020). These systems promise savings by letting one know what is going on at his/her home with the data collected by relevant hardware. However, there is no intelligent decision-maker control mechanism at these systems that decides how to carry out the savings by managing the energy flow while maintaining once comfort considering their personal preferences.

When we start talking about intelligent decision-making mechanisms, the other major component, controlling, steps into account. If both components, monitoring and controlling, exist in an energy management system, the system would lie somewhere between Level 2 and Level 4 concerning its AI power. A more generalised classification of functional components could be seen in Fig. 6.14.

Then, what should we expect from an advanced energy management system overall? First, the system should provide a comprehensive, informative overview of the situation of the house. The household should have good insight into their energy usage. Second, the system should allow for control and manual automation over the appliances, but its main job should be making this function unnecessary. The reason is that a robust advanced energy management system is supposed to manage everything optimally without any human intervention. For that reason, the system should forecast the future consumptions, arrange itself to the preferences of the householders to satisfy the occupant's comfort needs, schedule appliances and loads to reduce costs and environmental impacts.

The household appliances connected to the HMS for energy management are shown in Fig. 6.15. Especially, the trend towards the broad adoption of Distributed Energy Resources such as photovoltaics, wind turbines, and energy storage systems makes the energy management systems vital for the future. On the other hand, the electricity distribution grid's Demand Response (DR) programs are becoming more widespread, and their adoption is expected to expand steadily over the years ahead. The DR programs aim to encourage consumers to pay more attention to renewables and reduce their electricity consumption during peak energy demand periods. Some programs even let the consumers generate income by selling the surplus of electricity back to the grid (Alfaverh et al. 2020). Today, a lot of research handles the demand response by creating intelligent systems using state-of-the-art machine learning and reinforcement learning algorithms (Lu et al. 2019). However, in this chapter, we will not dig more into the demand response side.

Except for the industry's basic systems like the products we mentioned before, a good

HMS Level	Function	Class
Level 0	Provides an interface to control heating/cooling, lighting, water heater, batteries etc. individually	Energy Management
	Allows the user to turn devices on and off via related control systems	
	For the regular water consumption in various activities such as washing clothes and utensils, taking a shower - a smart water recirculation system can be built which recycles the used water that will remove calcium and magnesium	Waste Management
	The integration of rainwater harvesting system, smart taps, water meters and water flow sensors will help conserve and resend the water into the smart home system for lesser wastage.	
	The compartment can perform the compaction of the waste so that its volume can be reduced by up to 10 times, even before collection	
	Live video stream on mobile app from indoor and outdoor of home.	Security
	Flashing of the lights periodically when nobody is home.	
	Fire detector	
	Gas leak detector	
	Spotlight that turns on when an unexpected person arrives	
	Auto-zoom and tracking in an unexpected situation	
	Sound system in every room in the house can be controlled. Different music can be played in different rooms.	Entertainment
	Display systems in each room can be controlled by the Home management system.	
	Whenever you want to swim, you can control the pool temperature.	
	Show the data from leak inspector on water meter	Water Consumption and Generation
	Leak detection sensors which will be attached to all the pipes at home that will send an alarm to the make motoring system when the tap is not in operation and cuts the supply for that particular tap.	
	Provides detailed information about water temperature, usage and status.	
	Handles the treatment process and reuse of waste water	
Level 1	Monitors electrical activity of appliances from the control panel, and the mobile app	Energy Management
	Demonstrates real-time information about the power consumption and energy production from solar panels and wind turbines	
	Shows how much the user is consuming electricity at any moment	
	Allows the user to turn all devices on and off remotely via mobile app or general control panel	
	Provides a dashboard that reports daily, weekly, and monthly consumption and generation statistics for the situational awareness	
	There will be a camera in the door eye as well as outside the house door. This second camer The second camera will perform auto zoom and tracking when there is a door pressure.	Security
	The entire library of content can be accessed from any device in the home.	Entertainment
	Chronic Disease Tracking	Healthcare
	Remote Patient Monitoring	
	Users can access real-time flow data, previously preferred water temperatures, past usage statistics, and receive leak alerts for slow or major water leak events.	Water Consumption and Generation
Level 2	Resolve water issues in real time from anywhere with registered solution or web portal	Water Consumption and Generation
	Notification of irrigation schedule and watering times according to the type of flowers in the house	
	Giving warnings in case of an unexpected increase in heart rate thanks to smart watch data.	Healthcare
	Notify the relatives of the individual's heart rate and call 112 if necessary, thanks to the smart watch used by the resident.	
	Allows manual automation and scheduling of the connected devices (heating/cooling systems, water heater, clothes washer, clothes dryer, dishwasher, refrigerator, lighting, plug loads and connected electronics, rooftop photovoltaics, home batteries etc.)	Energy Management
	Sends alerts and notifications for unusual or unsafe energy usages. For example detects if something is accidentally left on	
	It notifies about the expiry day that needs to be dumped to the dustbin via mobile app	Waste Management
	Tracking how full the garbage is. When the garbage became full, the HMS notify the households.	
	Giving the information about the product such as name, system will notify the user about what date and time that the product should be used before it gets rotten.	Food - Smart Agriculture
	When the bell rings while watching TV, it can reflect who is in front of the door onto the screen.	Customization
	Automatically turns on the TV channel when your favorite program starts.	Entertainment

Fig. 6.13 HMS functions

Level 3	Recognizes abnormal patterns of energy use, pointing towards potential problems of several appliances	Energy Management
	Identify modifications in consumption and make suggestions that are specific to the household for reducing costs and environmental impacts	
	Customizes itself to the preferences of the houseowners with its predictive model which is trained on the data from connected devices.	
	Predicts future energy consumption and generation while projecting how to satisfy occupant's comfort needs	
	Decides when to charge batteries and store the energy that is generated by solar panels and wind turbines by AI powered control system without human intervention	
	A typical smart heating/cooling sensor will monitor the external temperature through data from weather websites or from a sensor outside the house and tallies it with the body temperature from the data collected via wearable devices and creates an ideal temperature for the user and their family.	
	Operates Advanced Demand Response system to reduce costs while managing energy resources optimally	
	Forecasts indoor air-temperature for the efficient energy management	
	It could adjust itself to the preferences of the households by exploring the patterns of the activity of them.	Customization
	Clothes Suggestions based on the agenda of the person, and the weather	
	Dynamic sign language recognition for interactive access to HMS	
	If someone is asleep in the house, the system warns when anyone want to increase the volume.	
	HMS can recommend the best drinks for your activity based on your fridge or wine collection.	
	It could give tips as a personal assistant according to the need of the households for several activity plans	
	Items or products which are expired or used will get added in shopping list.	Food - Agriculture
	Fridge stock estimation and diet advice using RFID -(A Content-aware Fridge Based on RFID in Smart Home for Home-Healthcare)	
	Recipe suggestions compatible with the products in the refrigerator are provided.	
	The system collects the food information in the refrigerator and analyzes the user's preferences and gives information about the products that should / should not be bought.	
	Determines the type of seed to be planted in vertical agriculture design.	
	Determines the most suitable water amount, nutrients and temperature range for the healthier growth of the plant considering the needs and demands of the household.	
	A robotic system in vertical farming can be developed for harvesting or surveillance of the systems. The system detects the mature plant through the camera and harvests the product	
	The mechanical arm will plant the suitable seed to the empty pot, respect to the user's demands or needs.	
	Giving suggestions in case of an abnormal body activity	Healthcare
	When there are extraordinary situations such as falls or injuries at home or unusual behaviour of a person, it takes snapshots from the camera and sends a notification to the relatives of the individual.	
	Performs automated assessment of cognitive health	
	Initiating certain activities when a group of words are uttered/ Initiating certain actions through voice command. Contacting law enforcement in case of trespassing or upon pronunciation of certain keywords	Security
	Detects the person waiting in front of the door is whether a foreigner or not with face recognition technology	
	Artificial intelligence powered system optimizes the network so that the households could continue to their activities without any interruption.For example, while watching online movies, it could turn off other unnecessary online devices in the home.	Entertainment
	While the room is changing, the system can follow you and move audio and video to the new room.	
	If the user do not sit in front of the TV for a long time or fall asleep, the HMS turns off the TV automatically.	
	Changes the lighting and the ambiance depending on the activity of the households.	
	Automatically adjusts the seasonal preferred tap water, drinking water and shower temperature	Water Consumption and Generation

Fig. 6.13 (continued)

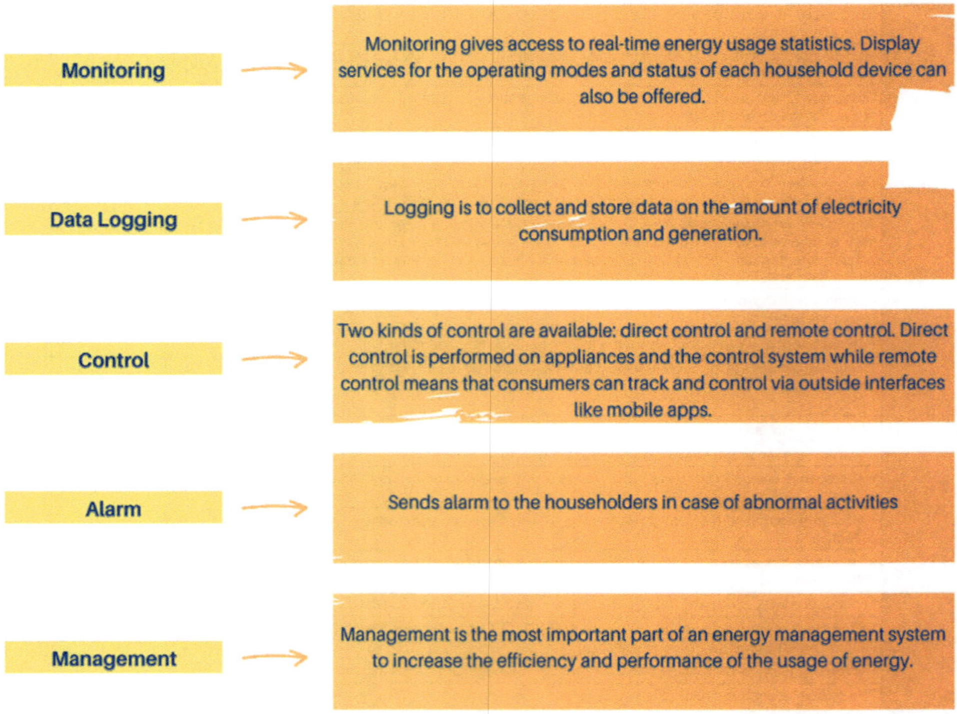

Fig. 6.14 Functions of the energy management system (Zhou et al. 2016)

Fig. 6.15 Energy intensive appliances at homes

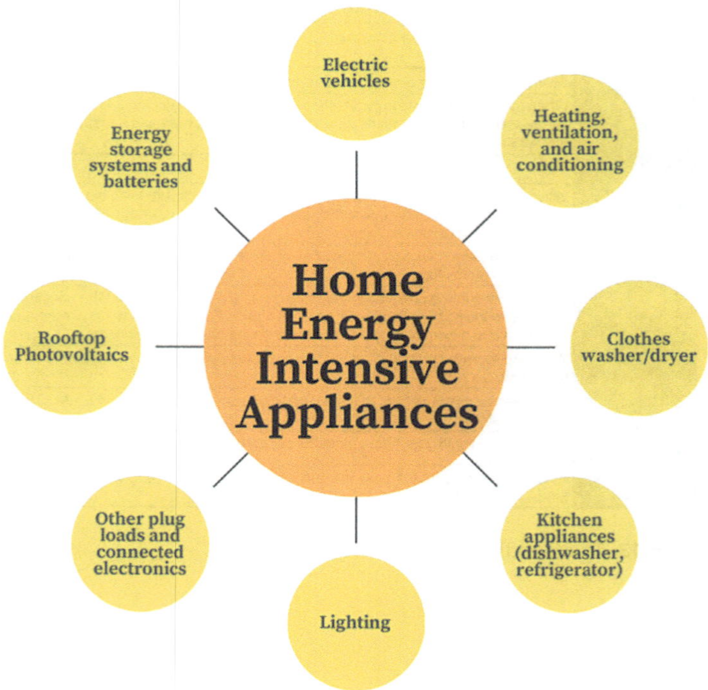

example of an advanced energy management system is the "foresee" of the National Renewable Energy Laboratory (NREL). Foresee received the R&D 100 award in 2018, thanks to its advanced capabilities compared to its equivalents. It is the first example of a near Level 3 system, but it is not a commercialised product yet. Foresee learns each home and its occupant's schedules and patterns to make highly accurate predictions of environmental impacts, comfort needs, energy costs, and future consumption by leveraging machine learning algorithms and advanced data analytics (NREL 2020). Foresee uses the Simple Multi-Attribute Rating Technique Exploiting Ranks (SMARTER) method to bring the user preferences to light. It uses the Expectation–Maximisation of Gaussian Mixture Models to learn the usage patterns. According to the researchers, the software promises up to 7.6% energy savings without requiring considerable behavioural changes (Jin et al. 2017).

Reinforcement Learning (RL) is recently getting more attention in the literature for designing energy management systems. We defined RL before in the previous sections. Especially after the integration with neural networks, RL has been used broadly in various fields by researchers. Today, Deep RL is the closest between other ML methods to the AI researchers' dream of achieving general intelligence.

RL algorithms could be classified into two subgroups: model-based and model-free algorithms. In model-based algorithms, the agent is expected to learn the environment's model by observing the environment's state changes. For example, Explicit Explore and Exploit, Prioritized Sweeping, and Dyna algorithms are designed with the model-based approach. In contrast, the model-free methods do not need to formulate the model of the environment. They aim to learn the optimal policy by trial and error. The most common examples of this kind of algorithms are Q Learning and State–Action–Reward–State-Action (SARSA) (Mason and Grijalva 2019). Today, the model-free approach is more prevalent in the research literature and product applications because they are more compatible with real-world problems and require less computational power. Q Learning, SARSA, and Actor-Critic are some of the foremost and fundamental RL algorithms. Although these algorithms are relatively old, in literature, there are still many studies utilising these methods. However, some severe advancements in reinforcement learning have paved the way for novel and robust algorithms. The most recent and state of the art Deep RL algorithms are Deep Q Network (DQN), Double Q Learning, Deep Deterministic Policy Gradient (DDPG), and Asynchronous Advantage Actor-Critic (A3C) algorithms. Deep Reinforcement Learning refers to the combination of RL methods with deep neural networks. These algorithms can speed up learning quite well, handle a broad set of environment states, and learn more complex policies than those look-up tables. Besides, the Multi-Agent Reinforcement Learning approach is also a well-known technique for multi-agent problems. A more general classification of RL algorithms is shown in Fig. 6.16.

RL algorithms significantly increase the energy efficiency of homes. In the literature, some studies report up to 20% savings using RL. Some even claim much more savings, but we should be careful while analysing the results. We have to make sure that simulations are highly realistic and the findings are reliable. We mentioned some studies for energy management with RL in the introduction. A detailed literature review on RL and Energy Management Systems can be found (Mason and Grijalva 2019).

Much of the recent research focus on DRL algorithms due to their success in solving complex problems. However, everything is not perfect, and there are some limitations to RL research. The vast majority of the studies are stuck in the simulation, and there is a lack of real-world implementations. This approach relies upon the accuracies of the simulators. Moreover, this is not unfavourable because many of the RL studies occur with simulations. Still, the future direction of the researches should move to the real world for more realistic consequences.

Innovation ecosystems like ODIH have a crucial role in building advanced technologies thanks to the environment they provide. ODIH

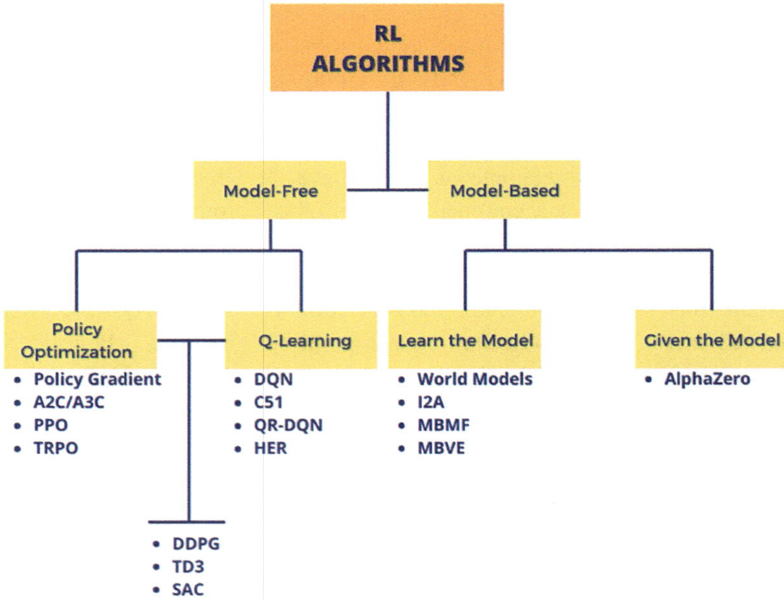

Fig. 6.16 A taxonomy of RL algorithms (OpenAI 2020)

will support the research and development process of energy management systems thanks to its open innovation environment. It will strengthen the development road to Level 3 and beyond by giving researchers and entrepreneurs a chance to test their systems. We think this is very important for filling the gap between simulations and the real-world.

6.3.1.2 Food & Agriculture

A. Food

Every living organism on earth needs the energy to maintain its vital functions, such as growth and reproduction. Living organisms can obtain this necessary energy only from one energy source: food. Providing essential nutrients from different types of food, such as carbohydrates, fats, and proteins, as in the case with animals and plants, is not just a survival fuel; it is also essential for protecting physical-mental health prevention of diseases in humans. However, food, which is invaluable for all living creatures, is at significant risk due to the problems that arise today, such as population growth and climate change.

According to the Food and Agriculture Organisation of The United Nations (FAO), climate change will cause extreme weather events in the short term, causing a change in temperature and precipitation patterns in the long term (Food and Agriculture Organization 2008). Consequently, results decrease access to the food, while the change in food production–distribution channels is observed. Like a chain reaction, these cause a wide range of events, from the deterioration of human health to the economic recession and the change in purchasing power (Food and Agriculture Organization 2008). Apart from that, research that The World Bank Development Research Group conducts shows that countries with high birth rates suffer from food shortages. For example, it is concluded that the Sub-Saharan African region's population will double by 2050 (Chen and Ravallion 2008). Undoubtedly, with more people living in the region, bigger the food crisis will begin to emerge. Another research revealed the situation with numerical data. According to Valin et al. (2014) as cited by (The Natural Resources Defense Council 2017), it is estimated that the population living on the earth by 2050 will demand 1.5 to 2 times more food than is needed in 2005.

Hence, as both studies show, uncontrolled population growth and climate change prevent people from accessing their most basic needs, namely "food" nowadays. However, in the near future, unless something is done as a solution for these problems, living beings on earth may be at significant risk of extinction due to famine and drought. Apart from the global problems mentioned, another big human-induced problem can be overcome if the necessary precautions are taken: food waste. According to the report prepared by NRDC, 40 per cent of food was wasted in the United States. (Rethink Food Waste through Economics and Data (ReFED) 2016) In other words, when America throws more than 400 lb of food annually, this can be expressed in a loss of $ 218 billion each year (Buzby et al. 2014). Another study found that less than a third of the food that is thrown away every day would be enough to feed 42 million Americans who face food insecurity (Coleman-Jensen 2016).

Consequently, simple personal precautions should be taken at home. Fortunately, an institution called the "Food and Agriculture Organisation" formed under the United Nations' umbrella, "Sustainable Development Goals" has been established. In this direction, studies are carried out to solve the problems, methods are developed, and solutions are offered in line with them.

As a solution, making living spaces smarter, wasted foods that are not under human control can be easily prevented. To do that, we aim to increase the efficiency in the area of food and prevent wastes thanks to the algorithms created in our Home Management System. These algorithms are listed below:

The first application is created to eliminate the overuse of foodstuffs. With the development of technology, people have started to buy more food than they need because they have a refrigerator to store foodstuffs without spoiling them safely. However, refrigerators start to fill up with more packaged foods, fruits, and vegetables than the actual need. This problem causes waste in food items. Recently, an application has been developed that provides consumption advice to the user before the food spoils, based on the food's average consumption time. Also, if the user ignored the consumption advice, it is the same function that gives a secondary warning that the product is spoiled after a certain period of time. Hence, the aim is to lower the food wastes that are forgotten and upper productivity. The function works in this way.

Some system requirements provide this system to be operated. In the first place, a computer that is interconnecting with the refrigerator is needed. This computing device has its own processor wherein the application is configured and a memory where the device can store all the data taken from the sensor. After that, a sensor identifies the food item and another sensor for obtaining the chemical gas level that belongs to the foods in the refrigerator. The measurement is made according to the threshold value of the chemical gas. If it exceeds a specific value, the system detects that the food item belonging to this measurement is spoiled and sends a notification to the user.

The second application is for those who do not know what to cook every day. Thanks to a system called Innit, the food in the refrigerator is efficiently evaluated. The system works as follows:

Using the sensors and cameras placed inside the refrigerator and with the contribution of the face recognition system, all the food in the refrigerator is identified. Even by using the application, it is possible to see inside the refrigerator and know which product it is. Besides, thanks to the camera that is placed on the kitchen ceiling, the system identifies when several carrots are placed on the cutting board and, the system offers a recipe suggestion by examining the database of recipes. Secondly, the algorithm developed to give the best cooking advice is provided with step-by-step cooking instructions.

Another application to be used in the Home Management System is to detect the most consumed foods by a refrigerator. When these "most consumed food items" run out, the system adds the product to the user's shopping list to be bought. According to the study, possible "food detection algorithms" are reviewed to extend the shelf life of food, increase efficiency, and decrease waste. The selection of the best method is essential. However, considering that there are different products in the refrigerator with entirely different shapes and sizes, this is not easy. Algorithms such as image recognition, RFID scanning methods, barcode and QR code scanning methods require manual control of the user, which is not practical. To choose the better algorithm, the best algorithm to be chosen is determined as the SSD target detection algorithm by comparing it with algorithms such as YOLO, SSD, R-CNN, Fast-R-CNN, and Faster-R-CNN, considering the actual internal state of the refrigerator (Gao et al. 2019, pp. 303–304). Figure 6.17 shows the resulting experimental result.

As a result, after performing various theoretical calculation and experimental results, they found out that an algorithm called "SSD 512," which is one of the types of the deep learning SSD algorithm, is the most logical method to choose (Gao et al. 2019, p. 308).

Thus, the refrigerator's product detection algorithm is obtained, the detection of the missing foodstuffs by using the image processing will be done, and automatically these items will be put on the shopping list.

B. Agriculture

Agriculture has been regarded as a system that can meet the food production need for decades. However, today agriculture has become a concept that includes the processing, marketing, and distribution of different food products such as forestry, poultry breeding, beekeeping, stock farming, and classical farming activities. Regardless of the type of agricultural activity, the difference in the physical characteristics of the area, such as weather conditions, geological elevation differences, and the agricultural area's width where the agricultural activity is carried out, is essential. Furthermore, it has become necessary to develop different methods due to the transformation of agricultural areas into settlements, loss of soil efficiency, and pollution,

Fig. 6.17 An experimental result depiction

which has become one of the effects of today's rapid population growth problem. For this reason, there has been a need to develop alternative agricultural practices such as aeroponics, hybrid seed technology, and hydroponics to obtain better quality and efficient products.

In ODIH, hydroponic or soilless agriculture has been chosen as the agricultural method. The soil-less farming method has been preferred since it does not take up much space and the plants grow faster with less water. Although the method is preferred as an effective farming method, many functions have been developed to increase the house's usefulness and efficiency.

The first function is a system that alerts when crops are harvested. While today's people cannot even meet basic needs such as cleaning and cooking when they come home from work, a warning system should be used to get high product quality and efficiency in a field such as food production. According to the study, the system is created using the "Fast R-CNN algorithm" one of the convolutional neural network methods to detect rotten apples. Using a convolutional neural network is explained as the success rate in image processing is much higher than traditional neural networks, although the training time is very long. Then, a Faster R-CNN approach based on Convolutional Neural Network architecture is used in the study. (Cömert et al. 2019) In the study, to decide whether the two images are the same or not algorithm, which consists of many layers, divides the image into its parts, and then a filter is applied on each part. As shown in Fig. 6.18, after the filter process, the picture becomes smaller and interprets the pixels obtained as a result of this process; decision making is started (Goodfellow et al. 2016).

Then, using the Food and Agriculture Organisation's food product database, a comparison is made depending on the real image belonging to the growing plant compared with the image defined in the desired product system. If the images match, a warning, "The food is ripe, please harvest!" is sent via the phone application.

The second function provides automatic control of the humidity, water, and temperature required for the plant established by the soilless farming method. Humidity and temperature are very much related to each other. Firstly, an increase in the temperature of the air increases the humidity level in the environment too. Since the plant requires this to thrive, high humidity can cause fungal pathogens to germinate and spread disease. Also, adjusting the water level correctly is the most crucial factor for plant survival. The main components of an automatic climate control system's should include the sensor to observe the critical conditions, a controller to adjust arrangements, and a wireless network to provide alerts and updates. In our Home Management System function operates thanks to the Wireless Sensor Labels and Kumo Sensors, which are products supported by Home Assistant, sensors monitor the temperature/humidity/soil moisture and water levels then, it warns the user when it detects temperature or humidity level lower or higher than it should be. The system, which can be seen in detail in Fig. 6.19, is included in HMS with Home Assistant, an open-source system. Integration can be done by the implementation called "wireless tag".

The third function is a system that closes the curtains when the plant does not need sunlight and opens when it does. To achieve this, we need two things:

1. A sensor that can detect daylight
2. An automatic arm system that can move the curtains
3. Object Detection Algorithm to identify the type of plant and determine the plant's light

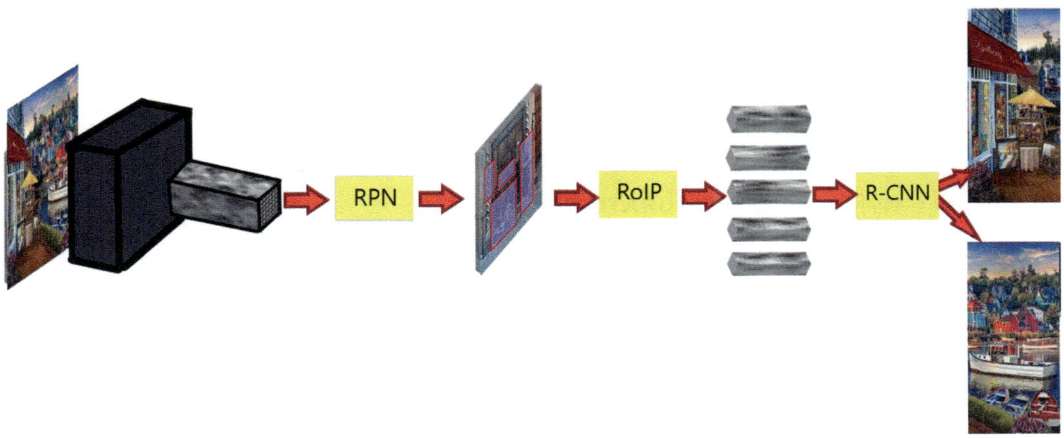

Fig. 6.18 The implementation steps of the faster R-CNN algorithm

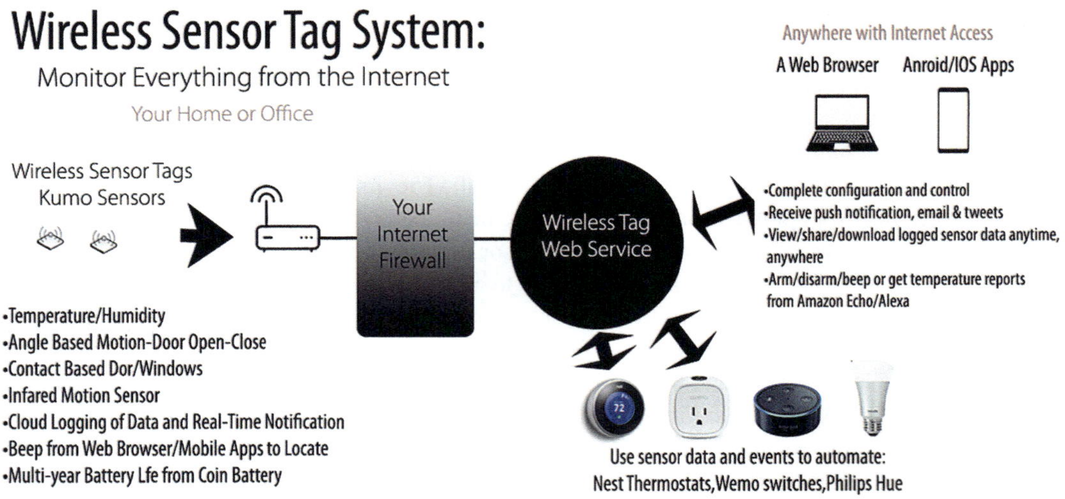

Fig. 6.19 The image shows the details of the system

requirement according to the features taken from the database.

The function mentioned is easy to integrate into our Home Management System. It is necessary to ensure that only the sensors from the brands supported by the Home Assistant System must be preferred. The system should also be configured with the configuration.yaml file, as shown in Fig. 6.20. Then all the arrangements are made automatically thanks to the application belonging to the device.

6.3.1.3 Water Consumption and Generation

Water is one of the basic life needs, and if it is consumed economically, it becomes advantageous in terms of reducing household expenditure and tackling climate change. In this respect, incorporating technology and digitalisation into our lives will increase our quality of life (such as getting rid of many workloads) and money savings.

A. Methods Applied to Improve Water Usage
   i. Automatic Meter Reading: For control of consumption, the meter's data using Wmbus and Wavenis technologies are sent to HMS for evaluation. Wireless M-Bus is a communication system formed by combining wireless communication structure and M-Bus communication protocol for meter reading operation (Texas Instruments 2020). Wavenis is a wireless data transmission technology that uses frequencies of 433/868/915 MHz and provides transmission up to 200 m indoors from a

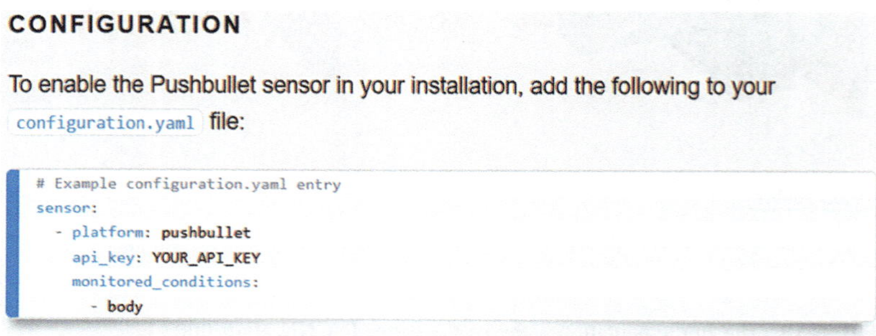
CONFIGURATION

To enable the Pushbullet sensor in your installation, add the following to your `configuration.yaml` file:

```
# Example configuration.yaml entry
sensor:
  - platform: pushbullet
    api_key: YOUR_API_KEY
    monitored_conditions:
      - body
```

Fig. 6.20 An example pushbullet sensor

distance of 1000 m in open space and up to a speed of 100 Kbps (Smart Energy International 2020).

ii. Leak Detection: Not all damage is visible. The leak should be detected quickly with leak detection sensors, and the leak's location should be seen. In possible risky leaks, the water in the house is cut off (Resideo 2020).

iii. Customer Portal: It aims to provide a portal where the user can find solutions to problems that have already been experienced or to identify possible problems and produce fast solutions accurately (Sensus 2020a).

iv. Demand Response: It is aimed to reduce production and purchasing costs by shifting the water for which energy is consumed (such as air conditioners, pumps, water heaters and coolers, and irrigation systems) to off-peak hours.

v. Pressure Regulation: For protection purposes, remote measurement and adjustment of pressure levels are aimed (Sensus 2020b).

vi. Alarms and Notifications: HMS warns the resident in case of problems in the house's water system, such as leaks, pressure, or temperature. If the resident is out, he/she receives an SMS.

vii. Dashboards: The properties that are aimed to be on the dashboard; monitoring, pressure management, and leak detection with that intelligent solution to help to conserve water and prevent leaks, configurable widgets organised in a grid, graphical display of the current and previous water consumption, headings create a shortcut to the information they represent. Figure 6.21 shows an example; it aims to show the water usage rates and usage distribution in the house (IOTSENS 2020).

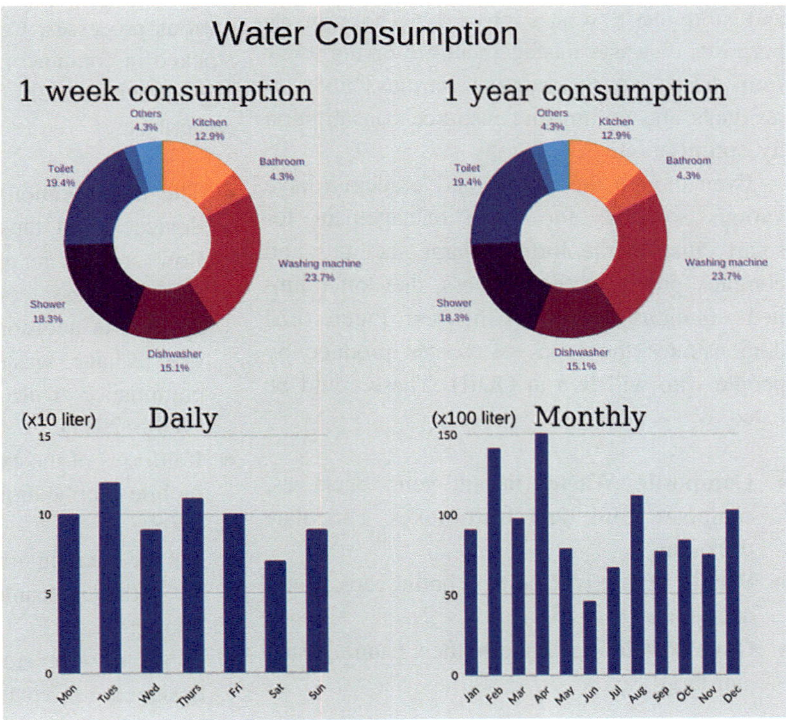

Fig. 6.21 Water consumption dashboard example

viii. Data Analytics: Data from the sensors are stored in a database to be compared later. This information will then be used to show the amount of water used in the house and to increase savings.

ix. Remotely Controlled Water: Plant species require varying amounts of water. It can simply be irrigated using a multi-channel water dispenser, and each fully automatically according to their specific needs. The water in some parts of the ODIH can be remotely controlled (e.g. filling the bathtub before coming home) (Eve Home 2020).

6.3.1.4 Waste Management

It is mentioned how to recycle domestic waste, water closet waste, agricultural waste, and left-over waste of ODIH without using AI and any automation in the "Water and Waste Management" part of this study. In this part, we will highlight the use of AI in solving the waste management issues of ODIH.

Like every house, ODIH will have some kind of residuals, and it will be required to separate and store these wastes properly. The primary purposes of waste management are taking away from home after separating various kinds of residuals and minimising resource consumption by optimisation.

Even though the market and academia have various solutions for waste management for smart cities in the form of large facilities and complex processes; nonetheless, they offer limited solutions for smart houses. Figure 6.22 demonstrates the types of wastes produce by people who will live in ODIH. These could be listed as:

- **Composite Waste** (instant soup packages, cardboard milk and fruit boxes, chocolate packaging)
- **Plastic Waste** (pet bottles, bottle caps, water demijohns)
- **Glass Waste** (beverage bottles, canning jars, jam jars)

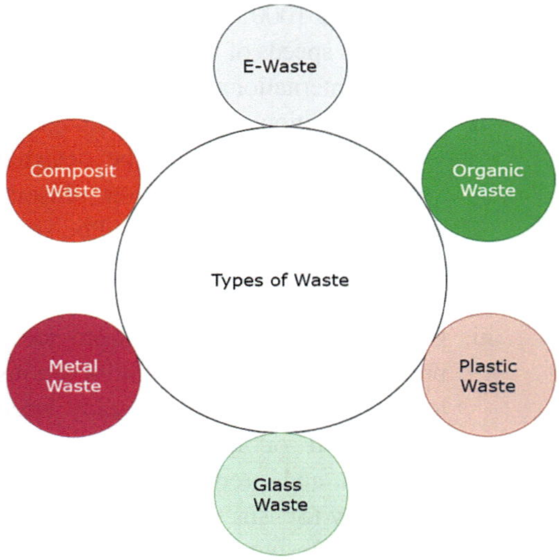

Fig. 6.22 Waste types

- **Metal Waste** (aluminium beverage cans, oil and tomato paste cans, canned foods)
- **Organic Waste** (fruit and vegetables)
- **E-Waste** (computers and phones).

Organic wastes are utilised for producing biogas that will be used in the kitchen stove after various processes. Conversely, other wastes are stocked in containers that have five layers properly. Other solutions for waste management and austerity are:

- The incorporation of a rainwater harvesting device, smart taps, water meters and water flow sensors helps maintain and return the water for less waste to the smart household system. In addition, this system tracks water use, leakage, waste, and potential demand in compliance with the consumption trends. (Willis 2011).
- It informs of the expiry date to be dumped via mobile applications to the dustbin. (Luo 2008).
- We are tracking how full the garbage is. When the garbage became full, the HMS notifies the households.
- A smart water recirculation device can be designed to recycle used water to remove

calcium and magnesium that constitute hard water for the daily use of water in different activities, such as washing clothes and utensils. (Nguyen et al. 2015).

- When any garbage is thrown away into the dustbin, the dustbin compresses automatically to create a freer space.
- The integrated system separates the wastes by their class, stocks them in distinct areas, and becomes full; it notifies the municipal waste collection centre.
- The waste management system can use human excrement to produce fertilizer.

6.3.1.5 Healthcare

Our HMS aims to monitor the safety of its occupants with lingering health issues. The targeted segment of this feature is not limited to elderly and disabled people. It also includes relatives and beloved ones. Our HMS aims to minimise distress that can be caused by any health issue that may arise. It is especially beneficial to those who live alone but are in constant danger of experiencing a health issue.

A. Chronic Disease Tracking

The smartwatch that the user wears collects data and sends it to HMS. HMS evaluates collected data and sends a notification via an app if a recurring issue exists.

B. Remote Patient Monitoring

Implemented system tracks and monitors residents within the house. Owners of the house can view the rooms remotely, which is especially useful for people who take care of elderly or disabled people. The inhabitant communicates with the system through voice inputs, which are processed through voice recognition. After the system is alerted, relatives or healthcare services are notified via smartphone apps and SMS messages.

C. Issuing Warnings in Case of an Unexpected Heart Rate Increase

Our HMS is capable of tracking patients' heart rates through smartwatch data. HMS will alert the patient in case of an unexpected increase in the heart rate. When the heart rate exceeds a predetermined value, HMS sends alert notifications via the app. If the situation persists or escalates, healthcare professionals and relatives are made aware of the situation via the app (healthcare personnel should use this system) and/or 112 is called with a voice mail that includes necessary emergency information.

D. Automated Alert Notification

The algorithm that manages the house also monitors its residents and their daily and weekly routines. In case of an abnormality, the system can automatically take pictures and alert relatives to confirm any threats. Relatives can confirm or deny the existence of a threat from the snapshot they receive on the HMS smartphone app. Should the relatives confirm the presence of danger, the system immediately notifies healthcare personnel (healthcare personnel should use this system). If there is no apparent danger, the system goes back to standby mode, where it continues to monitor its residents.

E. Observation and Feedback Function on Abnormal Body Activity

HMS tracks bodily functions and data via a smartwatch. HMS gives feedback and suggestions based on collected data and medical information via the app. Medical information and procedures are evaluated through the built-in machine learning algorithm to give an accurate suggestion.

F. Pillo

Pillo is a medical assistant IoT solution that is integrated into the users' home. The AI-supported system makes it possible for a personalised solution that learns and adapts itself to its user's needs.

As seen in Fig. 6.23, Pillo assists its user by waking up and proactively alerting the user when it is time for a medication dose. This makes it easier for the user to keep up with health regimens. Furthermore, Pillo allows professional

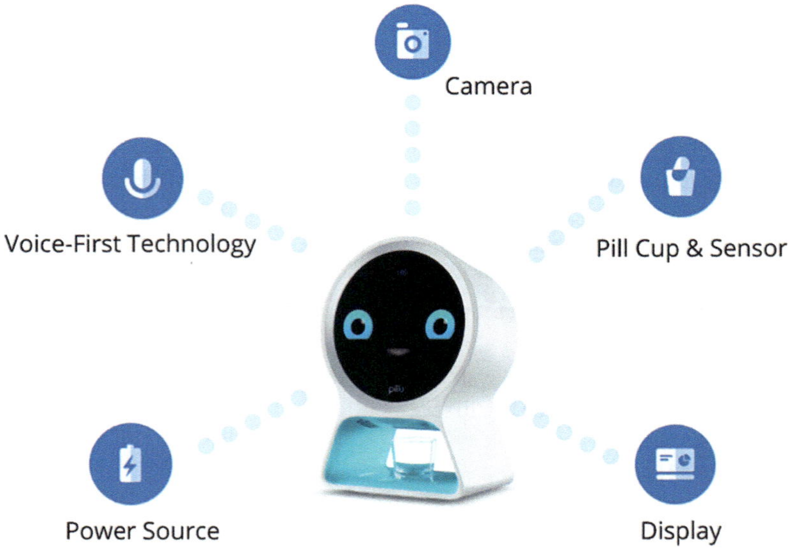

Fig. 6.23 The basic functions of Pillo

medical personnel and family members to monitor loved ones remotely. Through cameras and biometric recognition technology, Pillo makes it possible to ensure that the correct user is using the right medication. The user experience can be further personalised for the needs of the user. (Pillo Health 2020).

Pillo assists in the loading of stored medication and can detect when the drug is removed. Through voice commands and on-screen visual instructions, the user can sort pills, set timers, and not worry about making mistakes. Pillo stores up to 4 weeks of medication and can track refill times by alerting the user. Facial recognition and PIN entry processes, as well as the usage of sensors that confirm the delivery of medicines, allow a safe environment for the dispensary of medication. It provides consumers access to critical statistics, integrated health information, common causes of disease, and side effects of treatment.

The mobile app connects its users with their loved ones, who can track their medication schedule anytime as well as their health data. The mobile app also notifies them and health personnel of any missed doses or responses to check-in questions. Pillo seeks out and speaks to its user in the room about their health care needs. AI analyses continuous data streams to connect to users on a personal level and engage with them frequently.

Through the Software Development Kit (SDK), Pillo can be further optimized for the customer's unique needs. Pillo is a HIPAA-compliant platform that uses end-to-end encryption to protect patient data. The Home Assistant feature of Pillo also provides information on weather, clock, and general knowledge. Pillo makes it possible for healthcare professionals to do voice surveys en masse or specific segments, proactively deliver spoken messages, deploy photo or video content, and establish trigger alerts. Through its online web portal, Pillo allows monitoring and control of smart devices connected to it. Pillo can connect with PERS providers and alert them when a user triggers an emergency alert.

6.3.1.6 Customisation/Entertainment

The importance of home entertainment was understood during the pandemic. The home entertainment system enables the management, display, and automation of all digital entertainment systems in the house. By aggregating the entertainment systems into a master management system, it could be avoided from the hassle of unnecessary hardware such as remote control. This system enriches the home entertainment experience. The entire content collection can be accessed from any digital entertainment device in the home. When the room is changed, the system can follow the resident with sensors and move audio and video to the new room. The resident can play different music or videos in other places according to their mood. HMS can display and adjust the bath or pool temperature. The system can automatically turn off unused entertainment systems, thus increasing energy efficiency. HMS can alter the ambience and other parameters, such as network, in the house according to appliances' activities. Most of the home management systems available on the market are merely managing and displaying entertainment vehicles. In ODIH, there is a potential to develop more advanced systems. Unlike home management systems in the market, when the project reaches Level 3, it can give entertainment suggestions based on the information gained from appliances' lifestyle. At this level, HMS can use all data in the house and AI algorithms. It gives beverage/food suggestions based on the fridge or wine collection. It will also give clothes or tool suggestions based on activity and outdoor information such as weather information.

Customisation is one of the essential features of a smart home. It affects all smart home systems. Thanks to customisation, human interaction is minimised, and residents' comfort is maximised. A standard smart home can be too inefficient, so the whole house must be customised for the households. In this way, the efficiency of a smart home can be increased. Most products on the market have limited customisable features. These products offer standard solutions to all customers, which reduces the system's efficiency and cannot use the data collected at home. However, using artificial intelligence algorithms, the Home Management System has the potential to be customised according to the lives of the residents. The system learns the residents' lifestyles and arranges the whole system in the house accordingly. For example, the temperature system can be adjusted and the doors locked before the residents go to sleep. When the residents wake up, the curtains can be opened automatically, and the smart coffee machine can be operated. Unlimited scenarios can be produced with the customisation system. Thanks to the customisation systems, anomalies in the house can be understood by the system. In this way, it contributes to the security system. HMS will be more customisable in ODIH, and as the project levels increase, the effect of artificial intelligence increases. The system can adapt itself to the residents' calendar and give suggestions about calendar events. Artificial intelligence algorithms can adapt to residents' changing lifestyles by always following the life pattern at home. It can also help the energy management system as it adapts to the lifestyle of residents. Thanks to the customisation system, the system can remind the residents of the things they have forgotten. Suggestion algorithms currently used can be integrated into this system, and the system learns their tastes and can give them movie and music recommendations. It is straightforward to add new external devices to the system. The HMS software we use works well with devices from many brands.

6.3.1.7 Security

Our HMS aims to protect both the residents and the integrity of the house. Our HMS is designed to keep the house safe from interior and exterior threats and hazards. These features range from simple fire and gas detection to advanced facial recognition of guests or intruders and depend on the chosen level of HMS service. A variety of sensors are used to detect movement, gas, fire as well as other hazards. Sensors activate other responsory devices to react to the ongoing event. These events can be but are not limited to gas leaks, water leaks, fire outbreaks, and intruder alerts. HMS can take reactive actions such as alerting fire departments, law enforcement, residents, and their relatives.

A. August Wi-Fi Smart Lock

August Smart Lock is a smart lock system that communicates via Bluetooth and Wi-Fi. Security is provided in the August Home application using Bluetooth Energy and Transport Layer Security (TLS) technology. As seen in Fig. 6.24, it provides security in many areas, such as smart alerts and DoorSense. One of the security

Fig. 6.24 The security features of august smart lock

examples for this product is asking the user to verify his e-mail and phone number via the mobile application. Thus, additional security is provided with the two-step verification system. Thanks to the device's biometric authentication feature, the user can check the door lock while away from home. This is done by fingerprint or face authentication through August Home. Additionally, August Smart locks use the DoorSense sensor to securely inform the user whether the door is locked. It is compatible with the Home Assistant we will use in our HMS and can be easily integrated into the Configration.yaml file (August, 2020).

B. Reolink Cameras

Reolink cameras have 2560 × 1440 high resolution, provide better home security, and 4-megapixel live video streaming. When Fig. 6.25 is examined, the camera features such as Outdoor Waterproof Design, Movement Clip-on Micro SD Card, Remote Access and Control, Dual-Band Wi-Fi Strong Network Signals appear. Reolink wireless cameras operate in both the 2.4 GHz and 5 GHz bands. That is why it is one of the first dual-band Wi-Fi security cameras in the world. Thanks to this feature, it guarantees that there will never be any signal loss. Also, strong network signals are obtained with 2T2R MIMO antennas even over long distances of 164 feet. In addition, the Reolink camera is very successful in night vision thanks to its 18 LED lights and infrared radiation. The camera can provide security up to 100 feet in the night vision range. The user can create motion detection sensitivity and motion detection zones. In this way, Reolink RLC-410 W analyses the image and sends a notification to the user through the application in case of a violation. It can also send a photo or 30-s video clip to the user via e-mail. Images can be saved via an FTP server, so an NVR is not required. Reolink RLC-410 W camera can be connected to Home Assistant thanks to its integration with ONVIF Protocol. Thus, PTZ features such as zooming in, zooming out the camera can be used with the Home Assistant application (Reolink 2020).

Fig. 6.25 The general features of reolink RLC-410 W camera

C. Desi Smart Line

Thanks to the motion sensors, a notification is received via Desi Smart Line when a thief enters the house.

D. POPP Smoke Detector

POPP Smoke Detector reliably detects fire and smoke. It alarms in case of emergency and sends an alarm signal to the Z-Wave network, so all other smoke detectors and sirens within range can be alarmed. In other words, POPP Smoke Detector connects to HMS with Z-wave protocol and sends a notification to the user in case of gas leak or fire (Popp 2020).

E. Cyber Security

In addition to the house's physical protection, cybersecurity is of great importance for the user. It is crucial to keep the system updated regularly and configure secrets to ensure the Home Assistant's safety. Since configuration.yaml is a file that can be read by anyone, private information can be easily removed from a configuration file using secret. This method makes it easy to keep track of API keys and passwords as they are collected in one place. Porting private information to the secrets.yaml file is very similar to the case of splitting the configuration. The secrets.yaml file is created in the Home Assistant. The API keys and password inputs in the configuration.yaml file is the same. Entries need to change with identifier and secret. Using the Home Assistant cloud for remote access is considered the easiest and most reliable method. Other options are to use Let's Encrypt or between TLS/SSL.

6.3.2 Building the Smart Hub

6.3.2.1 Comparison of Three Different Home Automation Systems

At the beginning of our journey of building an HMS for ODIH, we could not think of a more convenient way to develop our HMS other than utilising open-source projects. We thought that the philosophy behind the open-source model concurs with the Open Digital Innovation Hub's approach. Therefore, we analysed the open-source projects for home automation, and we ended up with the three most successful ones.

OpenHAB, Home Assistant, and Domoticz offer end-to-end solutions for managing a smart home. All of them have advantages and disadvantages according to their distinct features. Table 6.1 presents a comparison of these Home Management System projects comparatively, according to various characteristics in detail.

6.3.2.2 Home Assistant

The Home Assistant is a free and open-source home automation software that provides access to manipulate IoT devices from a mobile device or a computer globally. Moreover, it integrates different gadgets into one network. Inherently, Open Digital Innovation Hub (ODIH) has different types of smart devices such as a smart thermostat, smart lock, smart fridge, etc. Almost all of these devices belong to various brands; thus, each one has a different control application. Home Assistant is used in smart homes to merge all dissimilar appliances and provide authority to one controlling system.

A. Installation and Configuration

Home Assistant can be worked on different platforms such as Raspberry Pi, Tinkerboard, Odroid, Intel NUC, and virtual machines. Nonetheless, it is recommended that running Home Assistant on Raspberry Pi 4 is the best. Home Assistant image is downloaded from its website, and it is installed on Raspberry Pi 4 by using an SD card. Subsequently, Home Assistant scans home networks and identifies devices that

Table 6.1 Comparison of three different home management systems

	OpenHAB	Home assistant	Domoticz
Installation	Straightforward (about 20–40 min)	Very elementary	It is easy to install
Configuration	OpenHAB has various user interfaces to change settings, control resident devices and objects to create rules and maps. HAB-min offers a range of functions to now overlap with the Paper UI	If the residents have a few electrical types of equipment, it is easy for the system to discover all the devices in their house and add them to the UI	A significant percentage of the configuration can be done using a Web UI
Flexibility	OpenHab can be as flexible as the resident wishes it to be. It comes with a cost and, it is not the easiest system. The web UI supports fundamental applications. The power is still in the configuration files	Home assistant can cover most people's needs. The auto-discovery function works fairly well and the system makes a pretty decent job guessing the users' needs	It is very stable and it can do the basics but it is quite limited in terms of supported devices and configurations
Community	One of the best things about OpenHab is the community of users. The community of OpenHab is much crowded and experienced	Its community is growing very quickly, but most of the users are naïve	It has fewer users than Home Assistant and OpenHab
Pace of development	It is quite slow, but the developers add new IoT devices every day	The developers publish new releases every week with recent devices	The small group of developers is doing great work, and every commit is checked/reviewed by Gizmocuz. However, there is no control of the documentation, and the actual commit is not always tested very well

(continued)

Table 6.1 (continued)

	OpenHAB	Home assistant	Domoticz
Automation	XBase syntax is not easy to learn and to use. It's possible to install the JSR223 plugin. This plugin helps the user to write Jython rules in JavaScript	Yaml is worse at defining automation rules, too. But the residents can use AppDaemon to code the rules in Python	LUA Scripting is easy to learn. It is clean and powerful
Updates	It is done by using a command line	It can be done by the click of the button located in the hass.io	It is done by one click on the UI or manually on the command line
Add-Ons	It has 424 add-ons	It has 21 official and 46 community add-ons	It has 80 add-ons
Devices	It supports about 2004 devices	It supports about 1700 integrations	It is compatible with 516 devices
Interface	It has many options for UI, but there is a lot of duplication and too many similar options	It has an adequate default UI which is convenient for beginners, and it is more customizable than others	There are plenty of sparkling themes that can be used to make the system a little bit nicer; in every theme, the residents can customize the icons of a sensor or switch
Mobile application	It has both android and IOS apps	It supports android and IOS apps	It has both android and IOS apps
Notifications	It has many more integrations with the notification services page has over 30 supported platforms	The "notify" integration makes it possible to send notifications to a wide variety of platforms	It needs integrations to send notifications
Integrations with HomeKit, Alexa and Google home	OpenHAB integration with home kit is relatively straightforward with the only requirement is to edit a couple of lines in the configuration lines	Both Alexa and Google Home can be configured almost with the press of a button if the residents use Home Assistant Cloud. The only drawback is that it is not a free service, currently costing $5 a month	It supports HomeKit, Alexa and Google home integrations
Stability	In terms of software updates, each of the OpenHAB system updates goes through a strict inspection. Also, before being a part of the OpenHAB system, new IoT devices should be proven in terms of capability. As a result, the system is stable	Although the system is updated frequently, it is a less stable system as it does not go through the strict control phase as OpenHAB does	Since there are limited devices and configuration options in the Domoticz, the system is quite stable

can be integrated into the system and demonstrate them on User Interface (David 2017). As seen in Fig. 6.26, Home Assistant has a basic and straightforward user interface.

In Home Assistant, YAML syntax is used for configuration. It is a complicated syntax to be used. On the other hand, it is very good at expressing complex structures. For integrations that are intended to be used in Home Assistant, code is added to the configuration.yaml file to specify its settings. This is especially true for integrations that cannot yet be configured via the user interface. Examples of configuration.yaml are available on GitHub.

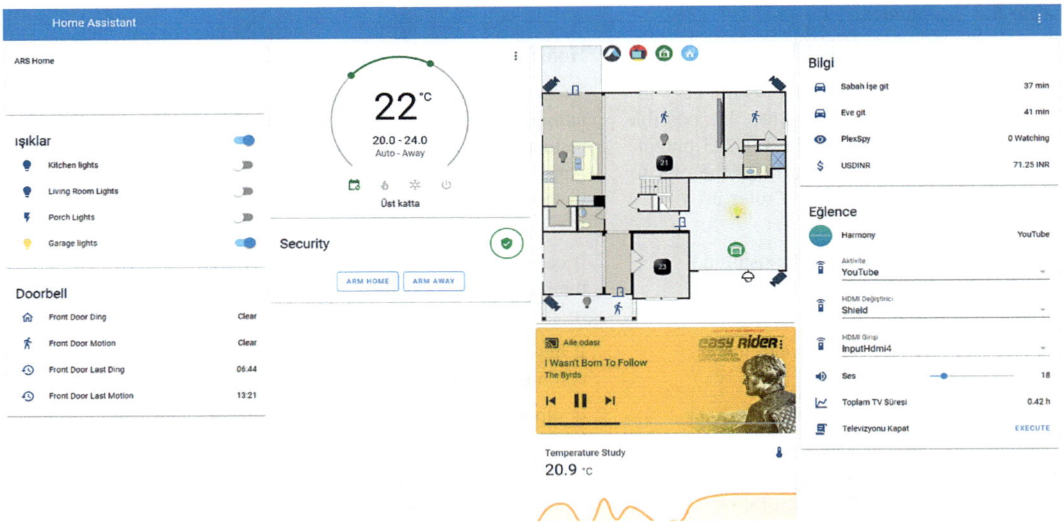

Fig. 6.26 User interface of home assistant

B. Features

Home Assistant is open source software which means its codes are editable by third parties. The source code of Home Assistant, which is written in Python language, is available on GitHub. If anyone wants to add new features or edit some of the features, they can do this via reaching the source code. Thanks to this great feature, entrepreneurs can integrate their product to the Home Assistant's system to test their HMS solutions. Besides, Home Assistant has an up-to-date and crowded community that allows users to ask their questions to other users and share their experiments with everyone.

C. Automation

The working principle of automation is divided into three bundles:

Trigger: It is the primary condition and initiates automation.

Conditions: These are conditions that limit the functioning of the function after the trigger is provided.

Action: This is the activity that is expected to take place after all conditions are met.

Example:

Automation for rainy days.

Turn on the light in the living room when it starts raining, someone is home, and it is afternoon or later.

As demonstrated in Fig. 6.27.

Trigger: when it starts raining.

Conditions: someone is home, and it is afternoon or later.

Action: turn on a light in the living room.

The automation rules are in constant communication with the Home Assistant. The change of current state within ODIH is used as the source of Triggers.

D. Integrations

Home Assistant is an open-source Home Management System software that supports many brands. It currently has 1700 integrations and continues to increase every day. It supports a broad spectrum from well-known brands such as Samsung, Google, and Xiaomi to brands one has never heard of. It ranks tenth on GitHub in terms of the number of contributors, and for this reason, newly released devices can be quickly integrated (Octoverse 2020). The system supports many different protocols, so the system can combine devices with different protocols and work in harmony. With UPnP and other discovery protocols, the system can automatically

```yaml
automation:
  - alias: 'Rainy Day'
    trigger:
      - platform: state
        entity_id: sensor.precip_intensity
        to: 'rain'
    condition:
      - condition: state
        entity_id: all
        state: 'home'
      - condition: time
        after: '14:00'
        before: '23:00'
    action:
      service: light.turn_on
      entity_id: light.couch_lamp
```

Fig. 6.27 Automation examples from home assistant

find some devices in the local network (Dague 2017). In Fig. 6.28, the system automatically detects Philips Hue and shows the configure screen. Thanks to mobile phone integration into the system, HMS can track the residents' location and adjust the house accordingly. With the Android and IOS applications, the house can be managed with mobile phones. The system can also work with devices using Bluetooth and Wi-Fi to track which room the resident is in the house. The system can also support smart home hubs such as ASUSWRT, Home Assistant IOS, Google Nest, and manage them. It also integrates with many cloud systems such as Google Cloud, Apple iCloud, and Home Assistant Cloud. Thanks to voice assistant support such as Amazon Alexa, Google Assistant, the smart home can also be managed with voice commands. If the

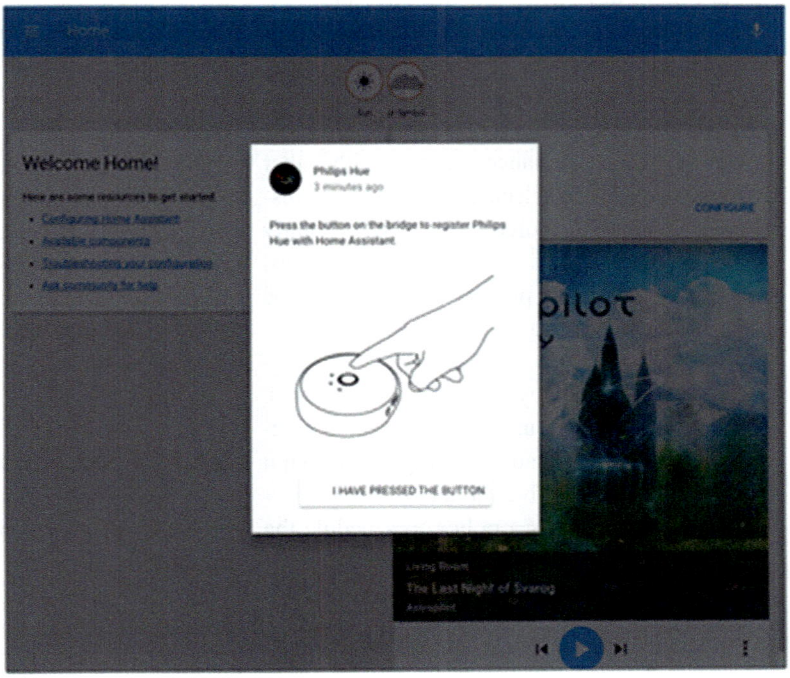

Fig. 6.28 Philips Hue configuration screen

residents have an unsupported device, they can introduce their device to the system by using MQTT, HTTP, and some programming and adjustments (Dague 2017). If the residents have trouble with the new integration, they can apply for community support. Since it is a system that is open to new integrations, start-ups can test their products in the ODIH ecosystem.

6.4 Results

This section will introduce the advantages and implementation processes of three cases where our HMS profoundly impacts smart homes: energy management, food & agriculture, and water. We will dig into the steps that bring the home to efficiency and sustainability with AI solutions. We will also demonstrate the effects and supremacies of HMS-powered homes compared to today's traditional homes.

6.4.1 Energy Management

Under the sustainable development paradigm, home energy management systems will have a crucial role in advancing the efficiency and reliability of energy consumption in future cities. The heating & cooling systems, water heater, lighting, and refrigerator account for most of the energy consumption in a typical house. Efficient control of these appliances, along with other components such as electric vehicles and distributed energy resources, can significantly reduce electricity bills and carbon emissions. Shortly, the operation process of HMS will be similar to the following:

- The data received by smart meters, sensors, and meteorology is used for forecasting electricity generation and consumption through the system's predictive model.
- The model generates a policy to schedule the shiftable appliances (washer, dryer, dishwasher, etc.) and optimise the controllable appliances (HVAC, water heater, lighting, refrigerator, etc.) considering householders' preferences.
- The model also provides a strategy to manage the house's energy flow based on electricity price and generation predictions.
- Furthermore, based on this strategy, HMS can sell the surplus of the generated electricity back to the grid.

According to the calculations done in Chap. 3, ODIH could save 425 trees and prevent the emission of more than 9000 kg CO_2 every year only by itself without energy management. By leveraging AI-powered energy management systems, these benefits can be increased further. According to a study, 59% of savings are achieved in a simulated smart home by managing the photovoltaics, energy storage system, air conditioner, and electric vehicle using an RL method called Double Deep Q Network (Liu et al. 2020). The time-of-use tariff is used in electricity price data for demand response. The researchers employed three hidden layers with 100, 512, and 50 neurons, respectively, for the neural network architecture, and the training operation of the whole network was kept simple. The DDQN training algorithm used in the study can be seen in Appendix. Despite some limitations like computational power and the need for further research to develop more robust architectures that utilize more advanced RL algorithms such as PPO (Proximal Policy Optimisation), TD3 (Twin Delayed DDPG), SAC (Soft Actor-Critic), C51 (Categorical 51-Atom DQN) etcetera, the results demonstrate DRL's potential in energy management. A more general review of energy management with novel DRL techniques can be seen in the study (Yu et al. 2020).

6.4.2 Food and Agriculture

Agriculture and food have never lost their importance since the development of humanity. Today, these two basic human needs are at significant risk due to factors such as rapid population growth, climate change, and the destruction of agricultural lands. According to the report published by the United Nations, 690 million people that corresponds to 8.9% of the world's population, were malnourished in 2019. Furthermore, if factors grow and today's habits continue, this number is expected to reach more than 840 million people by 2030 (United Nations 2020). So, with the growth of the mentioned risk factors, it can be expected hunger to increase even more in the future. Also, according to the data in the report, the number of hungry people in the world reached 821 million in 2017. This means that one out of every nine people is "hungry" (World Health Organization 2018). If no solution is found quickly, it is obvious that the number will increase. To prevent this, a project called "Sustainable Development Goals" tries to offer everyone a life with increased quality. In the project, there are 17 goals of which "Zero Hunger" and "Responsible Consumption and Production" draw attention. Because they closely affect the economy, health, education, equality, and social development. Hence, our mission is to perceive these goals as personal and social responsibility, sustainable agriculture and healthy food will be produced thanks to the AI-based functions created within ODIH. As a consequence, in this part of the study, how to make more efficient agriculture and reduce food waste will be mentioned.

In terms of agriculture:

To consume healthier and safer foods, a smart farming area is created within ODIH. A product called Vahaa is used to save space, reduce water and energy use, and finally provide more easily obtainable products to the user. Accordingly, the data of the lettuce plant produced by the product as a sample vegetable can be seen below:

First of all, lettuce is a cool climate vegetable that needs humid weather conditions and is partially cold-resistant. It can be grown in all regions in Turkey because the vegetation period is short. The best temperature for growing lettuce is 16-21 °C. Withstands lower temperatures are at 6–10 leaf cycles. It is seeded early before reaching its standard size at high temperatures. Lettuce is suitable for growing in all types of soil. However, factors such as soil type and the water-holding ability of the soil affect the product quality. Growing lettuce in a row in the same field causes diseases peculiar to lettuce to settle in the soil. Therefore, if the same field is to be used, crop rotation should be applied, and green fertilisation should be made every 2–3 years. Lettuce production is done by direct seed planting or seedling (DenizBank Finansal Hizmetler Grubu 2012).

To grow lettuce seedlings, seeds are placed in seedling pads on the sterilised mortar. It can also be planted in flower pots or plastic bags using the same mortar. Sprinkling is recommended as an irrigation method (T.C. Tarım ve Orman Bakanlığı Adana İl Tarım ve Orman Müdürlüğü 2002).

Fertilisation is one of the operations performed in lettuce cultivation. Before fertilisation, soil analysis is made, and fertilizer type and amount are determined accordingly. It is recommended that the field manure to be put into the soil before planting. Another process in lettuce is leaf binding. It is the process of tying the plant to the ends of the leaves with raffia or packing rubber in the core binding stage. In this way, the spread of the leaves is prevented and leaves that form the seeds of the desired colour and quality are obtained. When planted in spring, lettuce needs 55–70 days to mature. However, ripening is accelerated in the warm season in lettuce. It goes up to 140 days in the autumn period. Harvesting is done before the leaves are combed and before the seed stage. The harvest is done by hand cutting the plants one by one. The lettuce is cut from the part of the stem closest to the ground, far below (DenizBank Finansal Hizmetler Grubu 2012).

While lettuce cultivation is so difficult and challenging in classical soil agriculture, thanks to Vahaa, the harvest time is reduced from 70 days to approximately 26 days, and the individual can harvest faster with less effort. Also, the resident can grow fresh and chemical-free lettuce throughout the year. The user can check the light through the mobile application, see information such as the amount of water in the water tank, and the application sends a notification to the user when water needs to be added. The application informs the user when it is harvest time.

Considering the quality, quantity, and the area owned by the product offered by the Vahaa, it is obvious that the product is extremely useful. On the other hand, artificial intelligence is essential for the development of the system and for making functions more useful. Vahaa does not decide the harvest time of the growing product because it does not contain any artificial intelligence in its operation (Vahaa 2020). The overall production time in any soilless agriculture has been calculated and determined in advance. It only sends a notification according to the specified time. However, it should be borne in mind that environmental conditions can differ from one place to another. So by considering this, there is a need to develop another function based on artificial intelligence. As mentioned in the methodology section, the development of the product is monitored by cameras and sensors. Then, the "Faster R-CNN" algorithm, which is one of the algorithms of machine learning and the food database, is used. As a result, the function decides whether the image that is obtained belongs to a ripe fruit/vegetable or not. As can be seen, this function, which includes all possible environmental conditions in the decision-making process, will provide more efficient production. The "lack of decision-makers" in the function of the Vahaa will probably result in food waste as it does not take into account the changes in the humidity, temperature, and light conditions of the environment.

Other functions that are created in this field are, climate control system and the lighting control system. The climate control system provides automated control of the water, temperature, and humidity by some sensor data. Thus, while the amount of energy consumed as a result of manual adjustments for temperature, humidity, and water is high, the energy consumed will be much lower thanks to the automatic control system.

In terms of food:

The first application in the methodology part is the anti-food waste function. Based on the average consumption time of the food, an application has been developed that gives consumption advice to the user before the food spoils. In addition, the function, which gives a secondary warning that the product whose consumption recommendation is ignored, spoils after a certain period of time. It has been simply observed that the function increases efficiency by reducing forgotten food waste.

The second application to be used in ODIH is to determine the most consumed foods by the user that is put in the refrigerator. When these favourite foods are exhausted, the system adds the product to the user's shopping list for purchase. The benefit provided by this application is to increase usability by ensuring the required or preferred products are in continuous use. Also, since there will be no products that are not included in the shopping list, there will be no such thing as going to buy forgotten products again. Thus, both energy and time are saved.

6.4.3 Water Management

In this part of the study, we will focus on how to reduce the water consumption of ODIH by applying different methods (smart feedback, suggestions, social comparison, leak detection).

The Water Management System of ODIH provides a reduction of water consumption for the residents. The Water Management System realises this diminution by sending smart feedback and giving suggestions to the residents, and using leak detection mechanisms.

The literature indicates that smart feedback can decrease water consumption between 3% (Petersen et al. 2007) and 53.4% (Willis 2010) in terms of efficacy in curbing water consumption. In all studies, including positive feedback results, the mean decrease in consumption was 19.6% (Sønderlunda et al. 2014).

Furthermore, in another study conducted in South-east Queensland, Australia, Fielding et al. recruited 221 households and measured the impact of feeding customer-tailored information from utility-specific smart-meters (Fielding et al. 2013). The experimental groups consist of an information-only group, an information and social comparison group, and a feedback group. The information-only group received post-card information about how to save water in various situations of daily life (e.g. washing full loads, avoiding rinsing dishes, turning off the tap when cleaning teeth), and fixing leaks. The information and the social comparison group were provided with post-card information about how to save water and how much water consumes other households. Finally, the feedback group received a report about overall consumption and the consumption in specific parts of their homes besides information about how to save water. The study revealed that the feedback group consumed 15.5L (roughly 9.6%) less water than the other groups.

According to Fielding's study, it is clear that giving suggestions, sending smart feedback, and providing social comparison data to the residents decreases water consumption dramatically. One of the main purposes of the Water Management System is reducing water depletion of ODIH by using these methods integrated into of Home Management System.

The Home Management System cannot send post-card; nevertheless, it can send smart suggestions about how to save water to the residents. The residents who use the mobile application of HMS can receive these suggestions automatically. For example, while one of the residents is brushing his teeth in the bathroom, Home Management System sends the information about saving water to his smartwatch or his smartphone.

As shown in Fig. 6.29, using a few lines of code Home Management System can send a notification to every resident's smartphones. Figure 6.30 shows a notification screenshot which is about water-saving from an android phone sent by Home Management System.

Besides, the residents can access the comparison data consist of how much water is consumed by other households by using APIs or a social app.

When all these are taken into account, it is estimated that sending recommendations and providing social comparison information to the residents may reduce water consumption by about 9.6%, similar to the study in Australia (Fielding et al. 2013). That means a home integrated with HMS will consume less water than a home without HMS.

As mentioned in the water generation part of this study, the water usage for two adults will be approximately 355-L per day. Estimated water savings will be about 34.08 L which will be achieved by smart feedback and giving suggestions to the resident by HMS.

Leaks make up the majority of water losses in a home. Leaks in pipes cause serious financial and environmental damage. The properties of the pipeline are one of the variables that affect the volume of water lost. Significant losses occur when pipes burst or a connection is suddenly broken. Small losses are caused by leaky connections, fittings, service lines, and connections. Different variables are considered separately by leak detection and repair companies to calculate

Fig. 6.29 A sample of notification code in yaml format

```yaml
automation:
  - alias: Notify Mobile app
    trigger:
      ...
    action:
      service: notify.ALL_DEVICES
      data:
        message: "Turn off the water while brushing your teeth!"
        title: "Save water!"
```

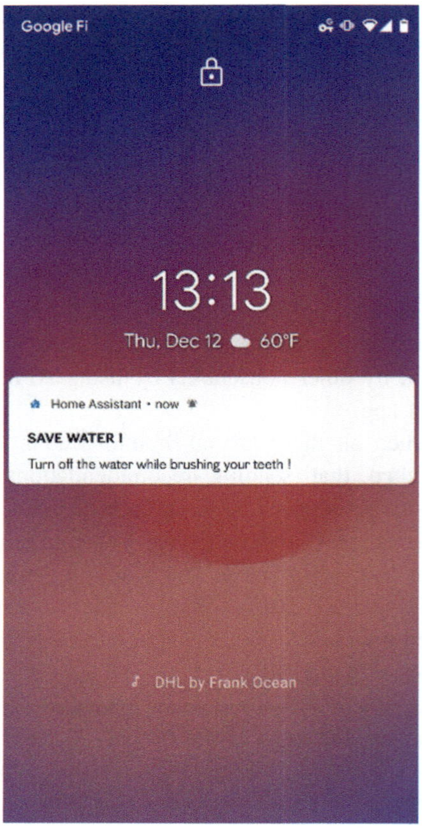

Fig. 6.30 A notification sent by the home management system

the volume of water lost. These are pressure in the water distribution network, type of soil (whether it shows leakage), detection time (time until the loss is detected), repair time (the speed at which the loss is corrected). Due to a large number of variables, it is not possible to obtain general numerical data for leaks (Farley 2001).

Table 6.2 shows the daily losses of possible leaks in the house. Earlier, we calculated that the daily water consumption of two people is around 350 L. Assuming that there is a total loss of 183 L, it can be determined that the leakage can cause both financial damage and damage to the environment can be great considering the decreasing water resources day by day. According to the United States Environmental Protection Agency (EPA), domestic leaks can reach about 4 trillion litres per year, and this equates to the annual water consumption of more than 11 million households (WaterSense 2017).

Physical losses are divided into leaks in the transmission and distribution network and leaks and overflow in the reservoir walls. Causes of leaks in transmission and distribution networks; Burst pipes (sudden rupture of a pipe section or connection) are recognised as leaky connections, fittings, service lines, and connections. Leaks and overflows from reservoir walls; leaks in old masonry or concrete walls have been described as the failure of floating valves. Non-physical losses fall into four categories: overuse forecast, under-use forecast, water theft, and wasteful use. Possible reasons for the overconsumption forecast were identified as insufficient or no measurement options and inadequate calibration program for batch meters. The reasons for low consumption estimates were identified as incomplete registration of customers' meters, poor quality/misfunctioning meters, meters that do not work, insufficient instructions for maintenance/replacement of meters, and for meter reading, underestimation of free supplies or operational use (Farley 2001).

In ODIH, we aim to detect water leakages rapidly. Using Underwater Wireless Sensor Network (UWSN) Technology, leaks or explosions in the pipeline that may occur underwater can be detected quickly. UWSN offers real-time monitoring underwater. As shown in Fig. 6.31, UWSNs consist of sensor nodes, surface stations, and autonomous underwater vehicles (AUVs) connected to the network to perform common monitoring tasks (El-Rabaie and Alsharqawy 2015).

Table 6.2 Water leaks and consumption amounts (Kirkland City 2020)

Total consumption (for 2 people)	Toilet leaks 4 L/hour	Faucet leaks 1 drip/second	Showerhead leaks 10 drip/minute
350 L	96 L	34 L	53 L

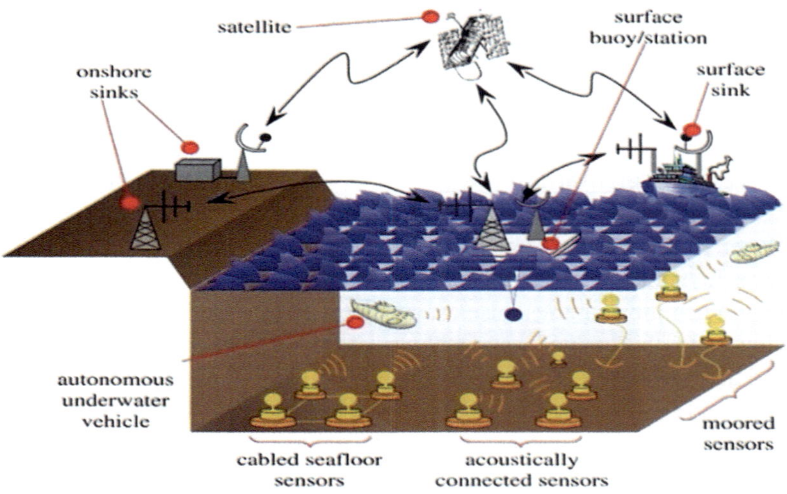

Fig. 6.31 Underwater wireless sensor network (El-Rabaie and Alsharqawy 2015)

By using UWSN, undesirable leaks will be detected easily; thus, water losses due to leakage will be reduced significantly. According to EPA; the average household's leaks can account for nearly 10,000 gallons (37,854 L) of water wasted every year (United States Environmental Protection Agency 2020). That means the water loss due to leaks per day is about 10,000/365 = ~27.397 gallons which are equal to 88.56 L. The water usage for two adults will be approximately 355-L per day; preventing unwanted water leaks will reduce the water consumption by about (88.56/355) * 100 = ~24.9%.

In summary, giving suggestions, smart feedback, and providing social comparison information to the resident by HMS will reduce the water consumption by about 9.6%, besides detecting the water leaks by using UWSN will decrease water consumption by about 24.9%. Consequently, we estimate that a home with HMS will consume around 34.5% less water than a home without HMS.

6.5 Discussion

This project aims to set a variety of levels for an HMS to be developed. As such, during this development process, some features will be progressively added. While initial levels of our HMS are currently attainable, later levels represent our futuristic vision. Some of these features may require advancements in other areas besides technology to be fully viable. Technological advancements, as well as social and juristic advancements, will allow the possibility and integration of such features. Below is our evaluation of possible improvements that can be made to enhance our HMS.

6.5.1 Energy Management

There are certain limitations in energy management systems at ODIH. The most important of these is the high-level hardware required by artificial intelligence algorithms. With the future hardware enhancements, the artificial intelligence algorithms we mentioned, such as deep reinforcement learning, can be trained better, and more effective results can be obtained. Academic and industrial development continues rapidly in the field of artificial intelligence. There will be many new and more effective algorithms in the future. In this way, energy management systems can be managed more effectively.

As the peer-to-peer energy trading systems become more popular, the energy trade will function more stable. As a result, energy and battery management systems will work more effectively. IoT devices in future smart homes will increase, and more devices will become controllable, and thanks to this, the amount of savings will increase with energy management systems.

6.5.2 Water Management

Water Management is one of the biggest problems in homes and even countries. No matter how much the countries raise awareness of their citizens for this, wasted water still persists to be a serious problem all around the world. In the planned HMS, the least user-dependent and most economical form of water management is aimed. In this context, it is envisaged that HMS will send notifications for excessive use and its awareness, show how much water is used in which area, and display statistics on a dashboard to compare past uses.

Another important issue is water leaks. The biggest problem with water leaks is that they are detected late. In this sense, automation systems developed to detect leaks can be used as the next step. In general, the activity status at home is determined by these sensors, and the first use statistics can be kept, and the next use can be compared with this information, and leakage can be detected (McCarthy and Kummer 2018). Another system calculates the maximum consumption amount of water by the household devices and assigns them to a threshold value. Then, it compares the water consumption of these devices, obtained with the help of sensors, with the threshold value. If the consumption of the device is higher than the threshold, the system sends a warning (Broniak et al. 2015).

6.5.3 Healthcare

Most of the problems in the healthcare module of our HMS originates from the fact that such applications of technologies are still in their infancy stage. This means that the journey that awaits our HMS will comprise of trial and error. Due to the nature of artificial intelligence, the algorithm must make mistakes to learn from the same mistakes and improve themselves. During this period, both the algorithms and the HMS must be manually monitored. Otherwise, unexpected and unwanted decision-making outputs may occur. For example, the HMS can acquire habits like falsely interpreting a situation and sending false notifications. The transition from a manually assisted state to a self-sufficient state may take some time to implement. During this time, HMS will not be fully automated and independent.

While all features are expected to be in the final product, some of these features may be implemented progressively. Certain features, such as establishing a connection with existing healthcare services, require implementation and coordination from third party entities. As such, a hindrance in the activation of these capabilities may be delayed. For the eventual activation of these features, the law must be compliant. Otherwise, some of these features may have to be deactivated in the final product to comply with the existing law.

6.5.4 Waste Management

Although waste processing requires large facilities and it is a complex process, this process may happen in a regular home in the future. Also, waste management will get its share from rapidly developing technology. For example, technological devices getting smaller and smaller every year; thus, machines and electronic systems used in waste management will be smaller than they are now. Besides, with advancing science and automation technologies, components for waste management systems will be cheaper so that these devices can be used in ordinary smart homes.

Thanks to all these developments, we should mention some actions that future homes may be capable of doing: they realise recycling process to produce new things from recycled raw materials, produce fertilizer using human excrement, annihilate toxic wastes safely, produce new papers from used papers.

Consequently, the waste management system may become an ordinary part of every smart home in the future with various actions that can be performed at home thanks to the development of technology.

6.5.5 Customisation and Entertainment

In the smart homes of the future collaborative intelligent entertainment systems can be designed. With future collaborative smart home systems, the residents can watch shared movies with their friends from separate houses. In this way, HMS goes beyond home and contributes to the whole entertainment life. Smart frames in the future can show pictures or artworks suitable for the environment. With technological developments such as VR and hologram, home entertainment can be enriched very well. With developments such as smart mirrors, residents can try the clothes recommended by the system on them. Artificial intelligence algorithms can increase their effectiveness with the development of hardware and software in the future. At Level 4, artificial intelligence can give suggestions like a coach or a mentor, based on the information it collects from home. Besides, as their level increases, human interaction decreases, and efficiency can be maximised. The customisation and entertainment system has a flexible structure in the start-up ecosystem. Thanks to the data collected at home, start-ups can test their products by integrating their applications into the house. From this environment, innovative solutions and products can emerge that will make our home life more comfortable and increase efficiency. To sum up, thanks to technological advances and new ideas discovered by start-ups, there are no limits within home entertainment and customisation.

6.5.6 Policy Recommendation

Further research and technological advancements only are not enough for the efforts of a sustainable future. The other side of the coin is the strategies and policies that policymakers need to determine to support the academic and industrial endeavours and lead to reliable growth and maximise economic, environmental, and social benefits. The following summarises our strategic goal suggestions and policy recommendations:

- Support R&D departments that develop AI technology further to create more intelligent Home Management Systems.
- Encourage universities and scientists to do research in AI and Smart Homes/Buildings/Cities by increasing the funding on projects and research centres. This can be achieved by direct government funds or by providing tax exemptions to capital owners for funding such projects.
- Form institutions that lead and supervise AI research to make sure the safety and reliability of progress. This is very crucial for AI systems to be beneficial for all by also considering ethical concerns. These institutions should make long-term strategic plans for AI development by assessing the potential risks carefully.
- Assure the databases of institutions and companies are proper, available, and reliable.
- Set realistic goals for future years to enhance energy and water savings, and waste management to improve the efficiency of residential and commercial buildings.

6.6 Conclusion

In conclusion, artificial intelligence and technological breakthroughs allow the possibility of augmenting houses to assist daily human life. This study focuses on the usage of artificial intelligence and its subsidiary topics to improve the healthcare, agriculture, and entertainment experiences of inhabitants as well as further manage and optimise water and energy usage in smart homes. To achieve this goal, this study highlights potential uses of machine learning, deep learning, and reinforcement learning. Machine learning processes allow the machine to be trained to make decisions and optimise decision-making operations. Meanwhile, deep learning further improves upon this concept by introducing additional neural networks that enable a range of possible operations from object detection to real-time motion analysis problems. Finally, reinforcement learning forms perhaps the most useful application of artificial intelligence for an HMS where resource management is tracked and optimized by a learning computer that works on the principle of success and failure.

The homes we live in and the electronic devices we use are getting smarter every year. Thus, Home Management Systems (HMS), the brain of a smart home, will be more advanced in the future. We classified HMS into five levels in the methodology section: Level 0 with a limited number of devices, Level 1 where a connection with a large number of devices and manual management is provided by systems such as mobile applications, Level 2 where HMS has started to perform basic manual automation, Level 3 with smart systems powered by AI, and finally Level 4 where HMS can manage all daily tasks and act as a sage rather than an assistant for householders. We also detailed our HMS with nearly 70 functions in energy management, food, water management, waste management, healthcare, customisation/entertainment, and security. Since energy consumption in the world will increase by 50% in the next 50 years, it is crucial to use energy efficiently and sustainably. In our HMS, energy management in smart homes has become controllable thanks to sensors and actuators. Furthermore, significant advances have been made in energy management systems, thanks to state-of-the-art deep reinforcement learning algorithms. In the future, access to food will not be as easy as it is now due to uncontrolled population growth and climate change. However, food waste is increasing day by day. As a solution, applications that give consumption advice to the user were developed. Today, since the population distribution is not regular across provinces, agricultural areas have turned into settlements. Moreover, the decrease in soil fertility has led us to the hydroponic method. Thanks to the soilless farming method, plants develop with less water in smaller areas. Our HMS will use the soilless farming method and applications that give consumption advice to users. In homes, water is as essential as food and energy. Our HMS aims to minimise water waste and ensure water management by using methods such as Automatic Meter Reading, Leak Detection, Pressure Regulation, and Remotely controlled water. Another aim of ODIH is to minimise resource consumption by recycling waste such as plastic, glass, metal, organic, and electronic. Our HMS has also taken serious measures in terms of health monitoring. It aims to make users' health controllable by using Chronic Disease Tracking, Remote Patient Monitoring and AI-supported Pillo methods. In addition, our HMS can give a warning in case of an unexpected increase in the user's heart rate. In other words, when abnormal body activity occurs, it can send an observation and feedback function. It is also of great importance that our HMS protect us from internal and external dangers. Our HMS aims to ensure the security of the door, warns the cameras in unexpected situations, and detect hazards such as gas and fire. Finally, customisation in our HMS is just as important as other areas. Thanks to customisation, the comfort of the residents is maximised.

As the name suggests, 'Open Digital Innovation Hub', entrepreneurs who want to test their projects will able to set up and test their projects at ODIH, and then they will able to see the results. As the residents live in ODIH, their data will be stored and shared online. Crucial real-time production and consumption datasets will be created to be used for research and business purposes. Furthermore, new IoT devices will be integrated into the Home Management System so that ODIH will always be kept up-to-date.

Appendix

Appendix: DDQN Algorithm for home energy management (Liu et al. 2020).

1. input: D- buffer; W -initial network parameters, w^{-} -a copy of w
2. input: N_r- buffer maximum size; N_b-training batch size; N-target network replacement freq.
3. for: episode $e \in \{1,2,\ldots,M\}$ **do**
4. Initialize frame sequence $x \leftarrow ()$
5. for $t \in \{0,1,\ldots\}$ **do**
6. Set state $s \leftarrow x$, sample action $a \sim \pi_\beta$
7. Sample next frame x^t from environment $\mathcal{E}$ given (s,a) and receive the reward r ,and append x^t to x
8. if $|x| > N_f$ then delete the oldest frame $x_{t\min}$ from x

end

9. Set $s' \leftarrow x$, and add transition tuple (s,a,r,s') to D, replacing the oldest tuple if $|D| \geq N_r$
10. Sample a minibatch of N_b tuples (s,a,r,s') $\sim$ Unif(D)
11. Construct target values, one of each of the N_b tuples:
12. Define $a^{max}(s';w) = \operatorname{argmax}_{a'} Q(s',a';w)$
13. $y_j = \begin{cases} r & \text{if } s' \text{ is terminal} \\ r + \lambda Q(s', a^{max}(s',w); w^{-}) & \text{otherwise} \end{cases}$
14. Do a gradient descent step with loss $\| y_j - Q(s,a;w) \|^2$
15. Replace target parameters $w^{-} \leftarrow w$ every N^{-} step
16. end
17. end

References

Alfaverh F, Denaï M, Sun Y (2020) Demand response strategy based on reinforcement learning and fuzzy reasoning for home energy management. IEEE 8:39310–39321

Amazon (2020). Echo show (2nd Gen). Premium. [Online] Available at: https://www.amazon.com/All-new-Echo-Show-2nd-Gen/dp/B077SXWSRP

Buzby JC, Farah-Wells H, Hyman J (2014) The estimated amount, value, and calories of postharvest food losses at the retail and consumer levels in the United States, s.l. United States Department of Agriculture (USDA) Economic Research Service

Broniak JA, Beyerle MT, Brian JM, Bingham DC (2015) Energy manager—water leak detection. United States, Patent No. US 9,019,120 B2

Brynjolfsson E, Mitchell T (2017) What can machine learning do? Workforce implications. Science 358 (6370):1530–1534

Bini S (2018) Artificial intelligence, machine learning, deep learning, and cognitive computing: what do these terms mean and how will they impact health care? J Arthroplasty 33(8):2358–2361

Brilliant (2020) Brilliant, smart home control system. [Online] Available at: https://www.brilliant.tech/pages/smart-home-control-system

Chen S, Ravallion M (2008) The developing world is poorer than we thought, but no less successful in the fight against poverty. DC, The World Bank, Washington

Chen Y, Luo B (2012) S2A: secure smart household appliances. CODASPY 12(2)

Cai Z, Fan Q, Feris RS, Vasconcelos N (Oct 2016) A unified multi-scale deep convolutional neural network for fast object detection. In: European conference on computer vision. pp 354–370

Coleman-Jensen et al (2016) Household food security in the United States in 2015, s.l.: USDA Economic Service

Cömert O, Hekim M, Adem K (2019) Bruise detection in apples using faster R-CNN. Int J Eng Res Dev 11 (1):335–341

DenizBank Finansal Hizmetler Grubu (2012) Marul, maydanoz, tere ve roka yetiştiriciliği. Ankara, GIDA TARIM VE HAYVANCILIK

Dey N, Fong S, Song W, Cho K (2017) Forecasting energy consumption from smart home sensor network by deep learning. Springer, Singapore

Dague S (2017) Open source [Online] Available at: https://opensource.com/article/17/7/home-automation-primer. Accessed Oct 2020]

David (2017) Smart home blog. [Online] Available at: https://www.smarthomeblog.net/openhab-home-assistant-domoticz/#Configuration. Accessed Oct 2020

Du M, Li F, Zheng G, Srikumar V (2017) Deeplog: anomaly detection and diagnosis from system logs through deep learning. In: Proceedings of the 2017 ACM SIGSAC conference on computer and communications security. pp 1285–1298

El-Rabaie S, Alsharqawy MA (2015) Underwater wireless sensor networks (UWSN), architecture, routing protocols, simulation and modeling tools, localization, security issues and some novel trends

Eve Home (2020) Eve aqua smart water controller. [Online] Available at: https://www.evehome.com/en/eve-aqua. Accessed 2020

Farley M (2001) International comparisons. In: Leakage management and control. WHO, pp 19–22

Food and Agriculture Organization (2008) Climate change and food security: a framework document. Rome, food and agriculture organization of the united nations. p iii

Fielding KS, Spinks A, Russell S, McCrea R, Stewart R, Gardner J (2013) An experimental test of voluntary strategies to promote urban water. J Environ Manage 114:343–351

Fang H, Hu C (2014) Recognizing human activity in smart home using deep learning algorithm. Nanjing, IEEE

Fjelland R (2020). Why general artificial intelligence will not be realized. Nature 7(10)

Goodfellow I, Bengio Y, Courville AA (2016) Deep learning. The MIT Press

Guo X, Shen Z, Zhang Y, Wu T (2019) Review on the application of artificial intelligence in smart homes. Smart Cities 2(3):402–420

Gao X, Ding X, Hou R, Tao Y (2019) Research on food recognition of smart refrigerator based on SSD target detection algorithm. Wuhan, Hubei, China, Association for Computing Machinery

Huang J, Zhou W, Zhang Q, Li H, Li W (April 2018) Video-based sign language recognition without temporal segmentation. Conference on artificial intelligence

Homey (2020) Discover homey. [Online] Available at: https://homey.app/en-us/homey/

IOTSENS (2020) IoT sens smart water. [Online] Available at: https://www.iotsens.com/solution/smart-water/. Accessed 2020

Jin X, Baker K, Christensen D, Isley S (2017) Foresee: a user-centric home energy management system for energy efficiency and demand response. Appl Energy 205:1583–1595

Khalili AH, Wu C, Aghajan H (Sep 2009) Autonomous learning of user's preference. In: Behavior monitoring and interpretation workshop at German AI conference. p 12

Kanjo E, Younis EMG, Ang CS (2019) Deep learning analysis of mobile physiological, environmental and location sensor data for emotion detection. Inform Fusion Cilt 49:46–56

Kirio (2020) Kirio in action. [Online] Available at: https://www.mykirio.com/kirio-in-action

Kirkland City (2020) https://www.kirklandwa.gov/. [Online] Available at: https://www.kirklandwa.gov/depart/Public_Works/Utilities/Water/Water_Bill.htm. Accessed 12 Nov 2020

Luo S (2008) Smart fridges with multimedia capability for better nutrition and health. Int Symp Ubiquit Multimedia Comput

Lee S, Choi D-H (2019) Reinforcement learning-based energy management of smart home with rooftop solar photovoltaic system, energy storage system, and home appliances. Sensors 19(18):3937

Lu R, Hong SH, Yu M (2019) Demand response for home energy management using reinforcement learning and artificial neural network. IEEE 10(6):6629–6639

Liu Y, Zhang D, Gooi HB (2020) Optimization strategy based on deep reinforcement learning for home energy management. Chin Soc Elect Eng 6(3)

Matsugu M, Mori K, Mitari Y, Kaneda Y (2003) Subject independent facial expression recognition with robust face detection using a convolutional neural network. Neural Networks 16(5–6):555–559

Marantos C, Lamprakos CP, Tsoutsouras V, Siozios K, Soudris D (2018) Towards plug and play smart thermostats inspired by reinforcement learning. Turin, ACM 2018:39–44

McCarthy III, BA, Kummer D (2018). Detection and mitigation of water leaks with home automation. United States, Patent No. US 9, 912, 492 B2

Mason K, Grijalva S (2019) A review of reinforcement learning for autonomous building. Comput Electr Eng 78:300–312

Mutis I, Ambekar A, Joshi V (2020) Real-time space occupancy sensing and human motion analysis using deep learning for indoor air quality control. Autom Constr 116

Nguyen et al (2015) Intelligent autonomous system for residental water and use classification. Autoflow applied soft computing

Nguyen K, Fookes C, Ross A, Sridharan S (2017) Iris recognition with off-the-shelf CNN features: a deep learning perspective. IEEE Access 6:18848–18855

Nugroho H, Harmanto D, Al-Absi HRH (2018) On the development of smart home care: application of deep learning for pain detection. IEEE

Neurio (2020) Neurio home energy monitoring. [Online] Available at: https://www.neur.io/home-energy-monitoring/. Accessed Dec 2020

NREL (2020) National renewable energy laboratory. [Online] Available at: https://www.nrel.gov/buildings/foresee.html. Accessed Dec 2020

Octoverse (2020) Octoverse github page. [Online] Available at: https://octoverse.github.com/. Accessed Nov 2020

OpenAI (2020) About. [Online]Available at: https://openai.com/about/ [Accessed 4 6 2021]

Petersen JE, Shunturov V, Janda K, Platt G, Weinberger K (2007) Dormitory residents reduce electricity consumption when exposed to real-time. Int J Sustain Higher Educ 8:16–33

Peng Z, Li X, Yan F (2020) An adaptive deep learning model for smart home autonomous system. Vientiane, IEEE

Pillo Health (2020) Pillo Health. [Online] Available at: https://pillohealth.com [Accessed 2020]

Popp (2020) Popp. [Online] Available at: https://poppandco.com/z-wave/z-wave-smoke-sensor [Accessed 15 12 2020]

Powerley (2020) Powerley web site. [Online] Available at: https://powerley.com/solution/. Accessed Dec 2020

Reolink (2020) Reolink. [Online] Available at: https://reolink.com/product/rlc-410/ [Accessed 24 11 2020]

Ruelens F, Iacovella S, Claessens BJ, Belmans R (2015) Learning agent for a heat-pump thermostat with a set-back strategy using model-free reinforcement learning. Energies 8(8):8300–8318

Rethink Food Waste through Economics and Data (ReFED) (2016) A roadmap to reduce US food waste by 20%. Rethink Food Waste through Economics and Data (ReFED)

Resideo (2020) Damage kontrol. [Online] Available at: https://www.resideo.com/us/en/solutions/water/. Accessed 2020

Rusen K (2020) Medium. [Online] Available at: https://medium.com/@kadirusen/reinforcement-learninge-giriş-nedir-nerede-kullanılır-578b14ed416a. Accessed 22 June 2020

Sønderlund AL, Smith JR, Hutton C, Kapelan Z (2014) Using smart meters for household water consumption feedback

Sun Y, Liang D, Wang X, Tang X (2015) Deepid3: face recognition with very deep neural networks. arXiv preprint arXiv:1502.00873

Silver D, Huang A, Maddison CJ, Guez A, Sifre L, van den Driessche G, Schrittwieser J, Antonoglou I, Panneershelvam V, Lanctot M, Dieleman S, Grewe D, Nham J, Kalchbrenner N, Sutskever I, Lillicrap T, Leach M, Kavukcuoglu K, Graepel T, Hassabis D (2016) Mastering the game of go with deep neural networks and tree search. Nature 529:484–489

Sas (2018) What is deep learning?. [Online] Available at: https://www.sas.com/en_us/insights/analytics/deep-learning.html. Accessed 15 June 2020

Sutton RS, Barto AG (2018) Reinforcement learning: an introduction, 2nd edn. The MIT Press, Cambridge, London

Sensus (2020a) Customer portal. [Online] Available at: https://sensus.com/solutions/customer-portal/. Accessed 2020

Sensus (2020b) Pressure regulation. [Online] Available at: https://sensus.com/solutions/pressure-regulation/. Accessed 2020

Smart Energy International (2020) Wavenis technology-technology overview. [Online] Available at: https://www.smart-energy.com/top-stories/wavenis-technology-technology-overview/. Accessed 2020

Statista (2020) Statista's digital market outlook. [Online] Available at: https://www.statista.com/forecasts/887613/number-of-smart-homes-in-the-smart-home-market-worldwide. Accessed 2020

T.C. Tarım ve Orman Bakanlığı Adana İl Tarım ve Orman Müdürlüğü (2002) Marul yetiştiriciliği. [Online] Available at: https://adana.tarimorman.gov.tr/Belgeler/SUBELER/bitkisel_uretim_ve_bitki_sagligi_sube_mudurlugu/sebze_yetistiriciligi_ve_mucadelesi/Marul.pdf. Accessed 25 Nov 2020

The Natural Resources Defense Council (2017) Wasted: how America is losing up to 40% of its food from farm to fork to landfill. Natural Resources Defense Council

Texas Instruments (2020) Wireless M-bus protocol software. [Online] Available at: https://www.ti.com/tool/WMBUS. Accessed 2020

Tutorialspoint (2020) Machine learning. [Online] Available at: https://www.tutorialspoint.com/machine_learning_with_python/index.htm. Accessed 15 June 2020

U.S. Energy Information Administration (2019) Annual energy outlook 2019 with projections to 2050. Washington, DC, U.S. Department of Energy

United Nations (2020) The sustainable development goals report 2020. United Nations, New York

United States Environmental Protection Agency (2020) United States environmental protection agency web site. [Online] Available at: https://www.epa.gov/watersense/fix-leak-week. Accessed Nov 2020

Vahaa (2020) vahaa. [Online] Available at: https://www.vahaa.co

Vanek J (1974) Time spent in housework. Sci Am 231 (5):116–121

van Otterlo M, Wiering M (2012) Reinforcement learning and markov decision processes. In: Wiering M, van Otterlo M (eds) Reinforcement learning. Springer, Berlin, pp 3–42

Willis RM, Stewart RA, Panuwatwanich K, Jones S, Kyriakides A (2010) Alarming visual display monitors affecting shower end use water and energy conservation in Australian residential households. Resour Conserv Recycl 54:1117–1127

Willis R (2011) Residential potable and recycled waater and uses in a dual reticulated supply system. Elsevier Science Journal on Desalination

Wang Y, Velswamy K, Huang B (2017) A long-short term memory recurrent neural network based reinforcement learning controller for office heating ventilation and air conditioning systems. Processes 5(3):46

WaterSense (2017) epa.gov. [Online] Available at: https://19january2017snapshot.epa.gov/www3/watersense/pubs/fixleak.html. Accessed 12 Nov 2020

Wei T, Wang Y, Zhu Q (2017) Deep reinforcement learning for building have control. Austin, IEEE, p 22

Wu Y, Hassner T, Kim K, Medioni G, Natarajan P (2017) Facial landmark detection with tweaked convolutional neural networks. IEEE Trans Pattern Analysis Machine IntellIgence 40(12):3067–3074

World Health Organization (2018) World Health Organization. [Online] Available at: https://www.who.int/news/item/11-09-2018-global-hunger-continues-to-rise-new-un-report-says. Accessed 6 Dec 2020

WeiWei Z, Wei L (2019) A deep reinforcement learning based human behavior prediction approach. 2019 International conference on robots and intelligent system (ICRIS), pp 59–62

WIRED (2019) How AI powered LG ThinQ points to a more advanced future smart home. [Online] Available at: https://www.lg.com/global/lg-thinq/news/future-of-smart-home

Xu A, Liu Z, Guo Y, Sinha V, Akkiraju R (2017) A new chatbot for customer service on social media. pp 3506–3510

Yang W, Jiachun Z (2018) Real-time face detection based on YOLO. IEEE, pp 221–224

Yu L et al (2020) Deep reinforcement learning for smart building energy management: a survey. arXiv e-prints, arXiv:2008.05074

Zhou B et al (2016) Smart home energy management systems: concept, configurations, and scheduling strategies. Renew Sustain Energy Rev 61:30–40

Zhang Z et al (2018) A deep reinforcement learning approach to using whole building energy model for hvac optimal control. ASHRAE/IBPSA, Chicago

Zhong C, Gursoy MC, Velipasalar S (2019) Deep actor-critic reinforcement learning for anomaly detection. In: 2019 IEEE global communications conference (GLOBECOM), pp 1–6

7 Demand Response and Smart Charging

Abstract

Soon, our living spaces will become a virtual energy cluster of the energy system with the increase and smart use of distributed energy assets such as electric vehicles, heat pumps, roof-top photovoltaics, and battery systems. The Open Digital Innovation Hub (ODIH) project aims to govern this hybrid architecture and its energy components and smart management systems. This study presents a broad perspective from the basic charger types to the revealed protocols as a complete component of the smart charging system in general and the comprehensive strategies and specific techniques that provide the smart charging solutions. A prominent contribution of this study with the ODIH project's implementation is that this chapter has an extensive level of detail that can be used as a reference while creating the relevant regulations. In this context, we have made critical recommendations for regulating smart charging and demand response functionalities consisting of electric vehicles, and residential prosumers penetrated grid systems. Our ultimate goal is to show that ODIH can be an active player in Demand Response through viable Smart Charging algorithms.

The author would like to acknowledge the help and contributions of Doğukan Aycı, Bora Batuhan İşgör, Ferhat Öğüt, Recep Tuna, Faruk Şahinoğlu, and Ulaş Özen in completing of this chapter.

7.1 Introduction

The electrification process of vehicles has been gaining momentum with the overcoming of the economic and technical barriers in the last 20 years. In 2018, more than 5.1 million electric vehicles were on the way, and this number is growing almost exponentially every year (IRENA 2019). With the introduction of electric vehicles in our lives, people have become familiar with a new routine: Charging of Electric Vehicles. Since the operation of Plug-in Hybrid and Electric Vehicles (PHEVs) are based on the battery system as an energy source, charging is a basic requirement for these vehicles. The International Renewable Energy Agency (IRENA) states that until the end of 2018, there are around 5.2 million charge points worldwide (2017 is 44% of this number) (IRENA 2019). Figure 7.1 illustrates the number of chargers and charger types between 2013 and 2018.

Turkey is just in the initial state of the e-mobility revolution, with approximately one thousand actively used EVs. However, this number is expected to increase exponentially every year due to the Domestic Electric Vehicle (TOGG project) initiative and the increase in vehicle ownership rate. According to the study of SHURA, it is predicted that 55% of the total passenger car sales will be electric and hybrid vehicles in 2030 and that it will represent 10% of the entire vehicle stock of electric vehicles. In this way, the total number of electric vehicles

S. Küfeoğlu, *The Home of the Future*, Sustainable Development Goals Series,
https://doi.org/10.1007/978-3-030-75093-0_7

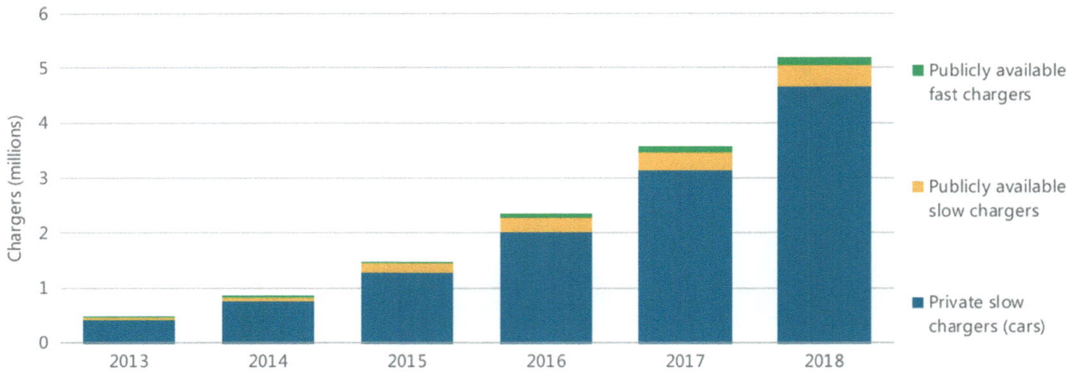

Fig. 7.1 Total charging stations around the world (IRENA 2019)

used in Turkey will reach 2.5 million at the end of 2030 (SHURA 2019).

7.1.1 Basics of EV Charging

According to data previously mentioned, electric vehicles and charging processes build a new habit in people's daily lives in the next ten years. To explain this change, it will be convenient to talk about Electric Vehicles Charging briefly. Electric vehicles can be considered as massive mobile batteries. The charging of an electric vehicle is quite different from the charging of any commonly used power tool due to its capacity and thus time, power, and energy requirements.

The power demanded by a charger during the process may be higher than the power demanded by the whole household. There are various power infrastructures in EV Charging. EV Service Equipment (EVSE) may supply energy in different forms (AC or DC), different voltage, and power levels. These can be categorised as.

7.1.1.1 AC Connectors

Type 1: Slow charger. Prominent in the US and Japan. Can support up to 7 kW. It only supports a single phase.

Type 2: Slow charger. Prominent in Europe and the rest of the world. Can support up to 22 kW. It also supports the three-phase.

7.1.1.2 DC Connectors

CHAdeMO: Used mostly by Japanese manufacturers. It has 50 kW power rating.

CCS1: Combination of Type 1 charger with two pins on the bottom for faster DC charging. Popular in North America as fast chargers.

CCS2: Combination of Type 2 charger with two pins on the bottom for faster DC charging. Popular in Europe as fast chargers. Can support up to 350 kW.

Tesla Supercharger: Proprietary Tesla supercharger. It can only be used with Tesla models at Tesla stations. It cannot be used for other models, even with an adaptor, since it requires authentication of the vehicle to be a Tesla vehicle (Lilly 2020). Table 7.1 illustrates the main charger types of EVs.

In EV charging, there are three levels according to the power rate, phase number, and output voltage.

In Level 1 charging, a standard single phase 16 A outlet is used, and this is the slowest method.

Level 2 charging is a faster and commonly used method for public facilities. Charging might take 4–8 h with a three-phase outlet with output power up to 22 kW.

On the other hand, EV's can be charged in only 20 min by using **Level 3 DC Fast Charging**, which may reach up to 400 kW currently (Shareefa et al. 2016; Gong and Rangaraju 2018).

Another classification is called Charging Modes. The concept of 'mode' refers to the charging technique in terms of its charging capacity, communication type, safety practices (IEC 2017; Skouras et al. 2019).

In **Mode-1 Charging**, a regular 230 V AC socket is used. This charging method lacks

Table 7.1 Main charger types of EVs (Lilly 2020)

Current type	**N. America**	**Japan**	**EU and the rest of the markets**	**China**	**All Markets except for EU**
AC					
Plug name	**J1772 (Type 1)**	**J1772 (Type 2)**	**Mennekes (Type 2)**	**GB/T**	**Tesla**
DC				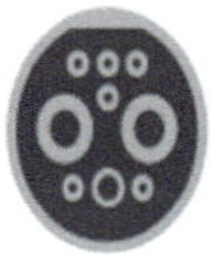	
Plug name	**CCS1**	**CHAdeMO**	**CCS2**	**GB/T**	**Tesla**

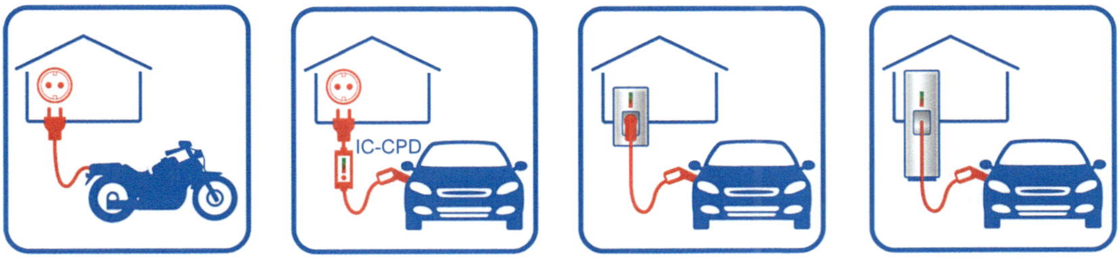

Fig. 7.2 Charging modes (IEC 2017)

communication, hence safety. Therefore, in Mode 1 the charging capacity is limited to a maximum of 2.3 kW in the scope of the IEC 61851-1 standard.

Mode-2 Charging provides an In-Wire Control Box for safety. In this mode, charging equipment can control charging capacity: the maximum charging capacity 7.4 kW 1-phase, 32A socket, or 22 kW 3-phase 32A supply.

In Mode-3, there is interoperation between the charging station and the connected EV, which coordinates the appropriate charging power. Public mode 3 charging stations offer 11 kW, 22 kW, or even 43 kW in most cases.

In Mode-4 Charging, AC to DC rectification takes place in EVSE. Hence, energy is directly transmitted to battery systems bypassing the dedicated charger. The delivered DC charging power usually ranges from 50 to 180 kW, and the charging current can reach 400 A. Figure 7.2 presents typical charging mode illustrations, Mode 1 to Mode 4 from left to right.

	Unit Power Rate / Vehicle Model	3.7 kVA	7.4 kVA	11 kVA	22 kVA	22 kVA
Full Charge Time	*Renault Fluence* 22kWh	6 hrs.	6 hrs.	6 hrs.	6 hrs.	6 hrs.
	Renault Zoe 22kWh	6 hrs.	3 hrs.	2 hrs.	1 hrs.	1 hrs.
	BMW i3 19kWh	5 hrs.	3 hrs.	3 hrs.	3 hrs.	3 hrs.
	Tesla Model S 60kWh	16 hrs.	8 hrs.	6 hrs.	3 hrs.	3 hrs.
Technical Specs	*Voltage*	1 Phase 230±23V AC		3 Phase 400±40V AC		3 Phase 400±40V AC
	Cables	1 Phase + Neutral + Ground		3 Phase + Neutral + Ground		3 Phase + Neutral + Ground
	Current	16 A	32 A	16 A	32 A	32 A
	Frequency	(50 ±0,5) Hz		(50 ±0,5) Hz		(50 ±0,5) Hz
	Plug Type	Type 2		Type 2		Type 2
	Charging Mode	Mode 3 (EN/IEC 61851-1)		Mode 3 (EN/IEC 61851-1)		Mode 3 (EN/IEC 61851-1)
Hardware	*Number of Plugs*	1		1		2
	Satellite Module	N/A		N/A		Available (Optional)
	RFID Card Reader	N/A		N/A		Available
	LCD Screen	N/A		N/A		Available
Safety	*Overflow Protection*	20 A	32 A	20 A	32 A	32 A
	Leakage Protection	30 mA (RCD)		30 mA (RCD)		30 mA (RCD)
	Env. Protection	IP 54		IP 54		IP 54

Fig. 7.3 An example of charging stations models is on sale in Turkey (eSarj 2019)

In the Turkish market, a company called eSarj provides numerous charging station options. Figure 7.3 shows the products and their technical details (eSarj 2019).

7.1.2 High EV Penetration Scenarios and Coordination Methodologies

EV charging will bring additional pressure on the distribution grid. With increasing EV uptake, the total power demand from the grid will increase significantly (Lillebo et al. 2019). Therefore, it is essential to understand the possible effects of electric vehicles on the distribution network, given concerns about additional network costs.

If charging is carried out uncoordinated;

i. Overloads may occur in the distribution system
ii. Energy unit price rises
iii. Due to the loads on the lines, the energy quality decreases, and frequency and voltage fluctuations are experienced.

These deteriorations may affect not only network operators but also end-users. Consumers may experience equipment failures due to power quality problems. In addition, energy price increases will be inevitable. Despite these trivial adverse outcomes, we should also mention that EV charging might also have a positive impact on distribution charges. Kufeoglu and Pollitt (2019) show that the distribution load drops approximately 2% with every 5% rise in EV uptake in the UK (Kufeoğlu and Pollitt 2019).

To overcome the adverse effects of EV penetration and to provide infrastructure for maximum EV penetration, different charging approaches have been developed. They can be classified into four groups (Rajakaruna and Shahnia 2015):

7.1.2.1 Dump Charging

In this approach, EVs have freely operated within their charging schedules without restrictions or incentives. EVs are such a king of power loads, like any other appliance. The charging process automatically starts by the time EVs plug-in and finish their charging when the battery has %100

State of Charge (SoC). There is no control, no interruption, or no power regulation in this model. Consequently, this model is not preferable since it has disruptive effects on the distribution network.

7.1.2.2 Multiple Tariff Policy

As in Dump Charging, EV owners used multiple tariff policy are free to charge their vehicles in any time period. The main difference is that energy prices are variable according to time or demand level. Therefore, EV owners are encouraged to shift their EV loads in peak hours to charge their EVs in low demand times. However, it is not an active method of control, and it is useful when electric vehicle users change their behaviour according to tariff dynamics.

7.1.2.3 Smart (Coordinated) Charging

Smart Charging is a sophisticated control model based on hieratical coordination between Distribution System Operators (DSOs) and EV Service Equipment (EVSE). In this model, the charging process is operated depending on the price, command, or demand signals coming from the system operator. Thanks to data-driven control architecture, goals such as high per cent renewable usage or minimal prices can be achieved through developed algorithms. Although smart charging provides many solutions and benefits, it requires sophisticated information and communication infrastructure.

7.1.2.4 Vehicle to Everything (V2X)

EV Charging can also be classified according to its power flow. EV Chargers can be unidirectional or bidirectional, depending on their architectures. As mentioned before, EVs can be regarded as mobile energy storage units. It is possible that EV battery systems can be used for not only powering the vehicle; they can also be used for power to the grid or a building. V2X applications are quite popular research areas nowadays since they provide significant advantages for smart grid applications and solutions for overloading (Soares et al. 2015). Table 7.2 summarises smart charging classifications presented by IRENA.

Table 7.2 Types of smart charging (IRENA 2019)

Type of application	Control methodology	Use cases	Technological readiness
Time of use policy without active control	There is no active control, based on financial incentives	Demand-side management with financial incentives. Encourage EV users to shift their energy into low demand intervals	Already exist
Basic on/off control	On/off capability with Connected EVSE	Enables charging scheduling and load shedding. Basic method for demand response	Accessible in market and it is in regulations
Adjustable charging power (VIG)	EV charging load adjustment capability	Frequency regulation and ancillary services by using active power demand control	Limited practice within the scope of but partially market regulations
Bidirectional charging between grid and vehicle (V2G)	Both EV and EVSE are capable of bidirectional power flow. EV can be regarded as mobile energy storage unit	Advanced frequency and voltage regulations, active and reactive power support as a power quality service. Economical energy optimization studies	Medium (generally in testing period)
Bidirectional charging between vehicle and all (V2X)	Two-way energy transaction is possible not only between EV and Grid also other appliances	Smart home energy management, virtual power plants, back-up supply	Medium (generally in testing period)
EV Charging according to dynamic pricing	EV Charging operation is scheduled according to dynamic tariff	Economic load dispatch, demand response services	Low (generally not in regulations)

7.1.3 Smart Charging Opportunities

Vehicle charging is not only a consumer need, but it is a new type of business field with commercial value. Therefore, it is essential to choose the right chargers and position them in the correct location. Real-time monitoring and control of charging equipment should be used in commercial or public spaces. Traceable and controllable charging systems have significant advantages. First, real-time monitoring of the charging processes contributes to the efficient use of the stations. Identifying user habits and data-driven correct positioning of the EVSE's will increase the benefit from the station. In addition, vehicle charging loads will constitute a large part of the electricity costs of buildings and institutions. Therefore, monitoring is also crucial for the control of commercial relations with customers or consumers (Green Flux 2019).

On the other hand, coordinated charging methods are available to increase renewable energy usage. Electric vehicles can be charged at night with electricity produced and stored during the day with rooftop solar systems. Many charging optimisation strategies can provide system flexibility for controlled various renewable energy sources to integration by grids. Many academic papers have reviewed smart charging coordination with renewable energy integration to cope with fluctuating renewable generation by adding technological and financial constraints to the optimisations model. Their investigation proves that while the individual EVs charging is anyhow satisfied, derangement of the system is diminishable by almost up to 44% with respect to the uncoordinated case. (Schuller 2015).

7.1.4 Demand Side Management via Smart Charging

Another phenomenon worthy of mentioning is Demand Side Management (DSM). Demand-side participation is a balance mechanism where Supply—Demand balance is not only the responsibility of the supplier, but the consumers are also actively engaged in it. They gain economic benefits by shifting their flexible loads during the peak loads. On the other hand, Demand Response (DR) is the consumer's reaction to changing prices or price signals. While the method provides support in the supply and demand management of the distribution network, it enables the users to gain financial benefits (Zhao et al. 2013).

Consumers ready to return incentive-based signals are referred to as active consumers in the literature. This type of consumers may exist like a giant consumer such as an industry or a commercial sector customer or a small consumer such as a residential customer. The demands of local customers, whose consumption level is very low compared to the average traded sizes in the electricity markets, can be met by an agent similar to the retailers. In addition, with governable resources, this tool, which is often called aggregator, can undertake new roles that provide simultaneous control of consumer devices with the developing technological know-how. Small customers will need the services of an aggregator agent that collects the flexibility provided by multiple prosumers and builds or pools to be purchased as a single resource in an active demand capacity. EV Charging Operators called 'EV aggregator' are a kind of aggregator for demand-side management services by using EV loads (Balram et al. 2015).

Electric vehicles are one of the most suitable energy loads for Demand Side participation. During the period when the network is overloaded, or there is frequency, voltage fluctuations, the system can terminate the charging process and reduce the load (Zhao et al. 2013). Charging equipment must be communicable and controllable. Also, there is a need for a platform to detect the price or demand participation signals coming from Network Operators (DSO) or system aggregators and provide load control (Palensky and Dietrich 2011). Smart tariffs are essential to achieve Demand Response. Different kinds of smart tariffs can be seen in Table 7.3.

While some methods are commonly used, some are still at a testing stage. To implement these smart tariffs, both DSOs and customers should be convinced and educated both

Table 7.3 Smart tariff designs

Tariff	Feature
Static time-of-use pricing (TOUP)	The electricity prices are fixed in long terms but the 24-h day is divided into certain intervals and has different prices
Critical day pricing (CDP)	For the days when balancing is determined, a price is determined in order to encourage day-ahead load reduction
Direct load control (DLC)	Some of the loads are able to remotely controlled for demand reduction
Critical peak pricing (CPP)	For the time intervals when balancing is highly required, operators set high prices to encourage users to shift or reduce their loads
Peak time rebates (PTR)	A pricing method that rewards the customer for peak energy reduction
Real-time pricing (RTP)	According to the real-time electricity market, the tariff is variable on an hourly basis or even in shorter time intervals
Dynamic demand (DD)	Prices are determined with time series according to consumption profiles
Market clearing price (MCP)	Hourly energy price balanced according to total supply and demand

economically and technically. Also, data security and privacy are critical issues to highlight. From a futuristic perspective, it will be possible for self-sustaining homes to make a profit by charging their batteries at off-peak times, restoring them to load at the peak time of another day. The learning ability of the various components enables the smart grid ecosystem to have mentioned intelligent characteristics. Since users in DR build their strategies using future load and price predictions, these need to be obtained through robust learning techniques by using historical data such as prices and load profiles. Smart charging problems can be addressed in two simple stages of the decision-making perspective:

- forecasting the possible responses of environmental and behavioural variables (e.g., electricity cost, driving habits)
- turn these predictions into actionable knowledge.

Scheduling, clustering, and forecasting are commonly used techniques to handle high incidence scenarios of EV usage. The key goals of those policies are to reduce the effect of charging on the delivery of electricity. To identify the highly used and recurring load profiles, clustering techniques have been applied to charging electric vehicles. It is possible to find some patterns and make predictions about daily charging behaviours of different kinds of charging profiles, such as residential or workspace charging. The certainty of the forecasting studies is vital for the creation of utility services and decision making. Coordinated EV charging will provide power grid resilience and ensure compatibility between demand and supply for electricity. Besides, it promotes G2V and V2G operations by optimising scheduling (Al-Ogailis et al. 2019).

7.1.5 Virtual Power Plants

Within the existing infrastructures of conventional central control systems, even Demand Response strategies will be insufficient in the face of future developments, and the need for distributed control systems will increase. The digitalisation of networks and integration of smart devices into the system offer solutions for the integration of renewable and distributed generation plants into the distribution network. The development of more efficient and cheaper communication technologies and cloud-based services enabled these technologies and services to be integrated within conventional systems. In this way, distribution network operators gained advanced traceability and faster and more

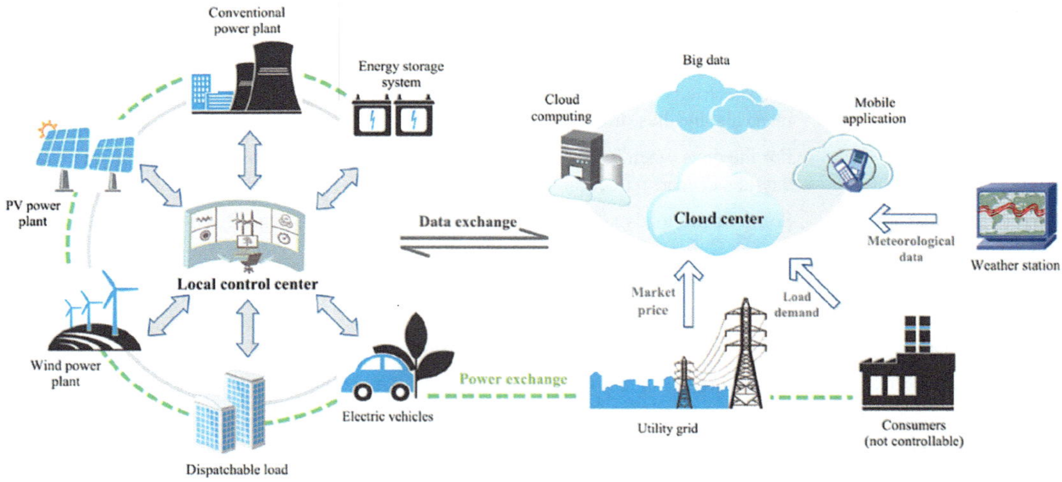

Fig. 7.4 Virtual power plant concept (Yu et al. 2019)

effective actions in their operations. On the other hand, many energy system components such as inverters, energy storage systems, smart home appliances, smart EV charging systems, HVAC loads are nowadays able to communicate and work in coordination.

Different types of energy storage systems, hybrid power plants, demand-side participation, smart charge–discharge methods, dynamic tariff structure, and many similar solutions have been developed to serve the goals of decarbonisation, high Renewable Energy Systems (RES) integration, grid stability, and market balance. Similarly, the concept of a Virtual Power Plant or called VPP, in which the smart communication infrastructure and the aforementioned network solutions are used for coordination, especially in distributed energy systems, has emerged (Molderink et al. 2010).

Virtual Power Plant (VPP) is a mechanism that enables countless distributed energy sources, storage, and consumer groups to operate in coordination through advanced communication infrastructure and decision-making mechanisms to act autonomously like a single digital power plant. In other words, VPP is a cluster of distributed generating units, displaceable loads, and energy storage systems that combine to function as a unique power plant. Each element of this cluster is in real-time and bidirectional communication with the relevant units and the required status, potential, etc. While transferring the information to the system, it can also organise its operations with the commands or pricing signals transmitted by the system (Saboori et al. 2011). Figure 7.4 presents the concept of a typical Virtual Power Plant.

Energy management operations are implemented through an advanced decision-making mechanism according to the operational potentials of the units in the VPP clusters, such as generation and load shifting, the operating conditions of the distribution network, and the determined targets. These targets are technical and economic targets such as low carbon emissions, high RES usage, minimisation of costs, high efficiency, or network flexibility. Real-time data collection from each asset and data processing is the key to achieve such goals. At last, system-related parameters such as distribution network capacities, its weak points are important in optimization operations (Mashhour and Tafreshi 2009).

7.1.6 Second Usage of Electric Vehicle Batteries

Recycling is an integral part of sustainable manufacturing since it prevents the reduction of natural resources and provides energy savings. By recycling and reusing, the amount of solid waste to be disposed of is reduced. It contributes to the economy and provides a clean environment for future generations. With electric vehicles' rise in popularity, EV battery recycling has become a lively field of attention. According to Lovell (2019), "The global e-mobility transition has an important task to find a solution to circular production retired EV batteries (Lovell 2019). The rapid adoption of intermittent solar and wind energy is increasing the demand for storage. Nonetheless, most energy storage systems are relatively expensive. There is a tremendous unmet need for affordable alternatives, especially for distributed energy production from rooftop solar installations on residential and commercial buildings. In the meantime, the prevalence of electric vehicles (EV) continues to expand, as does the amount of batteries that require replacement.

In conclusion, the figures show that the EV charging phenomenon is becoming more apparent in our daily lives. Though, if this increase is not appropriately managed, it will cause problems in the distribution system in the coming years. Therefore, Smart (Coordinated) EV Charging is essential to prevent negative impacts on the distribution network. Besides, it provides new opportunities in grid flexibility and high-level Distributed Energy Resource (DER) integration. Charging of EVs can also be shifted to periods where there is plentiful clean, renewable electricity generation. In this way, it also plays a vital role in global low carbon targets. Electric Vehicles provide a new role for consumers as part of a smarter and more flexible system. Consumers can benefit from financial incentives and avoid triggering unnecessary network reinforcement by becoming prosumer and smart home transformation. Tomorrow's smart homes and smart grids will undoubtedly be in coordination with electric vehicle charging operations. On the other hand, the e-mobility revolution will be an integral part of the sharing economy of smart cities.

7.2 Aim of the Study

By being fully smart and digitalised, the Open Digital Innovation Hub (ODIH) project will be the very core of future smart cities by providing a self-sustaining living environment. ODIH will be employing renewable energy sources, and it will be a mini-grid most of the time by being isolated from the grid. Nevertheless, domestic batteries will sometimes have an energy surplus and shortfall. In this study, we assume ODIH as an EV since they both have large batteries. We aim to constitute a smart charging environment for ODIH, depending on the charging algorithms. The purpose of this chapter is to:

- Design and construct physical infrastructure for bi-directional charging between ODIH and main grid
- Determine various smart charging algorithms to test and realise Demand Response
- Develop a hybrid energy management algorithm that coordinates charging and discharging operations in a self-sustaining and cost-effective way.

7.3 Methodology

In the scope of the Open Digital Innovation Hub Project, our main purpose is to develop a smart charging system that will interact with other home appliances and equipment such as batteries, inverters, controllers, or data sources such as forecasting and pricing signals. In this way, data-driven decision-making mechanisms will be developed by using machine learning and related tools, thanks to the interaction of data collected from other equipment and virtual sources. To realise the self-sustaining concept, it is critical to consume and store the generated energy in the most efficient way. Electric vehicles can be thought of as mobile battery units and play an essential role in storing the surplus generated renewable energy in this project. With decision-

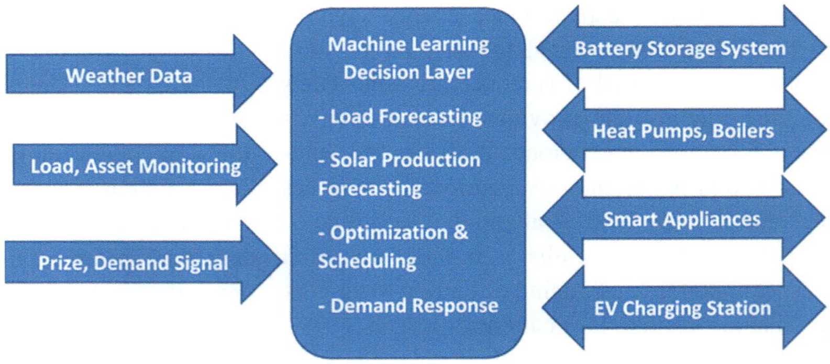

Fig. 7.5 ODIH energy and asset management system model

making mechanisms to be developed, production and consumption forecasts will be rendered in the day or different periods, and the energy generated through these estimations will be transferred to other appliance groups and storage groups. ODIH energy and asset management model is given in Fig. 7.5. As seen in the figure, there are some inputs (price signals, demand, weather data, etc.) from devices and web services. The application layer of the model consists of scheduling, forecasting, and decision-making functionalities. Thanks to these functions, applications such as demand response, production and consumption estimation, or load shifting are developed, and outputs to the assets such as on/off signals are generated. Figure 7.5 represents the model of ODIH Energy and Asset Management System.

7.3.1 Charging Station Selection

As mentioned in the previous section, choosing the right EV Charger depending on the application is essential. For an EV Supply Equipment (EVSE), there are several parameters such as connector type, power rate, power flow, or communication abilities. Different types of charging stations can respond better to different usage areas. For instance, although Type 1 EVSE is inefficient in commercial charging services with low power rates, they are satisfactory for residential usage. Besides, they are also desirable since they are cheaper and do not require additional infrastructure.

In line with the project objectives, the parameters and requirements for the charging station have been determined. A charging station to meet the project needs to consider the following:

i. One of the most important aspects is the communication ability with other devices and platforms. To achieve this, different communicating ports and protocols have been examined. A common protocol is Open Charge Point Protocol 1.6 V which will be mentioned in the later section. Also, some product provides Application Programming Interfaces (APIs). On the other hand, various wired & wireless communication ports such as Ethernet, Modbus, WIFI, GSM, or Bluetooth exist in different products. Other equipment in the project will also be effective for the selection to be made among them.
ii. The second parameter is the charging power rate. There are different charging levels in terms of its phase configuration and power rate. Since this project includes a residential application, and the main goal is to create a self-sufficient concept, a Level 1 charging station will suffice. In addition, this type of charging station will fit the technical needs and financial limits of the project, as they are easy to install and cheap.
iii. The third important issue is the smart charging abilities. It is expected in the project that the charging station can start, stop, and shift the charging process with remote signals transmitted from ODIH platform or user interfaces.

iv. Another expectation is the bi-directional charging ability. In Vehicle-to-Home (V2X) concept, an electric vehicle also has a discharging ability as with load management abilities. The electric vehicle will serve as the power source for the residence in the project.

v. Apart from technical points, there are also economic concerns. Therefore, a price-performance analysis will be made among the suitable products.

Various products that can fit the needs of the project are shown in Table 7.4.

7.3.2 Charging Station Connectivity

Connectivity is the key energiser of the coordinated charging. In the smart charging concept or within the scope of this study, the charging stations must be able to communicate with other equipment in the system and the cloud system. There are four communication layers in the system. They can also be seen in Fig. 7.6.

I. **Physical Layer** refers hardware part of the communication system, which consists of sensors, data receivers or transmitters, edge devices. The charging station is an element of the physical layer of this study.

II. **Networking Layer** is responsible for two-way data transfer among "the things" servers and users. To do this, **IoT gateways** are used for compiling the collected data converting it into digital form. It serves the same function as the modems in our homes in a more complex way. In this project, a dedicated IoT Gateway is responsible for collecting and transmitting edge devices data.

Table 7.4 EV charging stations from different makers

Product	Phase Conf.	Connector type	Power rate	Power flow	Communication port	Communication protocol
ABB Terra AC	Mono-phase or three-phase	Type 1 or Type 2	7.4 kW for 1phase or 22 kW for 3 phases	Unidirectional	Ethernet, Bluetooth, Wi-Fi, RS485, 4G	OCPP1.6, ABB ability API possible
Power Elektronik	Mono-phase or three-phase	Type 2	22 kW	Unidirectional	LAN, GSM, Wi-Fi (optional)	OCPP, IEC 61,851
Voltrun personal	Mono-phase or three-phase	Type 2	3.7, 7.4, 11, 22 kW	Unidirectional	Ethernet, GPRS	EN61851, Internet Platform
Schneider Evlink	Mono-phase or three-phase	Type 1 or Type 2	7.4 or 22 kW	Unidirectional	Ethernet, 3G, 4G	OCPP 1.5–1.6, Modbus
EVBOX Elvi	Mono-phase or three-phase	Type 1 or Type 2	3.7, 7.4, 11, 22 kW	Unidirectional	RS485, Wi-Fi, Bluetooth 4.0, 3G	OCPP 1.5 or 1.6
EnelX JuiceBox 40	Three-phase	American JSS	JSS	Unidirectional	Wi-Fi	Mobile App
Vestel EVC04	Mono-phase or three-phase	Type 1 or Type 2	7.4 kW max for single phase	Unidirectional	Ethernet, Wi-Fi, RS485, 4G	OCPP, Modbus
Delta AC Mini+	Monophase	Type 1 or Type 2	3.68 or 7.4 kW	Unidirectional	Ethernet—(standard), WLA N, 3G (optional)	OCPP

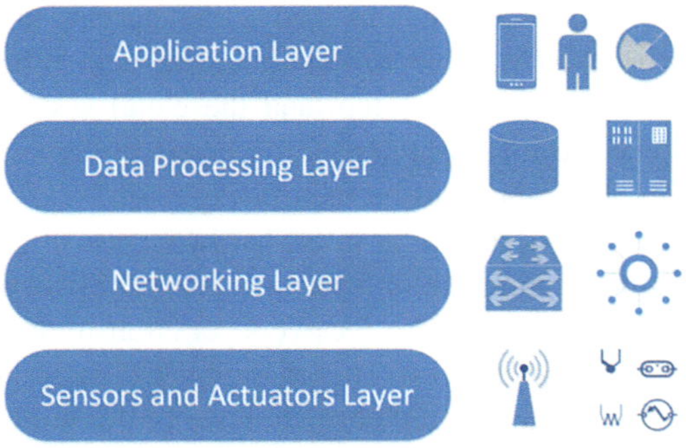

Fig. 7.6 Communication layers for the internet of things architecture

III. For more in-depth analysis and decisions, the whole data collected from home must be stored. Therefore, a **Cloud system**, or in other words, "**Data Storage and Processing Layer**" is essential for the study. There are various cloud service providers and end to end service platforms for this.

IV. **Application Layer** is the final layer of the IoT architecture. All of the computational functions, such as load management algorithms, mobile apps, or other user interfaces, are operated in this layer. Within the scope of the project, smart charging management functionalities are developed and operated in this layer.

As mentioned before, data exchange between devices and the cloud is a necessity for smart charging. Wired and wireless communication systems are available for use in smart home systems. Also, most of the communication ports are available in different EVSE products, and these can be seen in Table 7.4. It will be useful to explain the communication model over Open Systems Interconnection (OSI) layers. The OSI model is a world accepted model developed by the International Organization for Standardization that enables different systems to communicate using a common protocol, standard. Thanks to the OSI model, communication between computers and smart devices is operated in a world-accepted language. Figure 7.7 (Wikimedia 2015) illustrates the OSI model with the respective layers.

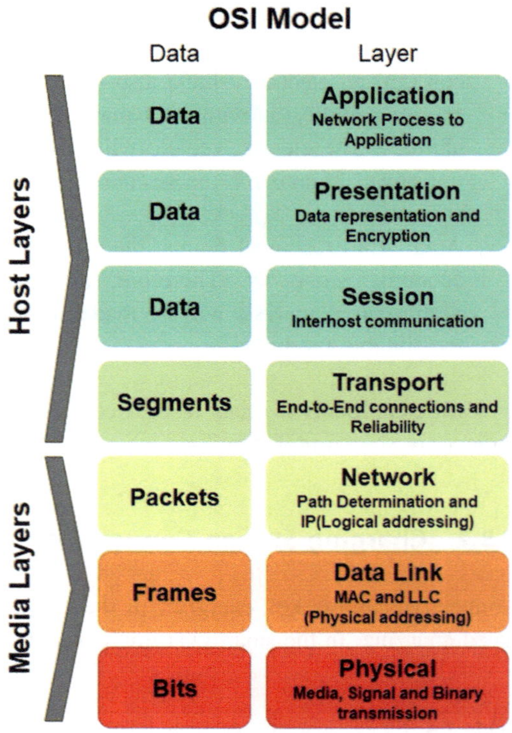

Fig. 7.7 OSI model and its layers (Wikimedia 2015)

In wired communication, RS485 and Ethernet are the common wired communication solutions between devices. **RS485** is located at the bottom layer of OSI layers and needs middleware to reach the internet. RS485 is commonly used in industrial devices such as solar inverters or motor drivers. **Ethernet** is also a physical data transfer system that uses twisted-pair or fibre optic wiring. Ethernet protocol is located at the second layer of OSI and convenient in local environments such as homes or buildings (Horyachyy 2017).

On the other hand, wireless technologies are getting more and more popular in IoT environments since they are easy to install and more capable to connect different devices. It is possible to classify wireless technologies by range, data rate, and power consumption. As a matter of fact, a single technology does not rule wireless networking. In most cases, systems that offer short-distance, low-power, low-bandwidth connectivity, free access, low service requirements, and safety standards are commonly used in home and indoor settings. ZigBee, Wi-Fi, Bluetooth, and Z-Wave are the most popular solutions for wireless smart home applications (Horyachyy 2017). Table 7.5 presents various home applications and corresponding recommendations for wireless communication technologies.

As can be seen in Table 7.5, products on the table are providing different kinds of communication technologies. In wired communication, Ethernet is the most popular medium and available in most of the products. In wireless communication, both Wi-Fi and GSM solutions are commonly available. In the scope of the ODIH project, there will be different kinds of devices and environments in the building. The charging station will be a part of coordinated and connected devices in the home area network (HAN). Hence, the technology to be chosen must be interoperable with other parts of the system.

On the other hand, wireless communication will be more logical as the project pilot is designed as an off-grid system. When we bring all these needs and benefits together, it is possible to say that Wi-Fi technology is the most suitable solution for smart charging.

7.3.3 Smart Charging Coordination via Charging Protocols

The state of charge and time of departure are critical pieces of knowledge to promote the implementation of smart charging. Standards for smart charging between EV and EVSE have been developed and are in operation to ensure that front-end communication and signalling procedures are interoperable at the base level. It is advised to concentrate on free protocols to use Original Equipment Manufacturers (OEMs) in the EV domain to get the charge rating (Elaadnl 2018).

The Open Charge Point Protocol (OCPP) is a publicly available and word-accepted protocol for interoperability between EVSE and a central EV charging management system, which is internationally recognized. This protocol is equivalent to mobile phones and telecommunications networks in terms of its methodology. Franc Buve and Joury de Reuver wrote the original edition. (Open Charge Alliance 2019).

The protocol is a project of a Dutch foundation, ELaadNL. The aim was to establish an open application protocol that would allow contact with the central management systems of the various providers between EV charging stations. There is a wide range of EV charging station manufacturers and central control services that use this protocol in operation around the world (Lohsen 2019). In addition, OCPP enables them to build a wide, open network using a number of different charging stations when only one operating system is needed (Greenlots 2018).

Another open-source protocol is **Open Charge Point Interface Protocol (OCPI)** that facilitates interoperability and communication between providers of mobility services, operators of charge points, and providers of navigation services. With OCPI, charging services become simple, standardized and harmonized. OCPI is responsible for connections between e-Mobility Service Providers (eMSP) with EV drivers as customers and Charge Point Operators (CPO) with Charge Point Management System (CPMS) control of charging points. In this model, CPMS is keeping track of the charging points status, communicating with external parties about their positions and charging sessions, and providing aggregators with an interface to monitor the charging stations' charging speed (Open Charge Alliance 2018).

The Open Smart Charging Protocol (OSCP) is another open accessed protocol for coordination between distribution system operators and charging station operators for demand management. The main feature of the OSCP is to bring Charging System Operators (CSO) forecasted capacity and flexibility. Besides, DSOs can send a request to CSO to increase and decrease their loads (European Commission 2020). In this way, CSO may regulate its charging demand according to DSO capacity, and incentives without expensive infrastructure are needed. Also, DSO's will able to increase the physical capacity of EV Loads by using this

Table 7.5 Comparison of wireless solutions for home area networks (Horyachyy 2017)

Smart home applications		Requirements							Recommended wireless technology
		Low power	Low cost	Security	Range	Topology	Network density	Throughput	
Home automation	Lighting	±	+	±	PAN/LAN	p2p, star, mesh	+	Low	ZigBee, BLE, Bluetooth, Z-Wave, ANT, WiFi HaLow, WiFi
	HVAC	±	±	±	PAN/LAN	p2p, star, mesh	±	Low	ZigBee, BLE, Bluetooth, Z-Wave, ANT, WiFi HaLow, WiFi
	Security	±	+	+	PAN/LAN	p2p, star, mesh	+	Low, upper medium	– ZigBee, BLE, Bluetooth, Z-Wave, WiFi HaLow, ANT, WiFi (low) – Bluetooth. WiFi (upper medium)
Energy management		–	±	+	LAN	p2p, star	±	Low	ZigBee, WiFi, WiFi HaLow, Bluetooth, Z-Wave, BLE, ANT
Entertainment		±	±	±	PAN/LAN	p2p, star	±	Upper medium, high	– Bluetooth, WiFi (upper medium) – WiFi (high)
Wearables		–	+	±	BAN/PAN	P2P, mesh	–	Low	BLE, ZigBee, Z-Wave, Bluetooth

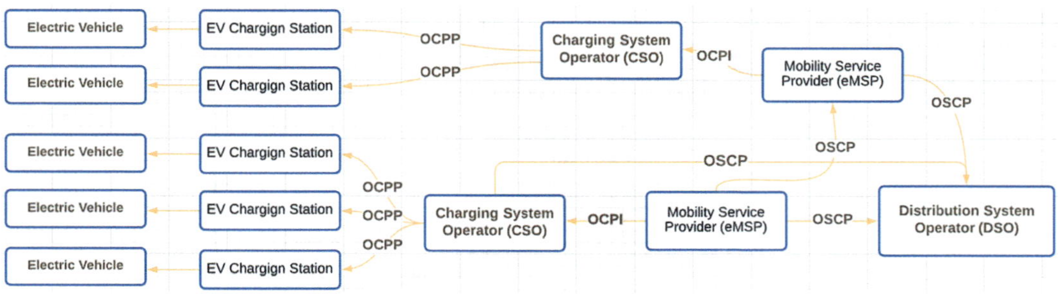

Fig. 7.8 Communication protocols between smart charging stakeholders

protocol (Portela et al. 2015). Figure 7.8 represents the communication protocols between smart charging stakeholders.

Our task in the Smart Charging hierarchy is to manage the charging process of ODIH in coordination with the decision-making mechanisms to be established. As a result, when these communication protocols are examined, it is seen that the protocol that will respond to our needs and tasks is the OCPP protocol between EVSE and CSO. The second alternative would be the Modbus communication protocol which is the most popular and open access industrial communication protocol in the field of process automation and SCADA. Some of the products, such as Vestel Product is capable of communicating with Modbus protocol.

7.3.4 Machine Learning Approaches for EV Charging Management

Many studies concentrated on an optimal charging schedule for smart charging management of an EV, arranging a sequence of serious agent's actions in an interval of bounded time arrival and departure time (connection time). This action simply consists of two-stage models. The first paradigm is the conventional time series forecasting future values, as mentioned earlier in this study, and the second step is an optimisation model (by adjusting the control variables). However, the robustness of these models is entirely related to the accuracy of the predictions. However, during the construction of these prediction analyses, historical data can be too complex, and it was mentioned that this process would be performed under the aggregator operation in smart grids.

At this point, Machine Learning (ML) methods became appropriate techniques. Stratification process, clustering, and classification, which come together within the ML, provide avoiding the redundancy of the data. ML algorithms and learning processes such as used dynamic programming have also achieved good results in optimising smart charging mechanisms in related studies. According to Lopez (2019), to establish a smart charging management system with ML, mainly four tasks should be carried out (Lopez 2019):

1. Constitution of an EV charging model using dynamic optimization that takes the electricity prices and consuming amounts of energy used as inputs.
2. Establishing an information system containing all the variables needed by the EV charging model.
3. In order to accelerate the operation of the model on the training, the most proper data selection is required to provide values close to the training error when the whole data using.
4. Developing a machine learning-based classification model that will decide whether to charge or not for EVs to guarantee optimality criterion.

Figure 7.9 illustrates the description of these four actions mentioned above.

Apart from this data collection, some of the required information for the model, such as location information, charging station

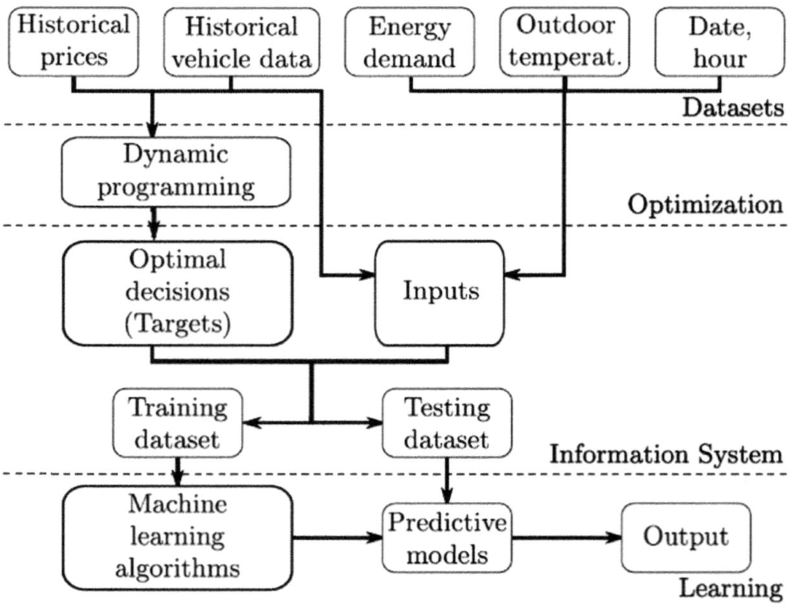

Fig. 7.9 Machine learning approach for EV charging coordination

availability, battery status of the EV, and energy need, should be provided by EV users personally. Therefore, whether user-based strategies will be allowed or not need to be decided so that there are alternatives in game-theoretic models in the literature (Robu et al. 2013). Regarding time series analysis, recent academic studies focused on Deep Learning-based approaches. (Ryu et al. 2016) proposed a Deep Neural Network (DNN) for forecasting short load period while (Mocanu et al. 2016) used the DNN method he presented for energy consumption prediction (Lopez 2019). In general, due to both the abundance of historical data and a dynamic decision-making process, the smart charging concept is highly convenient for ML models.

7.4 ODIH Hybrid Energy Management System Algorithm

Energy management can be broadly defined as the proactive, organised, and systematic management of energy use in a building or organisation to satisfy both environmental and economic requirements. ODIH Hybrid Energy Management System (HEMS) Algorithm is a smart charging prototype that controls and schedule energy transaction between distributed energy resources (DER), energy storage system (ESS), Smart Loads and the Grid Network. In the scope of the ODIH project, the main purpose is to create a self-sustaining building. Therefore, the algorithm tries to supply the building's whole energy need from renewable sources by scheduling charging and discharging periods of battery. However, total production and battery capacity are not enough always. Consequently, the hybrid system needs an external energy supply from the grid. In this case, the algorithm tries to match energy injection and price data in the most economical way. In short;

First Goal: Creating Self Sustaining HEMS Model that meets the energy it needs as self-sufficient as possible through its DER and ESS resources.

Second Goal: Developing a cost-effective HEMS Model that tracks Time of Use (ToU) and near dynamic market clearing price (PTF is equivalent tariff in Turkey) tariff and makes decisions about battery and grid charging and discharging operations.

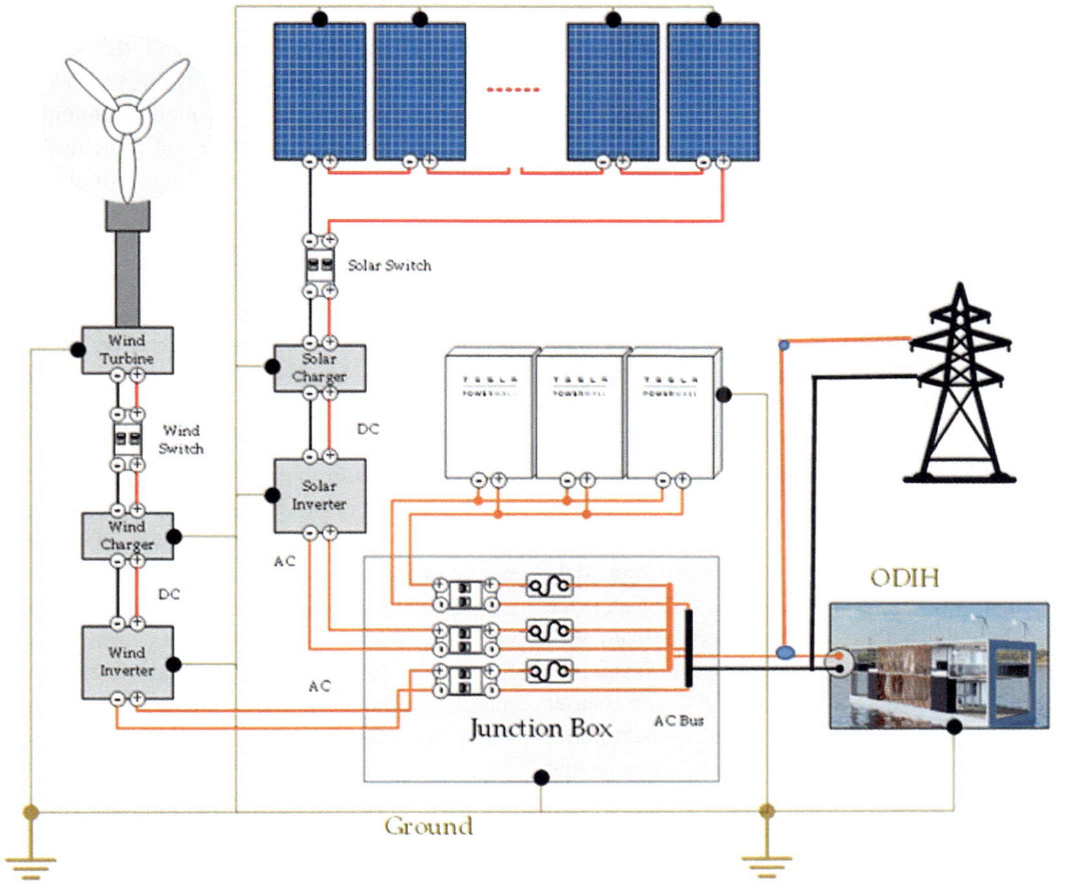

Fig. 7.10 ODIH hybrid energy management system

7.4.1 ODIH Hybrid Energy Management System Description

ODIH Hybrid Energy Management System from Sect. 7.4 can be seen in Fig. 7.10.

7.4.1.1 System Components

A. There will be solar panels on the roof of the building so that ODIH will generate its energy. However, energy production will not be entirely predictable and will depend on the weather. In addition to solar PV, there will be wind energy. Since the pilot area is a windy region, a small wind turbine was also placed in the house. Wind generation forecasting will also be done with analysis tools.

B. As an energy storage unit, three Tesla Power Walls will be used in HEMS. It stores the energy produced in solar panels up to a specific capacity and offers it to the use of the consumer whenever desired. Also, it acts as a backup source by activating the stored energy in case of power outages.

C. The third part is the smart loads. Smart loads can be explained as connected and controllable house appliances such as heat pumps or EV charging stations. When advanced algorithms are created, smart loads will be switched on, off, and shifted according to the load status.

D. In cases where the produced and stored energy is not sufficient, there will be a grid connection to meet the energy needs. As mentioned, grid supply is a worse case of the model, and care will be taken not to use it as much as possible.

7.4.2 Data Sources of HEMS Algorithm and Data Sample Methodology

The system determines real-time and the day before energy flow of battery and grid by processing the real-time and the production/consumption forecasting from the analysis in the flow diagram. Real-time data will be collected from devices such as energy analyser, inverters, and power walls. There is also production and consumption forecasting data obtained by estimation calculations. At last, price signals and whether data is collected from web services. Additionally, a dataset has been prepared to be used in HEMS models. The dataset contains 1-year scale production, consumption, temperature, and tariff data and their versions that have been taken from the necessary processes in synchronous columns with each other. The methodology for creating the data set will be Loads are divided explained while examining the data types. To list all data types;

7.4.2.1 Battery State of Charge (SoC) and Depth of Discharge (DoD)

State of Charge (SoC) is a ratio of present battery capacity over maximum battery capacity. It is vital to explain battery life. Another important term is the **Depth of Discharge (DoD)**. DoD is a ratio of capacity that has been discharged over maximum capacity. These two parameters are critical for battery charge and discharge operations.

7.4.2.2 Real-Time and Estimated Solar Production

While real-time production values are collected from solar inverters or smart meters, forecasting data were created with machine learning-based analysing tools. Two sources were used while preparing a one-year data sample: production history of a real Solar Plant in Ikitelli/Istanbul and solar production online analysing tools. The dataset is obtained from Istanbul Municipality Open Data Portal (IBB 2020) and it includes the hourly energy production of the İkitelli solar power plant in 2018 and 2019. The second data set gives the estimated monthly energy production calculated with sophisticated analysis tools. The calculation includes detailed parameters such as the number and model of panels to be found on the roof of the house, the angle of placement or the location of the house. To bring the calculated monthly production data to an hourly format, we obtained a ratio from the 7-month production data available to both production data. This is a ratio resulting from the installed power of the two power plants. Figure 7.11 presents the comparison of data obtained from analysing tools and normalised real production data from Ikitelli Solar Plant. The graph shows that the calculated monthly production data are quite consistent when compared to real-world data.

7.4.2.3 Real-Time and Estimated Wind Production

Wind energy production data will also be read in real-time via the relevant power converter or smart meter, just like solar data. To create the wind generation data, daily production estimates are made according to the installation site via web archive and analysis tools. Then, daily wind data were normalised according to the hourly average wind distribution and made hourly frequency. Figure 7.12 represents average daily production according to calculations.

7.4.2.4 House Demand

The energy load of the house will be collected in real-time via smart meters and smart devices. In addition, a consumption dataset has been obtained by distributing the energy loads of the equipment in the house properly. The energy consumption load profiles are divided into three categories for this data set to converge to the actual consumption. Base loads constitute the first category. The energy consumption of this group is fixed in specific time periods.

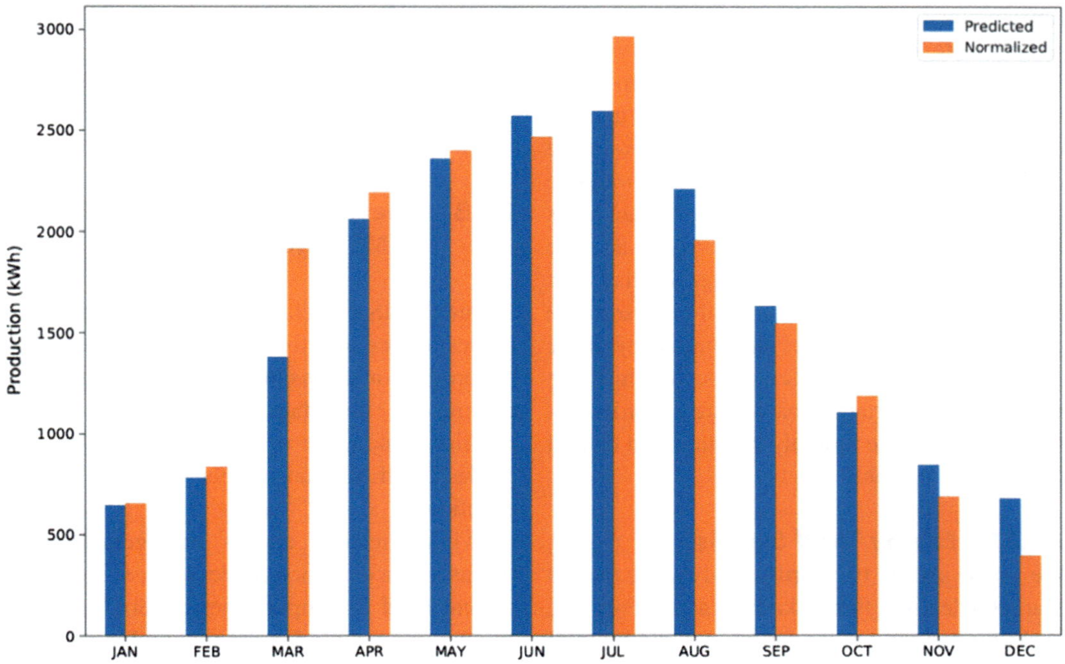

Fig. 7.11 Predicted and normalized monthly solar production

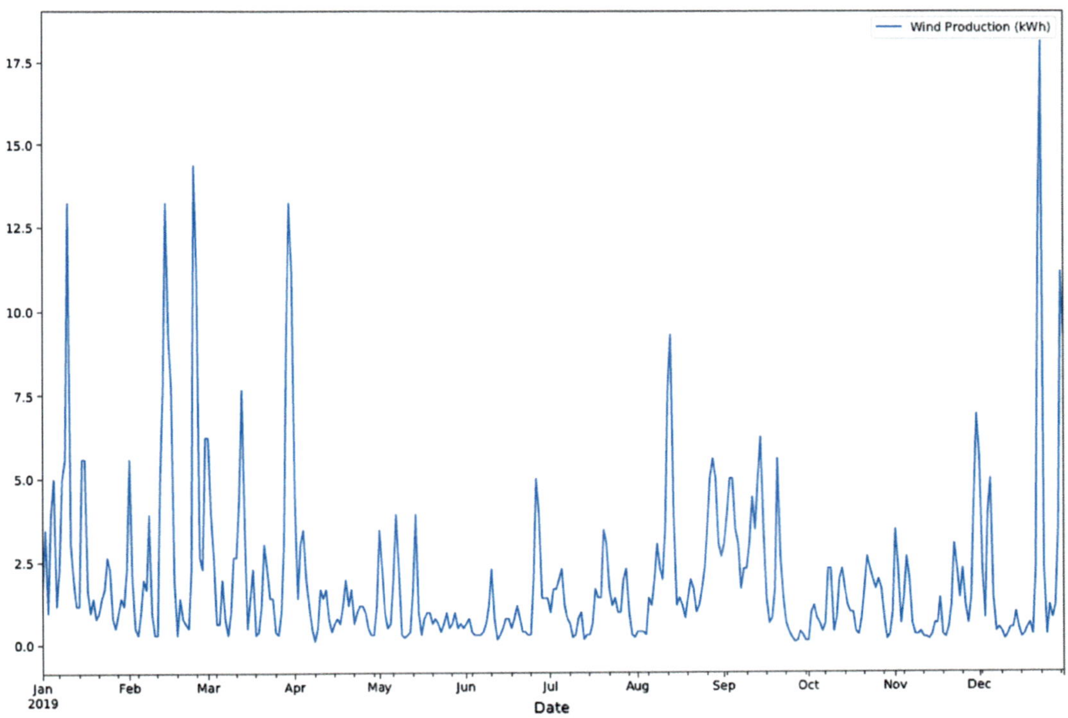

Fig. 7.12 Daily wind production

Refrigerators are examples of fixed loads. The second category includes dependent variable loads. The heat pump is a temperature-dependent variable, and the consumption will be normalised with the temperature data we will get from the web service. The last load group is variable loads. Loads such as toasters and kettles are loads that come into play at uncertain times and are difficult to predict. The advantage is that their usage time and the total energy they absorb are generally low. In this way, the consistency of the dataset will not be affected much while the remaining uncertain loads are distributed. A small part of the monthly consumption dataset can be seen in Table 7.6.

7.4.2.5 Energy Tariff Signals

We make use of two tariffs: a dynamic one is Market Clearing Price (known as PTF in the Turkish market) and a static time of use tariff (ToU). In short, the day-ahead tariff, known as PTF, is the reference electricity price formed by the matching of supply and demand in day-ahead electricity markets. With the consumption estimates collected on an hourly basis, TEİAŞ National Load Dispatch Center determines the amount of energy to be needed at the appropriate time of the day and makes the production planning. In the day-ahead market, the electricity reference price, that is, the market-clearing price (PTF), is determined by intersecting the hourly demand and supply. As can be seen in the table, it is lower at night hours when the energy demand of MCP is less than during the daytime when the demand is high. On the other hand, the Time of Use (ToU) tariff is the charging of electricity differently at certain times of the day. It is the most expensive hours as electricity usage makes a peak in the evening. At night, they are the cheapest hours because their electricity consumption is relatively low. Both of the price signals can be obtained from Epias web service. The market-clearing price tariff in a one-week period can be seen in Fig. 7.13.

As can be seen, the hourly price can reach twice the average value (near 300 TL for that week). In September, the maximum value reached 982 TL (Epias 2020).

7.4.2.6 Weather Data

Weather data are also obtained from web services with hourly frequency, and it will be used for estimating heating and cooling loads.

7.4.3 Operation Modes of ODIH HEMS Algorithm

In our HEMS model, there are several operating modes in case of different conditions. According to production, consumption, or energy price parameters, the management system makes decisions among the operation modes. Operation modes of HEMS can be seen in Table 7.7 with equations.

While creating energy management algorithms, some operation modes are assigned to the second level to simplify the model. The simultaneous operation of the battery and the network has been reduced in the created algorithm. As can be seen in the table, equations consist of solar, wind, grid, and battery powers.

P_{pv} Produced Solar Power.
P_{wind} Produced Wind Power.
P_{load} House Power Demand.
P_{batt-} Battery Charging Power.
P_{batt+} Battery Discharging Power.

The algorithm determines the choices and transitions between operating modes according to the previously mentioned self-sustainability goals. The key concept is that, to achieve self-consumption, produced energy from solar and wind should be used as much as possible. Priority is given to supplying the loads. In the case of surplus energy production or insufficient situations, the algorithm determines the required energy flow between battery to house or grid to the house.

While developing the HEMS algorithm, the products obtained from the literature review were also used. A study from Chekired et al. is an example of an energy management algorithm for a grid-connected-photovoltaic and battery storage system (Chekired et al. 2017). The key concept of this model is that to achieve self-

Table 7.6 Consumption breakdowns in ODIH

Devices/Months	J	F	M	A	M	J	J	A	S	O	N	D	Yearly total (kWh)
Fridge	53.0	53.0	53.0	53.0	53.0	53.0	53.0	53.0	53.0	53.0	53.0	53.0	636.0
Stove	135.0	135.0	135.0	135.0	135.0	135.0	135.0	135.0	135.0	135.0	135.0	135.0	1620.0
Oven	13.2	13.2	13.2	13.2	13.2	13.2	13.2	13.2	13.2	13.2	13.2	13.2	158.4
Range hood	0.9	0.9	0.9	0.9	0.9	0.9	0.9	0.9	0.9	0.9	0.9	0.9	10.8
Microwave oven	2.0	2.0	2.0	2.0	2.0	2.0	2.0	2.0	2.0	2.0	2.0	2.0	24.0
Dishwasher	12.0	12.0	12.0	12.0	12.0	12.0	12.0	12.0	12.0	12.0	12.0	12.0	144.0
Washer and dryer	43.5	43.5	43.5	43.5	43.5	43.5	43.5	43.5	43.5	43.5	43.5	43.5	522.0
Robot vacuum	1.8	1.8	1.8	1.8	1.8	1.8	1.8	1.8	1.8	1.8	1.8	1.8	21.6
Robot mop	1.5	1.5	1.5	1.5	1.5	1.5	1.5	1.5	1.5	1.5	1.5	1.5	18.0
TV	4.2	4.2	4.2	4.2	4.2	4.2	4.2	4.2	4.2	4.2	4.2	4.2	50.4
Kettle	6.6	6.6	6.6	6.6	6.6	6.6	6.6	6.6	6.6	6.6	6.6	6.6	79.2
Iron	7.8	7.8	7.8	7.8	3.5	3.5	3.5	3.5	7.8	7.8	7.8	7.8	76.4
Hair dryer	11.0	11.0	11.0	11.0	0.0	0.0	0.0	0.0	11.0	11.0	11.0	11.0	88.0
Toaster	13.0	13.0	13.0	13.0	13.0	13.0	13.0	13.0	13.0	13.0	13.0	13.0	156.0
Coffee maker	5.0	5.0	5.0	5.0	2.0	2.0	2.0	2.0	5.0	5.0	5.0	5.0	48.0
Modem–Router–Switch–Access Points–Computer	65.3	65.3	65.3	65.3	65.3	65.3	65.3	65.3	65.3	65.3	65.3	65.3	783.6
Phone charger	0.6	0.6	0.6	0.6	0.6	0.6	0.6	0.6	0.6	0.6	0.6	0.6	7.2
Laptop	48.0	48.0	48.0	48.0	48.0	48.0	48.0	48.0	48.0	48.0	48.0	48.0	576.0
Tablet	2.2	2.2	2.2	2.2	2.2	2.2	2.2	2.2	2.2	2.2	2.2	2.2	25.9
Desk lamp	1.2	1.2	1.2	1.2	1.2	1.2	1.2	1.2	1.2	1.2	1.2	1.2	14.4
Reoling cam outdoor	5.8	5.8	5.8	5.8	5.8	5.8	5.8	5.8	5.8	5.8	5.8	5.8	69.6
Burglar alarm	0.0	0.0	0.0	0.0	0.0	0.0	0.0	0.0	0.0	0.0	0.0	0.0	0.0
Smoke and CO alarm	0.0	0.0	0.0	0.0	0.0	0.0	0.0	0.0	0.0	0.0	0.0	0.0	0.0

(continued)

Table 7.6 (continued)

Devices/Months	J	F	M	A	M	J	J	A	S	O	N	D	Yearly total (kWh)
Motion sensor (water leak, temperature)	0.0	0.0	0.0	0.0	0.0	0.0	0.0	0.0	0.0	0.0	0.0	0.0	0.0
Door lock	0.0	0.0	0.0	0.0	0.0	0.0	0.0	0.0	0.0	0.0	0.0	0.0	0.0
Thermostat	0.0	0.0	0.0	0.0	0.0	0.0	0.0	0.0	0.0	0.0	0.0	0.0	0.0
LED bulb	37.8	32.4	32.4	32.4	27.0	27.0	27.0	27.0	27.0	32.4	37.8	37.8	378.0
Smart plug	0.0	0.0	0.0	0.0	0.0	0.0	0.0	0.0	0.0	0.0	0.0	0.0	0.0
Smart meter	1.8	1.8	1.8	1.8	1.8	1.8	1.8	1.8	1.8	1.8	1.8	1.8	21.6
Heat pump	1095.0	753.0	415.0	84.0	152.0	383.0	487.0	346.0	72.0	101.0	420.0	851.0	5159.0
Water from humidity	52.5	52.5	52.5	52.5	52.5	52.5	52.5	52.5	52.5	52.5	52.5	52.5	630.0
Reverse osmosis	s	0.0	0.0	15.4	26.4	32.0	33.7	29.5	15.4	0.0	0.0	0.0	152.4
Biogas reactor	132.0	132.0	132.0	132.0	132.0	132.0	132.0	132.0	132.0	132.0	132.0	132.0	1584.0
Agriculture lightning	59.4	59.4	59.4	59.4	59.4	59.4	59.4	59.4	59.4	59.4	59.4	59.4	712.8
Greywater	18.9	18.9	18.9	18.9	18.9	18.9	18.9	18.9	18.9	18.9	18.9	18.9	226.8
Rainwater	16.1	16.1	16.1	11.5	8.1	6.4	5.8	7.1	11.4	16.1	16.1	16.1	146.9
Hydrophore	22.0	22.0	22.0	22.0	22.0	22.0	22.0	22.0	22.0	22.0	22.0	22.0	264.0
Total (kWh)	1869.1	1521.7	1183.7	863.5	915.4	1150.3	1255.4	1111.5	846.0	869.7	1194.1	1625.1	14405.0

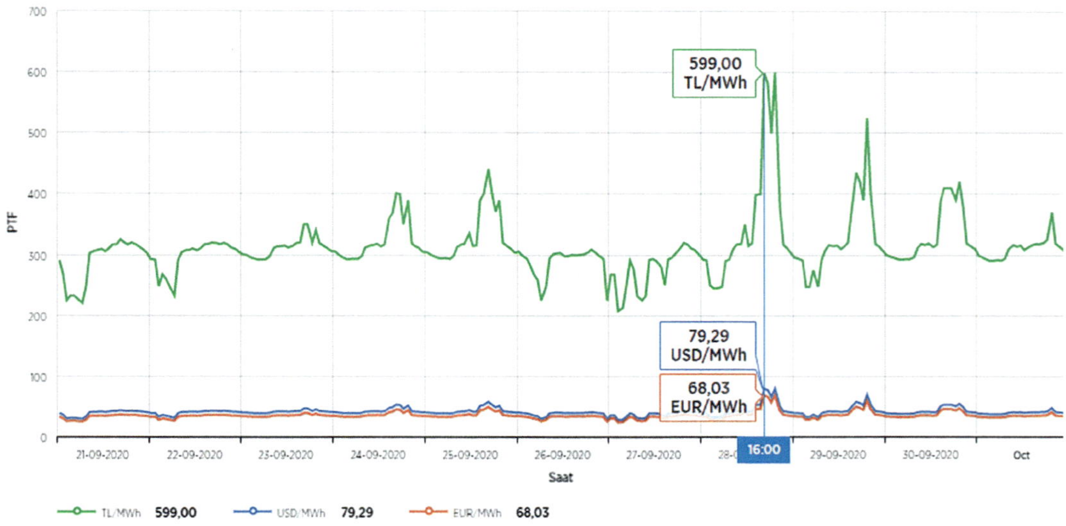

Fig. 7.13 Market clearing price with a one-week interval (Epias 2020)

consumption, the photovoltaic output should be used as much as possible. Priority is given to supplying the loads, then charging the batteries, and eventually feeding the excess energy into the grid. There are four operation modes depending on conditions. Although this model is parallel with ODIHs self-sustainable energy management purposes, it does not consist of any price based optimisation. Figure 7.14 shows the energy flow management algorithm.

Unlike this model, our study has cost optimisation functionalities that make energy management systems more profitable and shorten the redemption period. For this, the energy management system estimates the hourly total generation and consumption of the house and determines the energy surplus or gap. After that, using the algorithm, the charging and discharging schedules of the battery and grid are decided during the day. The algorithm optimises this scheduling process according to the current capacity of the battery and the price of the energy drawn from the grid. MCP tariff and ToU tariff, which are actively used in Turkey, are used as price data in the model.

The figure shows a HEMS algorithm customised in line with the dynamics of the MCP tariff. As mentioned, the MCP tariff is an energy tariff published a day before, with a different price for each hour of the 24-h time zone. With this feature, the MCP tariff is the model that is used in Turkey and closest to the dynamic tariff rates. As can be seen in the figure, when there is an energy gap in the daily forecast, the algorithm finds the time period when the battery system is sufficient until the target time, and the energy price is the cheapest. Thanks to this method, the system completes its energy gap with the lowest cost as soon as possible. Figure 7.15 presents ODIH HEMS Algorithm, which operates energy flow with optimisation of both energy production, energy consumption, battery capacity, and additionally day-ahead price signal.

In addition to the MCP model, in the model prepared according to the 3-time ToU tariff, cost optimisation has been developed not only for energy purchase from the grid but also for the energy to be transferred to the grid in case of excess energy. In this context, when the battery is not fully charged and it produces more than the demand power, the algorithm decides between charging the battery or exporting energy to the grid according to the time tariff. While making this decision, the algorithm acts under the self-sustaining goal, with priority over cost optimisation. For this, the high-priced time interval is not enough for the algorithm to make a decision to sell energy to the grid. The algorithm

Table 7.7 Operation modes of hybrid energy management system

Operating modes	Equation (P_{batt-} refers to charging)	Description	Involvement
Mode 1	$P_{pv} + P_{wind} = P_{load}$	PV and wind power feeds house load	Level 1
Mode 2	$P_{pv} + P_{wind} = P_{load} + P_{batt-}$	PV and wind power feeds house load and battery	Level 1
Mode 3	$P_{pv} + P_{wind} = P_{load} + P_{batt} \pm P_{grid}$	PV and wind power feeds house load, battery and grid	Level 2
Mode 4	$P_{pv} + P_{wind} = P_{load} + P_{grid}$	PV and wind power feeds house load and grid	Level 1
Mode 5	$P_{pv} + P_{wind} + P_{batt+} = P_{load}$	PV and wind power and battery feeds house load	Level 1
Mode 6	$P_{pv} + P_{wind} + P_{batt+} = P_{load} + P_{grid}$	PV and wind power and battery feeds house load and grid	Level 2
Mode 7	$P_{pv} + P_{wind} + P_{grid} = P_{load}$	PV and wind power and grid feeds house load	Level 1
Mode 8	$P_{pv} + P_{wind} + P_{grid} + P_{batt+} = P_{load}$	PV and wind power and battery and grid feeds house load	Level 2
Mode 9	$P_{pv} + P_{wind} + P_{grid} = P_{load} + P_{batt-}$	PV and wind power and grid feeds house load and battery	Level 1

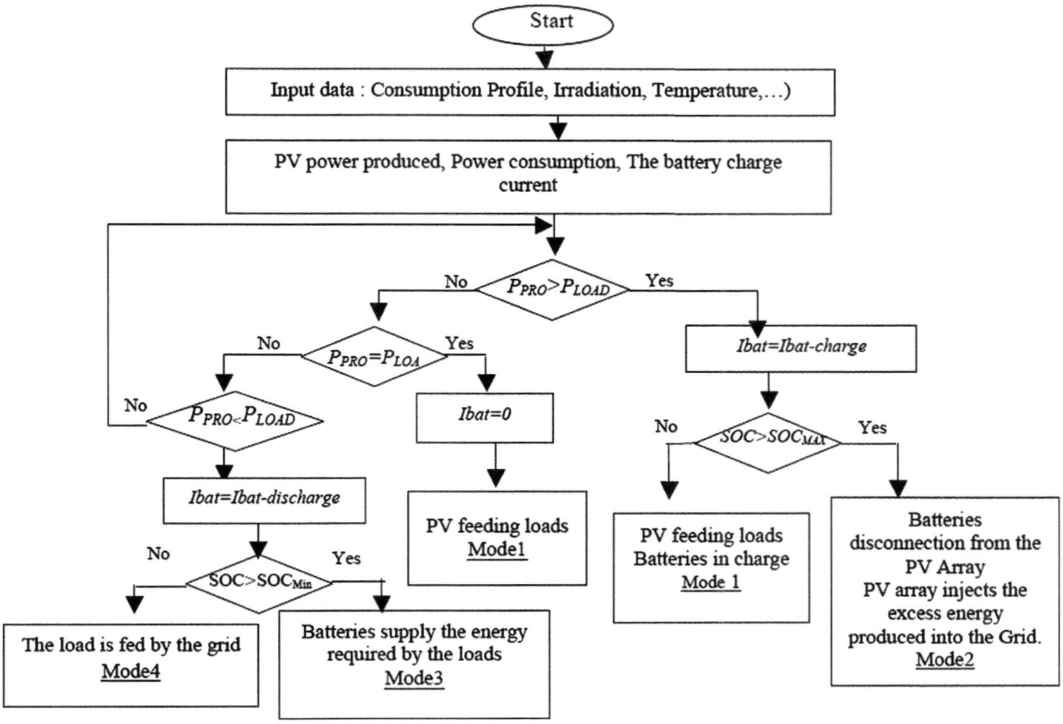

Fig. 7.14 Energy flow management algorithm (Chekired et al. 2017)

calculates the surplus energy for the rest of the day and makes a cost optimised decision about where to charge the battery and grid. Figure 7.16 presents the ODIH Energy Flow Management Algorithm optimised with ToU Tariff dynamics.

To summarise, in this study, hybrid energy management algorithms have been developed following the self-sustaining concept under cost optimisation targets. In the continuation of the study, the designed algorithms will be coded and trained through simulation programs over the created data sample.

7.5 Results

As a result of our comprehensive literature study on EV Charging, Demand Side Management, Distributed Generation, and Energy Management issues and the dataset we prepared, the outputs may be grouped under four titles.

7.5.1 Uncertainty and Imbalance in Energy Production and Consumption

Distributed generation systems are on the rise exponentially around the world, both with the incentives given in line with the carbon targets and being economically profitable. However, as mentioned in the study, the most significant barrier in the high integration of distributed and renewable energy systems is that these resources depend on parameters that cannot be directly controlled. In particular, seasonal conditions directly affect both energy production and the behaviours of heat pumps that make up a large percentage (34%) of energy consumption. Figure 7.17 presents heat pump consumption by months.

The heating and cooling demand is a parameter that is directly dependent on the outside temperature. Figure 7.17 shows the hourly average consumption of the heat pump by

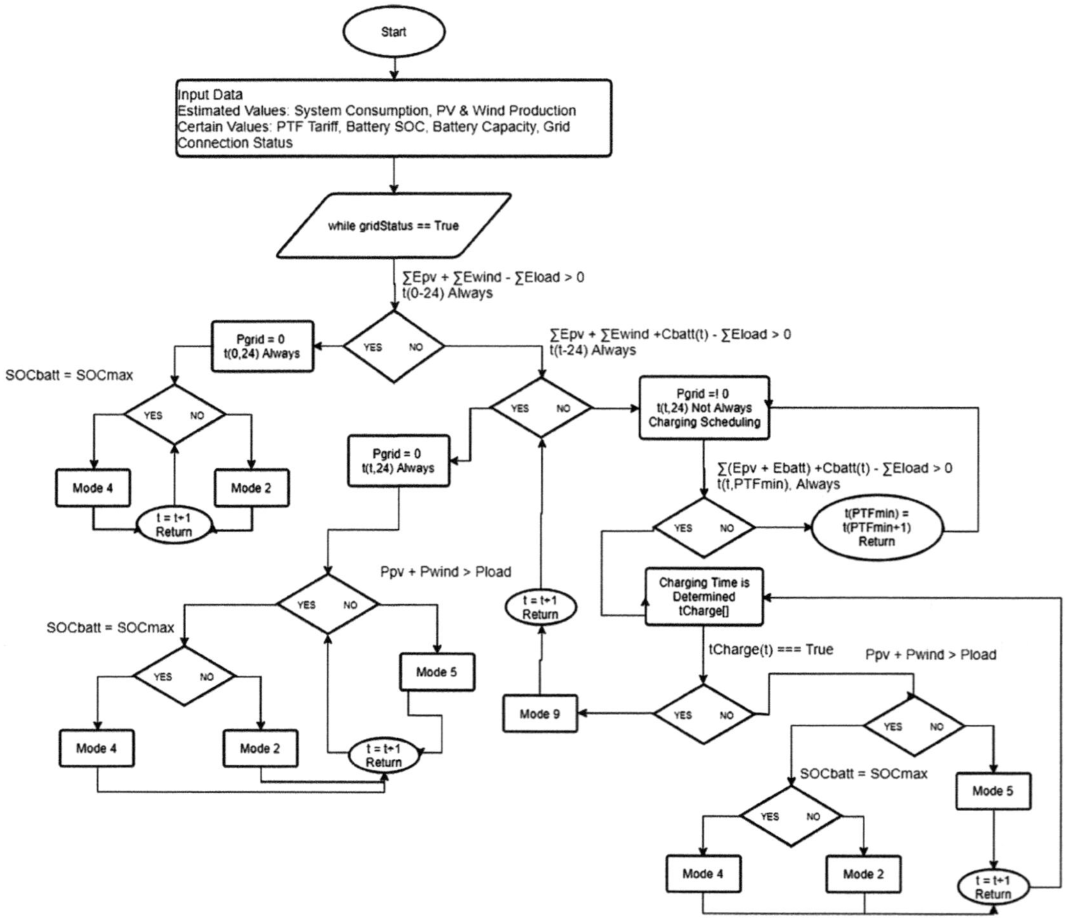

Fig. 7.15 ODIH HEMS algorithm according to the MCP Tariff

months. As can be seen, the heat pump energy consumption acts as a cooler in summer, and its load increases towards noon and decreases at night. On the other hand, the heating load, which is more linear in winter, decreases a little with the increase of temperature at noon.

Weather conditions also directly affect wind energy production. The energy obtained from wind turbines is quite uncertain with the change of wind speed on an hourly and daily basis. Figure 7.18 shows the energy planned to be obtained from wind turbines during the month of March. In addition, according to our calculations, the monthly maximum energy production can reach approximately ten times the average monthly energy production.

Solar energy is another source of unbalanced production. The energy production of the photovoltaic system, which will be the main energy source of the designed living area, increases and decreases seasonally in the long term. It can only produce during daytime hours due to the limited daily sunbathing hours and fluctuates due to temporary effects such as shading. PV production in May can be seen in Table 7.8.

7.5.2 Importance of Energy Storage

Fluctuations in energy production and consumption cannot be prevented due to uncontrollable seasonal parameters. This is an important

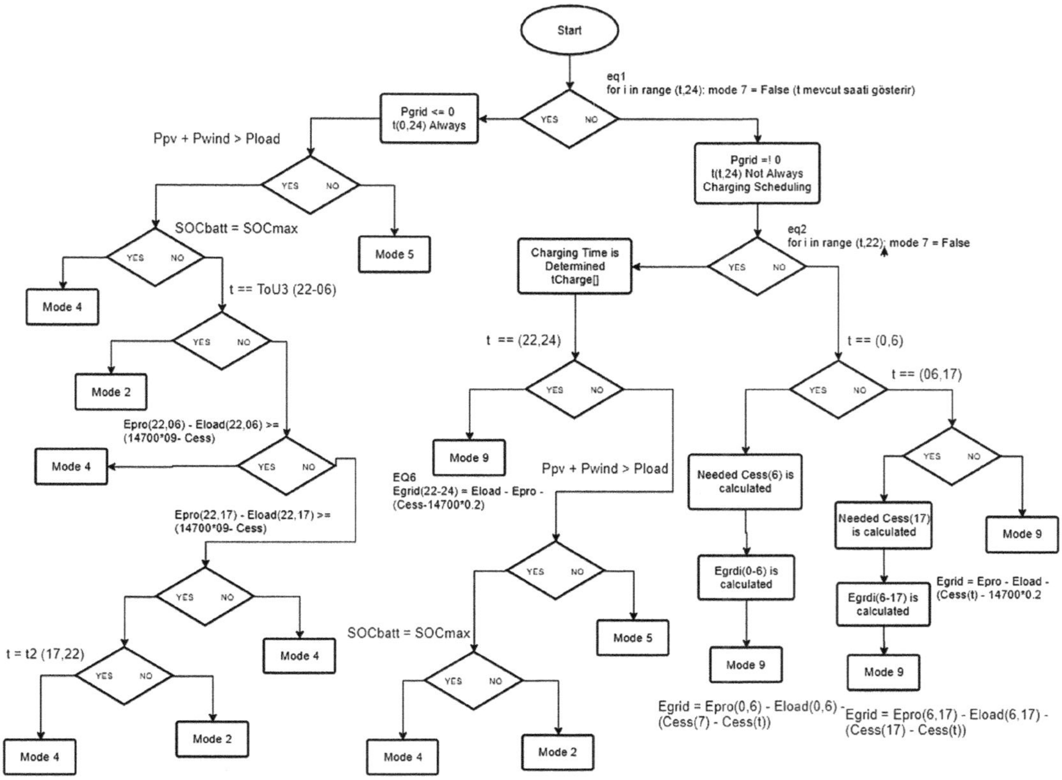

Fig. 7.16 ODIH HEMS algorithm according to the ToU Tariff

challenge for the energy management of both the decarbonized grids of the future and especially the self-sustainable living space in our research. Figure 7.19 shows the hourly average of total production and consumption in the summer and winter seasons. As seen in the figure, there is an energy surplus and gap at any time of the day. Energy storage systems offer an important solution for the management of imbalances. Storing excess energy during the day and feeding the system from the network in cases of energy deficit provide an energy solution to the self-sustainability goal of the living space.

7.5.3 Opportunities for Load Scheduling and Smart Charging

As can be seen in Fig. 7.19, there is always an energy surplus and gap between production and consumption. In addition to the support provided by the battery system in energy imbalance management, another support mechanism is the load shifting of house loads. In ODIH system, there are two options for load scheduling: EV Charging and Water Purifier Systems. According to our calculations, the water treatment system consumes 3.8 kWh of energy to produce a certain amount of water per day. Since it has its own clean water tank, it is possible to shift this energy load to the desired time period. On the other hand, electric vehicles can be considered as mobile batteries and are a very suitable source for demand-side scenarios. Thanks to the smart charging methodology mentioned in the previous sections, the system we designed can also use the electric vehicle in addition to the battery system in energy management. In addition, significant economic savings are achieved by shifting these loads to time periods where energy is cheap in time periods with an energy deficit.

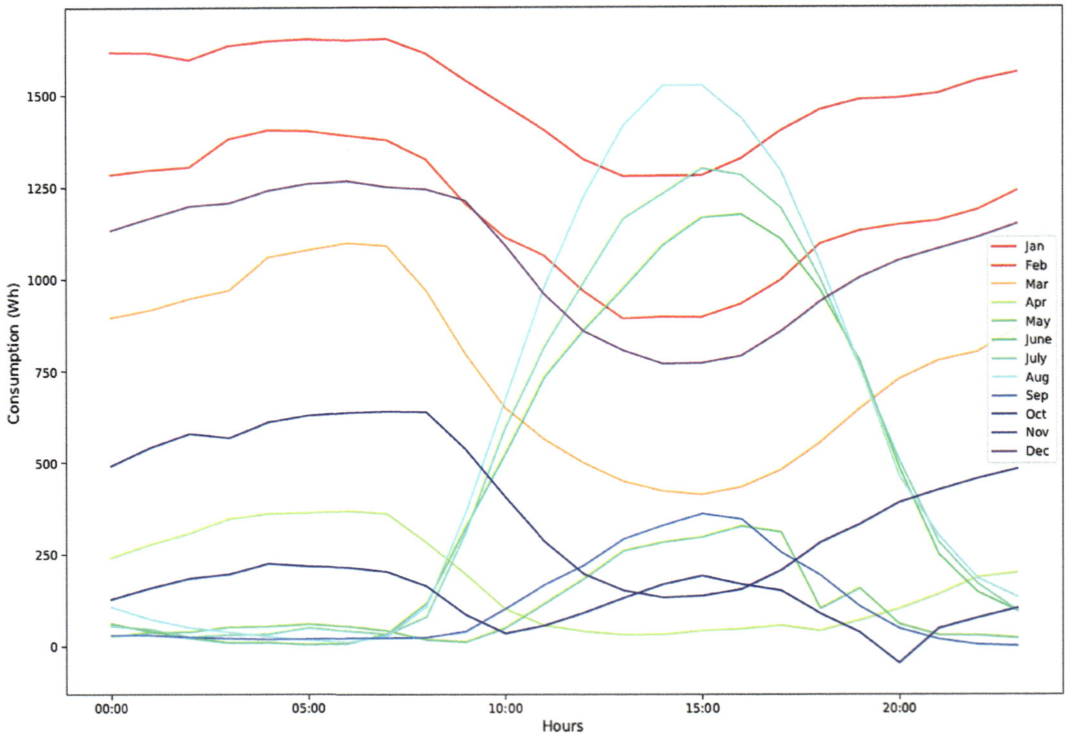

Fig. 7.17 Heat pump consumption by months

7.5.4 Advantages of Smart Energy Management Algorithms

Energy management is an obligation in both large-scale demand-side management applications in the network and micro-net systems such as ODIH. It is possible to mention 4 layers for integrated and successful energy management. These are:

A. **Enterprise Energy Transaction Layer** which makes energy management economically reasonable for all stakeholders (prosumers, EV owners, Distribution System Operators, etc.) of the system
B. **Energy management Assets Layer,** which includes energy generation units (PV, Wind Turbines, cogeneration, etc.), energy storage systems (batteries, thermal storage, etc.), and controllable loads (Smart EVSE, smart HVAC units, etc.)
C. **Control and Communication Layer** is responsible for the communication and control of all assets in the energy management system. All of the energy and economic transactions are operated with sophisticated handshaking mechanisms
D. The fourth layer that manages and optimizes all these systems is the layer where the algorithms are executed. The most important part of realizing and developing next-generation energy clusters such as ODIH is the development and execution of sophisticated energy management algorithms. Realization of many goals such as self-sustainability, decarbonisation, cost optimization or imbalance management is possible with the development of smart management mechanisms in addition to providing the specified economic and technical infrastructure.

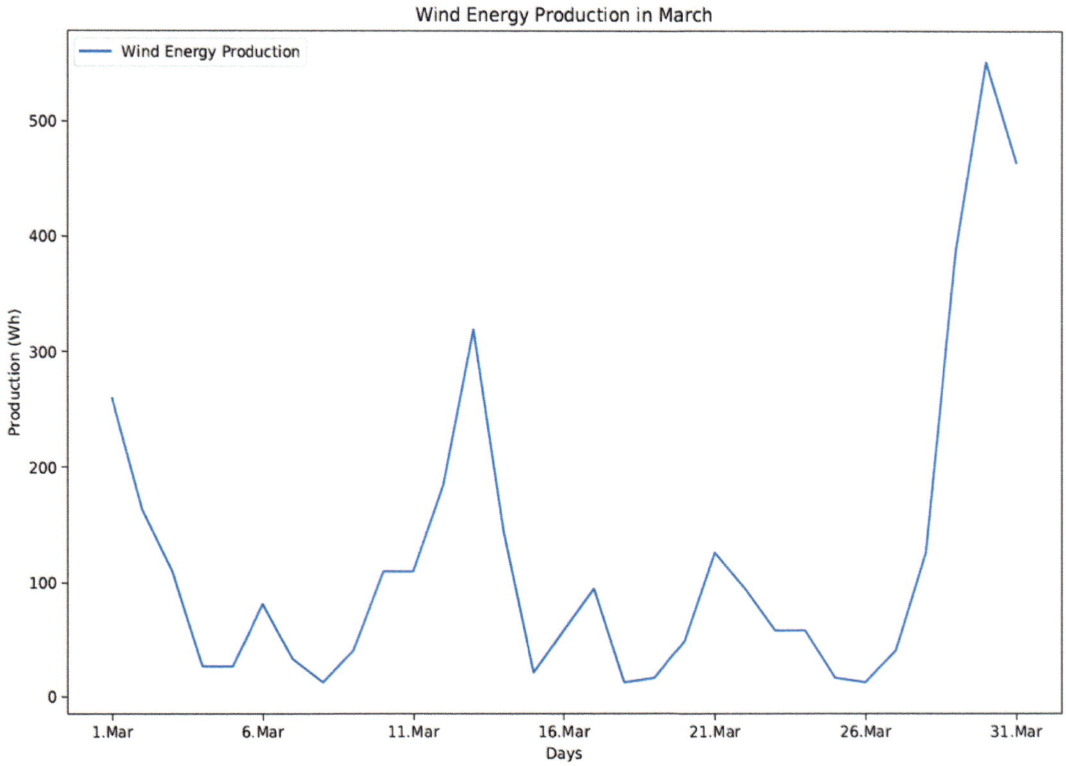

Fig. 7.18 Daily wind production in March

7.5.5 Tariffs for Demand Side Management

Turkey is in an initial state of both energy and e-mobility transformation. Thus, this is a perfect time to take actions for shaping this transaction. In the current situation, there is no tariff option for EV charging yet. Also, current tariffs for rooftop solar systems are needed for improvement. New tariffs suitable for smart city and smart grid concept must be produced in order to make our energy sources decentralized, decarbonized and democratized, to accelerate the realization of electrification of the transportation sector and to solve the problems that this transformation will create. Considering all these, we would like to propose novel tariffs to empower individuals and prosumers.

The first tariff we recommend is an **annual deal** for prosumers who do not have energy storage units. According to this tariff, if there is a production surplus in a month, that energy can be used for consumption in the following months. Production surplus electricity generated during the summer months can be used for offsetting during the winter months, thus maximizing the earnings for end users. Because it means that instead of purchasing electricity at the most expensive unit price (0.75 TL), it can use the electricity that it previously produced in its own solar power plant at the lowest price (0.10 TL). In the event that electricity distribution companies apply such a tariff, they will encourage widespread uptake of rooftop solar PV installations in urban areas.

The transition to electric vehicles should be supported by the government since the electrification of the transport sector has a huge impact on the mitigation of climate change. In order to promote e-Mobility, we recommend a **Critical Peak Pricing** tariff. In this tariff, EV owners can charge their EVs at even lower prices than the

Table 7.8 Heatmap of solar production in May

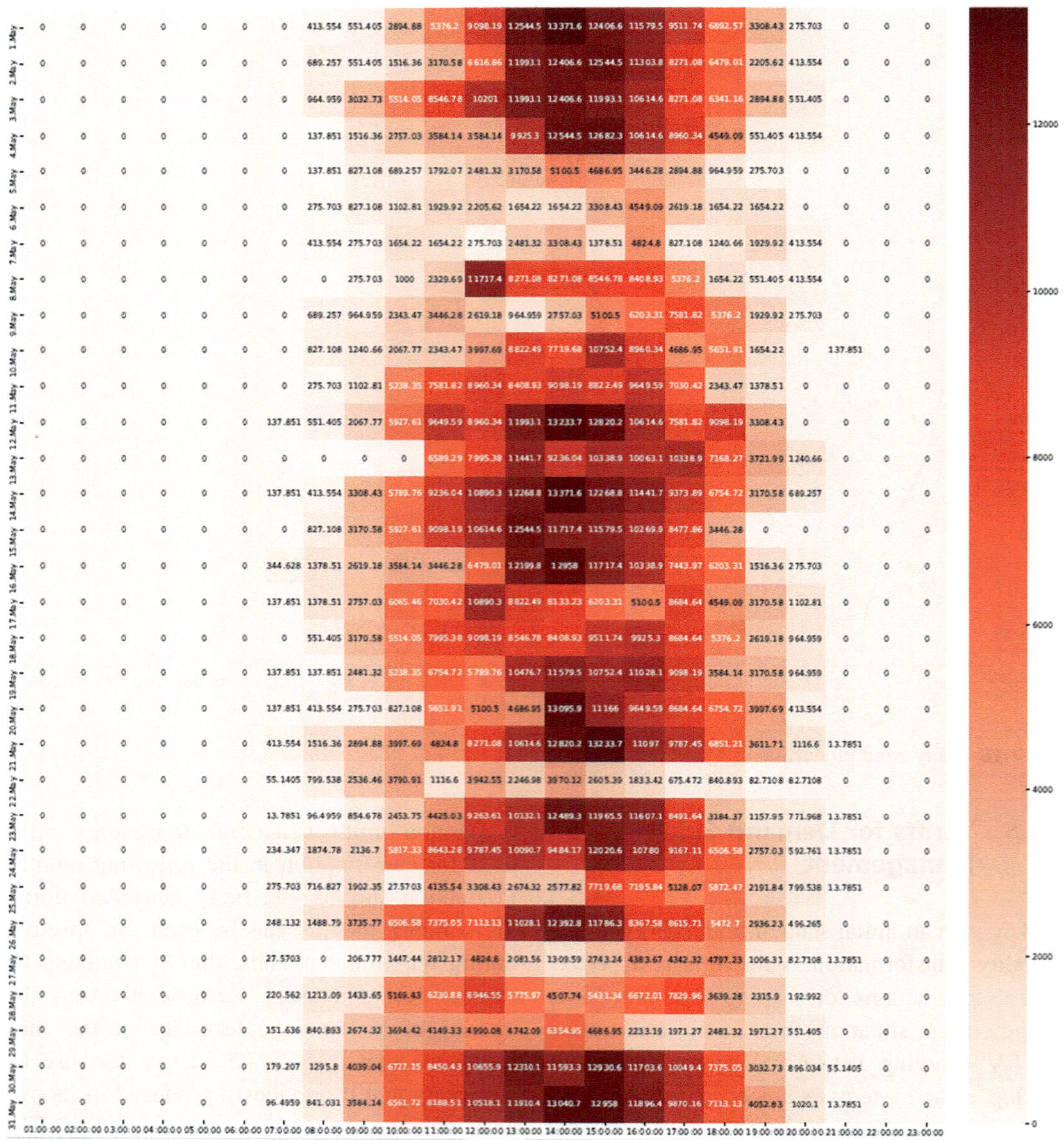

existing night tariff in return for the highest rate during peak times to prevent overloading. Also, it is possible to improve this tariff by covering demand response abilities. As mentioned before, EV loads are one of the most suitable consumption groups for load reducing, bidding, and scheduling options. Communication and aggregation, both technically and economically between EV Users, Charging Stations, and Distribution System Operators, enable sophisticated and flexible demand-side management operations. Another tariff recommendation is a Dynamic Demand or Dynamic Time of Use (ToU) tariff, which is perfect for electric vehicles, storage, heating, ventilating, and air conditioning (HVAC), or anyone who can shift their

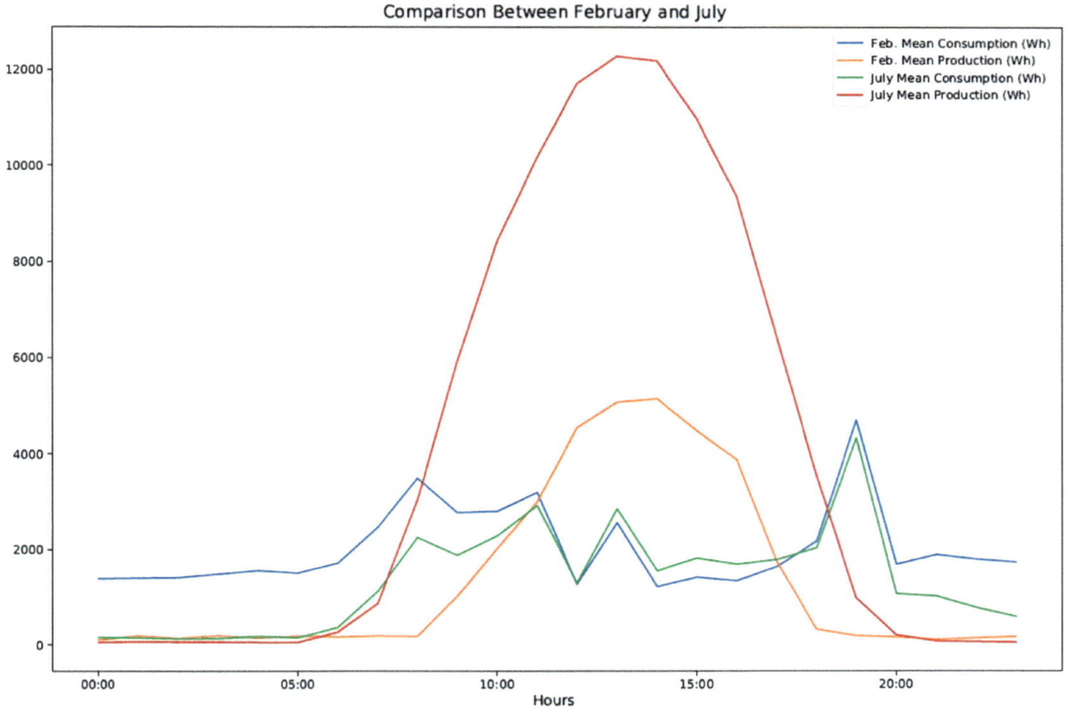

Fig. 7.19 Energy production and consumption behaviour in different seasons

electricity use outside of peak times. In this tariff, electricity price changes every 30 min according to the needs of the distribution system and the supply and demand balance. Consumers manage their assets and optimise energy transactions between battery, house, and grid in line with dynamic tariff prices. The proposed tariff encourages customers to participate in Demand Response programmes at the same time to gain economic benefit.

7.6 Discussion and Policy Recommendation

7.6.1 Empowering e-Mobility

It is now clear how EVs have shifted the current automobile industry. Beyond any doubt, it is an impressive success that TESLA is the herald of this new industry (Perkins and Murmann 2018), in addition to the progressive adoption of well-established vehicle manufacturers. It is obvious that the trend in user preferences will accelerate with values such as new technology options offered by the automotive industry. The development of business model innovations such as automobile connectivity will increase the interest in e-mobility, as technology giants take a good position in these areas. Another motivation for this transformation is the increased need for personal mobility services, especially with the COVID 19 pandemic period. Even though giants such as UBER and LYFT perform the already widespread shared electric vehicle service, new start-ups have also emerged in this field. The German government allocated € 300 million funds to promote the roll-out of rapid and standard recharging points and earmarked € 100 million to purchase EV for its own government fleet in 2019, an increase of 20% compared to last year (Federal Ministry for Economic Affairs and Energy of Germany 2020). Therefore, as of 2018, German manufacturers have achieved to put 29 different EV models on the market. E-mobility can further be empowered with the following recommendations:

I. Not just some of the developed countries, all countries should incentivise the purchase of EVs to make a shift from ICE cars to EVs gradually.
II. Electric mobility roadmaps should be created and executed.
III. E-fleet formations should be made in places such as government offices so that this will set an example for the public.
IV. Start-up companies such as e-car and e-scooter sharing should be supported both by the central authorities and municipalities.
V. More investments in R&D and incentives are needed for battery technologies as well as a second-life use case of batteries.

7.6.2 Smart Charging and Prosumers

Electric vehicles are the most suitable energy loads for Demand Side participation in the residential sector. The process takes place by stopping the charging process or reducing the load in the period when the network is overloaded or when there are fluctuations in frequency and voltage. In order to realise this, charging equipment must be communicable/controllable and in aggregation with decision-making mechanisms. And, of course, the necessary regulations must be introduced by the authorities. In this scope, we make the following recommendations:

Technology investments should be increased for bi-directional charging capable EVs and smart grid adaptations.

I. Coordination and joint operation of charging stations and prosumers should be encouraged.
II. V2G option vehicles should be integrated, and V2G market models should be developed and prosumers should be informed and trained for V2H and V2B business cases.
III. Smart meters are the enabling technology for smart charging and prosumer integration. Therefore, smart meter roll-out is necessary for countries that wish to take part in the flexibility markets and demand response schemes.

7.6.3 Developing Smart Tariffs for Prosumers and EV Owners

The net-metering tariff currently applied in Turkey is a monthly offsetting tariff. The meters record electricity consumption. The difference is evaluated monthly and is paid or billed in the middle of the following month (Enerji Piyasası Düzenleme Kurumu 2019). The consumed electricity unit price is around 0.75TL for residential users (EPDK 2020). With the use of electronic meters, residential users also get the opportunity to use a three-rate tariff. However, this tariff also offers users a limited optimisation opportunity. Prosumers can sell their surplus energy for a lower price. The production surplus is paid at a unit price of 0.10 TL. In this case, users might choose to establish a facility that meets their only needs since the capital expenditure of larger PV systems will naturally result in a longer return on investment times. In Turkey, the electrical energy obtained from solar power plants in the summer months can be almost twice the winter months (Global Solar Atlas 2020). On the other hand, due to heating, electricity consumption in the winter months is higher than the summer consumption. In this case, rooftop solar panel owners cannot meet their consumption during the winter months and draw electricity from the grid at high unit prices; in summer, they sell their surplus to the grid (Association of Mediterranean Energy Regulators 2018).

With the digitalisation of energy systems, each electricity service user should communicate not only with the network operator but also with other users connected to the same network. By the time the energy source and assets become distributed, the flow of energy is to be controlled in the merging of technologies, policies, and financial drivers. Besides prosumers, EVs, microgrids, and VPPs are the entities of this new prosumer active energy environment.

7.7 Conclusion

In this study, we present a broad perspective from the basic charger types to the revealed protocols as a complete component of the smart charging system in general, as well as the comprehensive strategies and specific techniques that provide the smart charging solutions. A prominent contribution of this study with the implementation of the ODIH project is that the report has an extensive level of detail that can be used as a reference while creating the relevant regulations in the country. In this context, we have made critical recommendations for regulating smart charging and demand response functionalities consists of electric vehicles and residential prosumers penetrated grid systems after drawing the big picture on the report. Our other contribution is to bring an explanatory solution to existing sustainability by the Open Digital Innovation Hub Hybrid Energy Management System Algorithm by cooperating a decentralized smart grid framework most of the time by being isolated from the grid. This algorithm provides the data-driven decisions for charging and discharging scheduling of ODIH assets within the fully dynamic environment in terms of grid tariff, state of home battery charge, and our renewable energy production amount. For prosumer entities like ours, which will be integrated into the grid bidirectionally, our algorithm is capable of also optimizing energy transactions with the grid in a cost-effective way, apart from conventional charging controllers. The algorithm was modelled by training from our estimation for production, consumption, and tariffs. To estimate our energy productions, firstly, daily productions were predicted according to our wind and solar capacities and technologies by using simulation tools represented in Chap. 3.1. These daily datasets were converted to hourly by normalizing corresponded to the same year's hourly production of the nearby facilities in the IBB database. On the other hand, the consumption amounts necessary for water, food, and energy in a sustainable home where at least two people can live permanently have been determined.

At the end of our simulations and comprehensive literature studies on EV Charging, Demand Side Management, Distributed Generation, and Energy Management issues and the dataset we prepared, we reached the goal of simulating the following topics. Uncertainty and imbalance in energy production and consumption, the importance of energy storage, opportunities of load scheduling and smart charging, advantages of smart energy management algorithms, and tariffs for demand-side management.

According to Figs. 7.17 and 7.18, our study actual consumption and production of energy within the analysed duration. It showed that weather conditions directly affect both of the simulated data.

Therefore, consumption and production may not be controlled as explained. Hence, we concluded these results to improve our energy storage within the needed periods according to Fig. 7.19. Apart from that, load scheduling, a kind of demand response activity, plays a considerable role to handle production uncertainty and demand-side management.

Last but not least, we developed a smart energy management algorithm which controls, manage, and schedule energy consumption and production of ODIH. As the advantages of smart energy management algorithms are listed in the result Sect. 7.5.4, we put our confidence in the algorithms to strive to achieve affordable and reliable energy for ODIH.

References

Al-Ogaili AS, Hashim TJT, Rahmat NA (2019) Review on scheduling, clustering, and forecasting strategies for controlling electric vehicle charging: challenges and recommaditions. IEEE Access 7:128353–128371

Association of Mediterranean Energy Regulators (2018) Support to the evaluation of a net metering system In: Palestine, s.l.: MEDREG

Balram P, Tuan LA, Tjernberg LB (2015) Centralized charging control of plug-in electric vehicles and effects on day-ahead electricity market price. In: Plug in electric vehicles in smart grids charging strategies. s.l.:Springer, Berlin, pp 267–298

Chekired F, et al. (2017) An energy flow management algorithm for a photovoltaic solar home. Energy Proc 111:934–943

Elaadnl (2018) EV related protocol study. [Online] Available at: https://www.elaad.nl/uploads/downloads/downloads_download/EV_related_protocol_study_v1.1.pdf. Accessed on 23 Sept 2020

Enerji Piyasası Düzenleme Kurumu (2019) Elektrik Piyasasında Lisanssız Elektrik Üretim yönetmeliği. Resmî Gazete,Ankara

EPDK (2020) 1/10/2020 Tarihinden Geçerli Elektrik Tarife Tabloları, s.l.: Resmî Gazete

Epias (2020) Exist transparency platform. [Online] Available at: https://seffaflik.epias.com.tr/transparency/piyasalar/gop/ptf.xhtml. Accessed on 24 Oct 2020

eSarj (2019) eSarj. [Online] Available at: https://esarj.com/en/products. Accessed 20 Aug 2020

European Commission (2020) [Online] Available at: https://ec.europa.eu/info/funding-tenders/opportunities/portal/screen/opportunities/horizon-results-platform/20914;needList=10,11,12

Federal Ministry for Economic Affairs and Energy of Germany (2020) Electric mobility in Germany, s.l.: s.n.

Global Solar Atlas (2020) PV Production for Istanbul. [Online] Available at: https://globalsolaratlas.info/detail?c=39.820107,27.851372,7&s=41.041006,28.763237&m=site

Gong X, Rangaraju J (2018) Taking charge of electric vehicles – both in the vehicle and on the grid. Texas instruments, p 13

Green Flux (2019). GreenFlux EV smart charging whitepaper. [Online] Available at: https://www.greenflux.com/whitepaper/. Accessed June 2020

Greenlots, 2018. Open vs. closed charging stations: advantages and disadvantages. [Online] Available at: https://www.openchargealliance.org/uploads/files/OCA-Open-Standards-White-Paper-compressed.pdf. Accessed Sept 2020

Horyachyy O (2017) Comparison of wireless communication technologies used in a smart home: analysis of wireless sensor node based on arduino in home automation scenario. Blekinge Institute of Technology—Master of Science in Electrical Engineering with emphasis on Telecommunication Systems, p 71

IBB (2020) IBB open data portal. [Online] Available at: https://data.ibb.gov.tr/dataset/ikitelli-gunes-enerjisi-santrali-elektrik-uretim-miktarlari. Accessed on 20 Sept 2020

IEC (2017) IEC 61851-1: electric vehicle conductive charging system—Part 1: General requirements, s.l.: IEC

IRENA (2019). Innovation outlook: smart charging for electric vehicles. International Renewable Energy Agency, Abu Dhabi

Kufeoğlu S, Pollitt MG (2019) The impact of PVs and EVs on domestic electricity network charges: a case study from Great Britain. Energy Policy 127:412–424

Lillebo M, Zaferanlouei S, Zecchino A, Farahmand H (2019) Impact of large-scale EV integration and fast chargers in a Norwegian LV grid. J Eng 18:5104–5108

Lilly C (2020) EV connector types. [Online] Available at: https://www.zap-map.com/charge-points/connectors-speeds/. Accessed on 1 Sept 2020

Lohsen (2019) Medium. [Online] Available at: https://medium.com/@joachim_21503/a-new-chapter-of-ev-smart-charging-the-era-of-ocpp-2-0-9f9f0da8962d. Accessed on 2 Sept 2020

Lopez KL (2019) A machine learning approach for the smart charging of. s.l., s.n.

Lovell J (2019) Storage: retirement home for old EV batteries?. [Online] Available at: https://www.energycouncil.com.au/analysis/storage-retirement-home-for-old-ev-batteries/. Accessed on 28 Aug 2020

Mashhour E, Tafreshi M (2009) The opportunities for future virtual power plant in the power market, a view point. Capri, s.n.

Mocanu E, Nguyen PH, Gıbescu M, Klıng WL (2016) Deep learning for estimating building energy consumption

Molderink A et al (2010) Management and control of domestic smart grid technology. IEEE Trans Smart Grid 1(2):109–119

Open Charge Alliance (2018) OCPI basics. [Online] Available at: https://www.nklnederland.com/projecten/onze-lopende-projecten/ocpi-the-independent-open-protocol-for-ev-roaming-/

Open Charge Alliance (2019) The importance of open protocols. [Online] Available at: https://www.openchargealliance.org/protocols/. Accessed on 12 Sept 2020

Palensky P, Dietrich D (2011) Demand side management: demand response, intelligent energy systems. IEEE Trans Industr Inf 7(3):381–388

Perkins G, Murmann JP (2018) What does the success of tesla mean for the future dynamics in the global automobile sector?. s.l., s.n.

Portela CM, Klapwijk P, Verheijen L, Boer Hd (2015) Open smart charging protocol. Lyon: s.n.

Rajakaruna S, Shahnia F (2015) Plug in electric vehicles in smart grids charging strategies. Springer, s.l.

Robu V, et al. (2013) An online mechanism for multi-unit demand and its application to plug-in hybrid electric vehicle charging. Journal of Artificial Intelligence Research, 48, pp.175–230

Ryu S, Noh J, KIM H (2016) Deep neural network based demand side short term load forecasting. basım yeri bilinmiyor:yazarı bilinmiyor

Saboori H, Mohammadi M, Taghe R (2011) Virtual power plant (VPP), definition, concept, components and types. Wuhan, IEEE

Schuller A (2015) Charging coordination paradigms of electric vehicles. Plug in electric vehicles in smart grids charging strategies. Springer, Singapore, pp 9–13

Shareefa H, Islam MM, Mohamed A (2016) A review of the stage-of-the-art charging technologies, placement methodologies, and impacts of electric vehicles. Renew Sustain Energy Rev

SHURA (2019) Türkiye ulaştırma sektörünün dönüşümü: Elektrikli araçların Türkiye dağıtım şebekesine etkileri. SHURA Enerji Dönüşümü Merkezi, İstanbul

Skouras TA et al (2019) Electrical vehicles: current state of the art, future challenges, and perspectives. Clean Technol 2(1):1–16

Soares FJ, Barbeiro PN, Gouveia C, Lopes JA (2015) Impacts of plug-in electric vehicles integration in distribution networks under different charging strategies. Plug in electric vehicles in smart grids charging strategies. Springer, Singapore, pp 89–102

Wikimedia (2015) Wikimedia. [Online] Available at: https://commons.wikimedia.org/wiki/File:Osi-model-jb.svg. Accessed Nov 2020

Yu S, Fang F, Liu Y, Liu J (2019) Uncertainties of virtual power plant: Problems and countermeasures. Appl Energy 239:454–470

Zhao J, Kucuksari S, Mazhari E, Son YJ (2013) Integrated analysis of high-penetration PV and PHEV with energy storage and demand response. Appl Energy 35–51

8 Blockchain Applications and Peer-To-Peer Tradings

Abstract

Blockchain can be considered as a very successful technique that can simplify the metering and billing systems for Peer-to-Peer (P2P) energy trading. Blockchain decentralised ledger technology's promising features indicate its feasibility to meet the prevailing challenges to current power structures, as we review in the Open Digital Innovation Hub (ODIH). This study examines the Blockchain technology, smart contracts, the application of P2P energy trading from hardware, software, regulatory perspectives within ODIH, a self-sustaining smart home project. We show promising results in terms of the future potential of decentralised structures offered by Blockchain technology, which includes a reliable and transparent trading infrastructure for P2P energy trade. In ODIH, we presented our proposed business architecture design and implementation results for Smart Home System based on the Ethereum Blockchain. We implemented a smart contract using Solidity language to handle and record the transactions between the homeowners and Ethereum miners. ODIH shows that the integration of Blockchain technology in the hybrid P2P electricity market has a positive impact.

8.1 Introduction

Digitalisation in the energy sector includes the creation and use of computerised information and processing of the full-size quantities of data that are generated at any levels of the energy supply chain. It guarantees a lot for each phase of the energy ecosystem: households, prosumers, distribution, transmission, generation, and retail and is often said as likely to result in a change of the energy system (Küfeoğlu et al. 2019). Digitalisation is helping to improve the safety, productivity, accessibility, and sustainability of energy systems around the world. In this part of the book, we will describe how we adapted this phenomenon to the overall study and how to establish ODIH as a financial entity. We will discuss respectively: Peer-to-Peer (P2P) energy trade, Blockchain applications, and smart contracts.

8.1.1 Peer-To-Peer Energy Trading

Today, the energy market is as centralised as it gets. Monopolies decide where and when to build power capacity, decide how to bridge the distance between generators and loads, and most importantly, the price of electricity. In addition,

The author would like to acknowledge the help and contributions of Ayşe Korkmaz, Esra Kılıç, Metehan Türkay, Ömer Faruk Çakmak, Taha Yasin Arslan, and Ulaş Erdoğan in completing of this chapter.

S. Küfeoğlu, *The Home of the Future*, Sustainable Development Goals Series,
https://doi.org/10.1007/978-3-030-75093-0_8

decentralized generation will reduce grid losses and conserve primary energy compared to massive centralized power plants providing electricity that flows down the voltage chain (Hu et al. 2017). P2P energy trade mechanism can support the decentralised electrical network.

Peer-to-Peer (P2P) energy trading is a business model that constitutes an online marketplace network where prosumers, producers and consumers meet directly without the need for an intermediary to exchange electricity (Tushar et al. 2020). The P2P electricity trading model was born after the rapid increase in distributed energy generation uptake. Supporting the prosumage through incentives is another motivating factor behind the model. In addition, by finding a better result compared to higher tariffs and lower buy-back prices compared to the P2P settlement, prosumers can directly exchange energy with others in order to obtain a win–win situation. P2P trading guarantees buyers' cost savings and sellers' benefit (Liu et al. 2019). Figure 8.1 illustrates the structure of the P2P trading model.

A Peer-to-Peer exchanging model can be set up among peers inside a nearby local area, just as for a bigger scope, among different gatherings, networks or areas. It can also keep the community resilient to outages in emergencies. P2P exchanging offers a marketplace for prosumers and consumers to exchange and purchase sustainable power, often at a superior cost, empowering the arrangement of the disseminated power age (Liu et al. 2019). Also, P2P exchanging permits consumers to have power over their power utilization and its cost, expanding adaptability in the framework. Additionally, the P2P power exchanging model makes environmentally friendly power more open, enabling buyers while utilizing their dispersed energy assets. P2P trading allows participants to employ more on-site renewable generation via support from their local communities. For an isolated prosumer, storage might be compulsory; however, in a community connected to each other, the amount of storage would be lower thanks to the P2P trading mechanism. Simultaneously, purchasers without sustainable production limit can profit straightforwardly from neighbourhood inexhaustible age through P2P power exchanging.

8.1.1.1 The Potential Impact on Energy Sector Transformation

Consumers without production limits can profit by an ease nearby renewable power supply. Two effective models, where P2P exchanging have brought about expense investment funds, are:

i. Power Ledger, an Australian P2P exchanging site, has saved a normal of USD 424 every year for its energy purchasers on yearly power charges and aided sunlight-based housetop framework proprietors twofold the investment funds they typically get from their sun-based plants (Kabessa 2017).

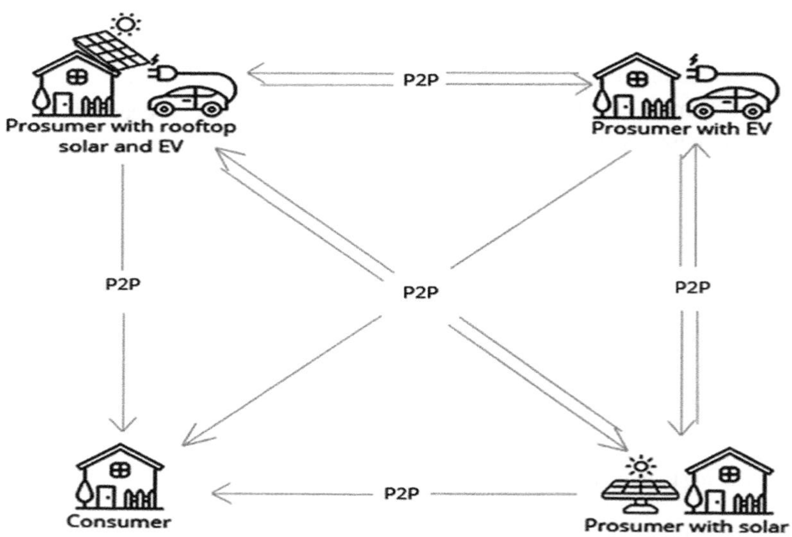

Fig. 8.1 The structure of the P2P trading model

ii. The New York-based energy start-up Drift, a P2P trading platform, has assisted buyers with saving 10% of their power costs contrasted with Con Edison, a nearby utility in the New York zone (CNBC 2017).

In our implementation, estimated values of Open Digital Innovation Hub (ODIH) are used to calculate the cost of generating electricity and compared with regular techniques with a standard cost per unit. In addition, when P2P energy is traded with Blockchain technology, the cost of electricity consumption will decrease at the consumer level, and PAR (Peak to average ratio) will decrease at the grid level (Yahaya et al. 2020). By using P2P energy trading, the peak to average ratio (PAR) and the cost of power is decreased, which benefits both the electricity grid and the energy users.

8.1.1.2 Where Is P2P Energy Trading Available at the Moment?

There are a few ventures on P2P energy exchanging around the world. The first recorded Peer-to-Peer energy exchange occurred in Brooklyn, New York, in 2016, when an occupant with solar panels offered some kWh of energy to his nearby resident over the Ethereum Blockchain (Infinite Energy 2020). Since the concept started circulating the globe, there have been various private preliminaries occurring in the US, Australia, and various places in the world.

Here are a couple of the organizations kicking things off inside the Peer-to-Peer trade all through the globe:

A. Brooklyn Microgrid

Brooklyn Microgrid is a local area energy market inward a microgrid. Thanks to the platform, peers can purchase and sell energy from one another with smart contracts utilizing Blockchain (Mengelkamp et al. 2018).

B. Centrica Pilot

Centrica Pilot is a platform to build up a nearby energy market in Cornwall, UK. They offer access to renewable generation, storage and reward their consumers for flexible demand. The members utilize the most recent computerized advances to interface with a virtual marketplace platform that will empower them to offer their adaptable energy ability to both the grid and the broad energy market. Centrica makes use of LO3's Blockchain-powered energy trading platform (Centrica 2018).>

C. LO3

LO3 is reforming the manner in which energy is shared and dispersed. The customers, households and businesses, trade energy on an auction market platform called Exergy. After a critical accomplishment in New York with the Brooklyn Microgrid preliminary, LO3 is currently chipping away at another venture in South Australia, cooperating with Yates Electrical Service (Infinite Energy 2020).

D. Power Ledger

Power Ledger is an Australian energy organization that uses a Blockchain-empowered P2P renewable energy exchanging site. The site encourages the purchasing and selling of environmentally friendly power progressively, empowering users with PV boards to exchange their electricity with their neighbours. To execute the innovation, the organization has just evolved associations with energy retailers, industry and neighbourhood governments, and as of now has 22 ventures in eight nations, including Australia, the United States, Italy and Thailand. In 2018, Power Ledger's innovation received Sir Richard Branson's worldwide Extreme Tech Challenge (Infinite Energy 2020).

E. Lumenaza

Lumenaza's "utility-in-a-box" energy network facilitates local, regional and national P2P sharing of energy and communities. The program ties electricity producers to customers, tracks demand and supply (e.g. by loading batteries), and provides the management of the balance group, collection, billing and visualization of energy flows. Enables energy societies to engage in the design of the electricity market (Lumenaza 2020).

F. SOLshare

SOLshare builds mini-grids on a small scale, linking local customers and allowing them to share local electricity. Consumers who have a solar panel mounted on their rooftops may provide those who do not have access to electricity with surplus capacity. A consistent power network is available throughout the location by supplying a mini-grid (UNFCCC 2020).

The errands of the pilot project incorporate the production of a P2P exchanging application, the planning of strategy rules for Colombian policymakers, and a business scale-up guide (UCL 2019).

H. SonnenCommunity

The SonnenCommunity enables individual consumers who use Sonnen's batteries to share self-produced renewable electricity. This surplus energy is not fed into the grid but into a virtual energy pool that, during periods when it does not generate, provides energy to other members of the society. The price of electricity is set at approximately USD 25 cents/kWh (EUR 23 cents/kWh). For the website, the monthly access fee is EUR 20 (Sonnen 2019).

8.1.1.3 How Can We Use P2P Energy Trade in the ODIH?

We have designed our project as a 'platform'. That is why we have defined the concept of 'identity' for prosumers. The identity structure, on the other hand, will contain specifications that are reduced to specific standards regarding the relevant prosumer and the energy it produces. Thanks to our digital identity structure, our data will be transparent.

As the platform also contains the identity structure of different prosumers connected to the same network, it will provide flexibility in buying and selling energy. Peer-to-Peer energy sharing helps customers to negotiate from whom they buy and to whom they sell electricity. The exchanging of energy is done over a safe platform, regularly utilizing innovation like Blockchain. Peer-to-Peer energy exchanging sites, for example, Blockchain, will allow clients to divide their abundance of energy between another and control how it is scattered through microgrids. Regardless of whether one doesn't have solar boards, they can buy energy in any case from others. Purchasers can decide to purchase sun-based energy from a neighbour or to help a nearby wind or solar homestead.

8.1.2 The New Trends of Future Energy Markets: Digitalisation, Decarbonisation, and Decentralisation

Around the world, energy markets are, as of now, encountering extremist advancements in their customary designs and activities, driven by endeavours to decentralize, digitalise, and decarbonise. Every one of these parts involves various perspectives, for example, atmosphere activities and energy strategies, which have emerged from the need to react to environmental change and secure the stockpile of energy in the midst of the development of worldwide interest for energy (Bompolakis 2020).

8.1.2.1 Digitalisation

Governments and business executives that keep up with the digitalisation approach in the energy industry increase the security, sustainability, reliability, and affordability of energy. Therefore, the energy sector around the world is now experiencing a rapid shift from their traditional frameworks to modern ones, which emphasise environmental, economic, and social perspectives. Many experts consider the Blockchain technology that comes as a result of digitalisation to be a disruptive force for the energy sector. Increased RES (Renewable Energy Systems) deployment and grid modernisation are allowed by the digitalisation of the power sector (Bompolakis 2020). The usage areas of digitalisation in the energy sector are shown in Fig. 8.2.

Purchasers using computerized arrangements are engaged with better administration of their energy creation and use. They can effectively partake in the framework by taking care of in their inordinate energy or finishing P2P energy

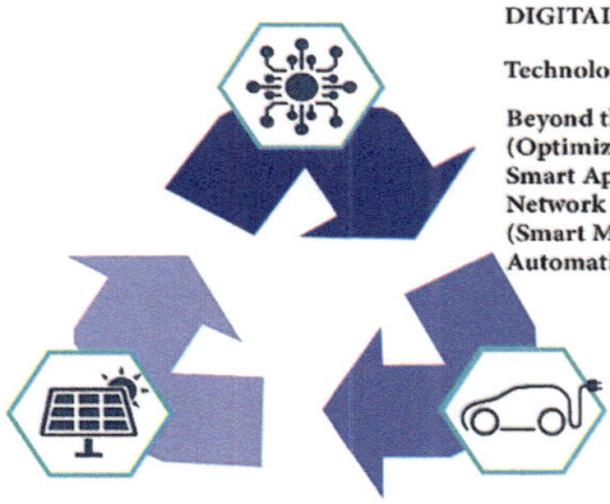

Fig. 8.2 The new trends of energy systems: digitalisation, decentralisation, electrification

exchanges (Livingston 2018). Digitalization of buildings may also reduce energy demand by about 10% by leveraging real-time data to boost operating performance (IEA 2017).

8.1.2.2 Decarbonisation

Decarbonisation represents the decreasing average carbon intensity of primary energy through the use of modern and sustainable energy sources over time. Decarbonisation in the energy sector means reducing carbon concentration (CO_2/kWh) per kWh (Silvestre et al. 2018). Advanced innovations can back the move to zero-carbon, circular, and versatile social orders. Global energy demand will increase by 2% annually by 2020, and by 2040 there will be a massive 55% demand increase; therefore, even more innovation is required. More than 1.2 billion individuals today need admittance to power, and in excess of 7 million passing happen yearly because of air contamination. As the energy request keeps on developing, more brilliant and greater local area driven advancement is needed to handle the expanding zap, energy effectiveness, and carbon expulsion difficulties to encourage proceeded with financial development and improve individuals' quality of life (IEA 2017).

8.1.2.3 Decentralisation

The energy sector is on the verge of digital transformation. Digitalisation is obscuring the distinction among the market. It likewise makes openings for customers to partake straightforwardly in the ongoing harmony among supply and demand, alongside the gradual zap of the energy lattice and the development of decentralized force sources (IEA 2017). A decentralised energy industry requires real-time data on energy production and usage to be available and accurate at different points on the grid. Figure 8.3 shows the difference between the traditional way of energy trading and decentralised energy trading.

Hardware infrastructure is required to capture the data in the first place, in addition to a software layer that is needed to handle and process all the data flow. This system will consist of a network of connected items, such as energy consumption of smart meters and sensors, which will provide details on the condition of the electricity grid. According to the statistics and predictions, such devices would be implemented more widely (DEEP 2018).

Most of today's Blockchain implementations in the energy sector have focused on the electricity industry, with more than half of them concentrating on decentralised energy exchange and funding energy projects (Bompolakis 2020). One of the critical aspects is the decentralised design of Blockchain networks. Unlike centralised systems, where a single entity makes decisions, the members operate Blockchain networks, and transaction records are copied in a decentralised manner to every participating node. This helps to prevent losses or modifications of data and makes malicious behaviour less immutable to the networks (Gates 2017).

Blockchain can deliver a wide variety of advantages in a commercial context, such as

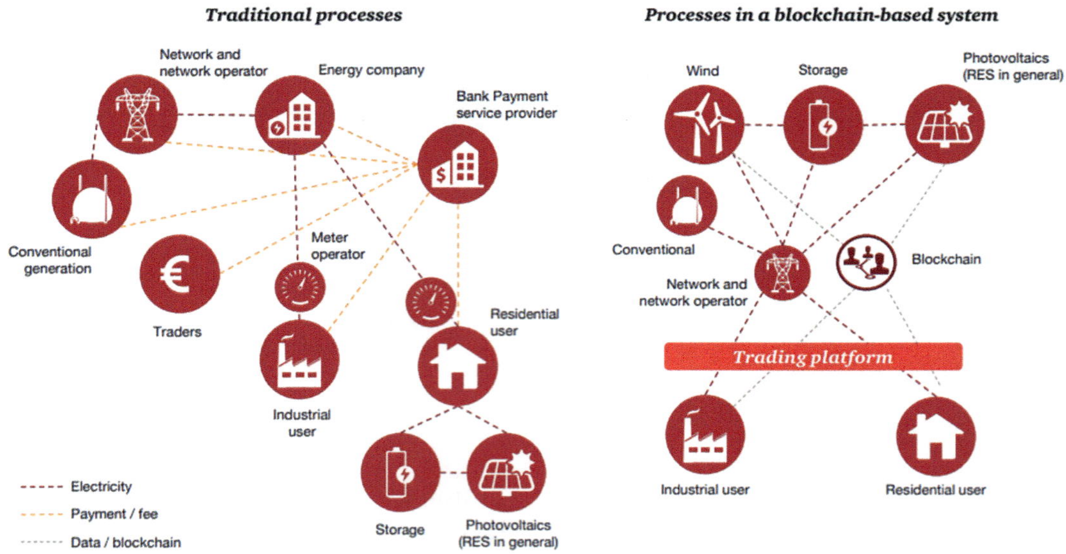

Fig. 8.3 Transformation of market structures on the introduction of a decentralised transaction model (PwC 2016)

decentralised wholesale power trading, enhancing the protection of the grid, and creating new business models. Utility businesses should also expect grid maintenance and related activities such as energy balancing, metering, and customer billing to be significantly automated (WEF 2017).

8.1.3 The Blockchain

Blockchain is an information structure that empowers to make a computerized record of information and divide it between an organization of free companions (Laurance 2017). The Blockchain uses cryptography innovation to permit every member of an organization to deal with the record safely without the requirement for a focal position to implement the standards (Laurance 2017). Eliminating central authority from the information base construction is perhaps the most unimportant qualities of Blockchain.

Blockchains make lasting records and chronicles of exchanges; however, nothing is perpetual. The perpetual quality of the record depends on the lastingness of the organization (Laurance 2017). With regards to blockchain, this implies that a huge piece of a blockchain local area should consent to change data and are urged not to change information.

If data is recorded in a Blockchain, it is very hard to alter or delete it. When the transactor wants to add a transaction and a record to the blockchain, users with validation control on the network verify the proposed transaction. Figure 8.4 simply shows the working principle of the Blockchain mechanism.

Each block contains some information, the hash of the block and the hash of the previous block. The information put away inside a block relies upon the kind of Blockchain. A block likewise has a hash, and it recognizes a block and the entirety of its substance. The hash of a solitary block is consistently special. When a block is made, its hash is being determined. In the event that the fingerprint of a block changes, it is not, at this point, the same block. The third component inside each block is the hash of the past block.

8.1.3.1 Why We Are Using Blockchain? How Does It Relate to P2P?

With blockchain set up, applications that beforehand could just go through a believed merchant would now be able to run decentralized without the requirement for focal permission.

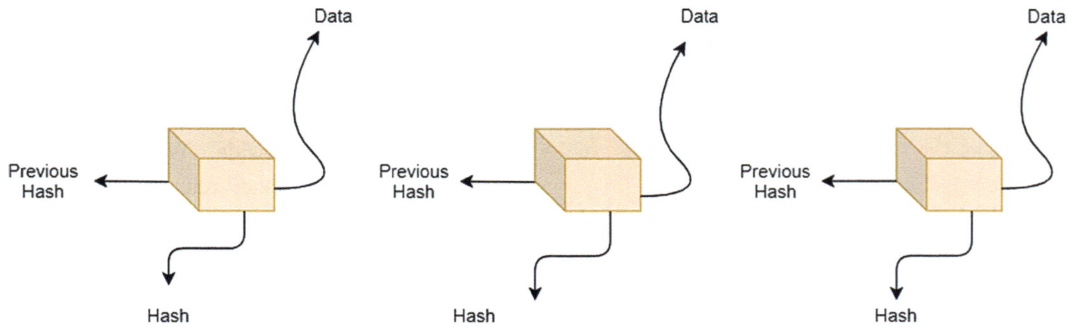

Fig. 8.4 Blockchain working mechanism

These applications would now be able to accomplish similar usefulness with a similar measure of assurance (Christidis and Devetsikiotis 2016). Blockchain gives us an information structure that can be reproduced and divided between individuals to empower secure, straightforward, and decentralized energy exchanging a P2P organization (Tushar et al. 2020).

8.1.3.2 Blockchain Applications

There are plenty of applications of Blockchain since the system is applicable to many sectors. Generally, Blockchain applications are divided into two separate parts. These are defined as non-financial and financial. Separately, we will see what practices in the energy sector can be highlighted regarding our topic.

A. Financial Applications

i. Cross-Border Transactions: Traditionally, getting cash across borders has been moderate and exorbitant on the grounds that frameworks, as a rule, travel through a few banks and clearings while in transit to the last objective of the instalment (Hydrogen 2019).

ii. Trade Finance Platforms: Another Blockchain innovation in the account is to exchange money to watch. Numerous banks use Blockchain to fabricate smart contract members' exchange stages (Hydrogen 2019).

iii. Clearing and Settlements: The precise accountabilities of Blockchain can one day make current clearing and settlement systems excess, bringing about quicker exchanges and decreased expenses for monetary establishments (Hydrogen 2019).

iv. Digital Identity Verification: Using Blockchain-enabled IDs, Blockchain helps banks and other monetary organizations to recognize citizens (Hydrogen 2019).

v. Credit Reporting: Credit reports affect the financials of consumers critically. Credit reporting based on Blockchain is more reliable than conventional server-based reporting (Hydrogen 2019).

B. Non-financial Applications

i. Supply Chain Management: For the situation of inventory network executives, a record should be kept of all products added or eliminated from the stock. Blockchain is the most suitable arrangement for this situation, as it permits the exchange record to be kept as a changeless record.

ii. Healthcare: Blockchain can help make records of patients' treatments and give specialists significant data by bringing the whole data on the web. Since information would be encoded, it is a lot of secure than straightforwardly putting away the information over a distributed storage.

iii. Voting: In this subject, it is a likelihood that the genuine information is messed with or controlled by abusing a weakness in the framework. In actuality, if Blockchain is utilized to deal with similar data, we can make a much more secure system to deal with a similar assignment.

C. Energy Sector Applications
   Blockchain technology has the potential to transform the energy sector.
   i. Grid Management: Blockchains can assist with decentralized organization, the board, versatility administrations or resource the executives. Blockchains can incorporate flexibility exchanging stages and enhance adaptable assets, which can prompt costly organization redesigns (Andoni et al. 2019).
   ii. Smart Grid Applications and Data Transfer: Blockchains have the potential for the use of smart devices for communication, data transmission or storage. (Andoni et al. 2019).
   iii. P2P Trading and Decentralised Energy: In this class, decentralized exchange microgrids, respective exchanges among prosumers and consumers, and business-to-business (B2B) energy exchange are referred to as potential use cases. (Andoni et al. 2019).
   iv. Metering, Billing, and Security: When coordinated with the metering infrastructure, Blockchains give a programmed charging chance to energy administrations for purchasers and dispersed generators. (Andoni et al. 2019).

8.1.4 Smart Contracts

8.1.4.1 Definition and History of Smart Contracts

We have been encountering the term "smart contract" for many years. Firstly, it was proposed by the computer scientist and cryptographer Nick Szabo as "a set of promises, specified in digital form, including the protocols within which the parties perform on these promises" (Szabo 1997). Szabo compared smart contracts with a vending machine. He describes a vending machine as a contract between the vendor and the buyer. The locked case of the vending machine protects the coins and products inside. When the buyer inserts the coin, the device releases the product. Obviously, beyond vending machines, smart contracts offer way more sorts of contracts to be stored digitally (Szabo 1996). Szabo thought that smart contracts could be substantially superior to paper contracts with their functions through the time implementations of verification and cryptography protocols. Nevertheless, they only started to make sense after the emergence of Blockchain technology with distributed ledgers and consensus protocols (Wang et al. 2019). A smart contract is a computer protocol that runs on Blockchain platforms, and they have a significant role in Blockchain. The function of smart contracts is to make sure that the contract made between two or more parties is correctly and safely executed by the consensus protocols. A smart contract can contain any kind of content that is ruled by its programming language (Luu et al. 2016). In Fig. 8.5, smart contracts are compared with traditional contracts.

The first cryptocurrency, Satoshi Nakamoto's Bitcoin, which was introduced in 2008, is the principal of modern smart contracts. Bitcoin, perhaps the most famous one, is good for coin transactions but has limited capability for coding. Five years after Bitcoin's publication, Vitalik Buterin introduced his own Blockchain platform, "Ethereum" which can serve smart contract needs comprehensively. Ethereum can host complex Blockchain applications and smart contracts. Ethereum is the most advanced platform to code one's own contracts. It allows users to create any kind of contract. Moreover, there are several smart contract platforms, and the smart contract adoption of Blockchain platforms is growing up day by day (Rosic 2018), (Kim and Ryu 2020).

Smart contracts are advanced software pieces that can be coded in various programming languages like Solidity, C++ , Golang, Python, JavaScript, and more (Nejjari 2018), (Mistry

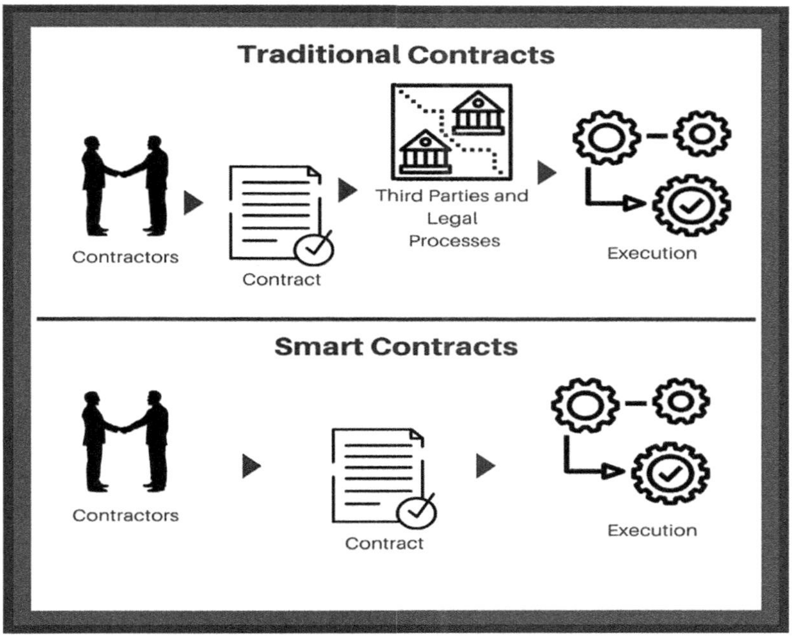

Fig. 8.5 Traditional contracts and smart contracts

et al. 2020). Modifying the software is impossible after being applied. Smart contracts can manage the reliability, trust, and process of the transaction, and they are under the guarantee of the smart contracts (Singh et al. 2019). The development process of smart contracts is at the beginning stages, and soon their use will increase. Many organisations will start to include Blockchain technologies and smart contracts within their operations (Kabra et al. 2020).

8.1.4.2 Benefits of Smart Contracts

The benefits of smart contracts go parallel with Blockchain. Smart contracts offer significant benefits for Peer-to-Peer transactions such as faster speed, lower transaction costs, increased accuracy, and, perhaps most importantly, security. Thus, traditional contracts are being replaced by smart contracts day by day. The nature of the Blockchain brings safety with inalterable ledgers and unique encryption, which makes smart contracts secure. The technology itself ensures the execution and the security of the contracts; therefore, it is possible to say that smart contracts are aspiring to be the substitute of a middleman (DevTeam.Space 2018). The features of smart contracts can be listed as:

A. Speed and accuracy: Fully digital and automated structure of smart contracts prevent time to waste and faults. Also, computer code is more exact than the legalisation process where traditional contracts are written in (Gopie 2018).
B. Trust: Smart contracts ensure that the transactions desired to be performed processed automatically in an encrypted framework with predetermined rules. Thus, it is ensured that no transaction or information is manipulated in line with personal interests (Gopie 2018).
C. Security: Encryption on the Blockchain transactions make smart contracts hard to hack. The data takes place in separate blocks, and the blocks are connected to previous blocks. It is impossible to change a block without changing all blocks (Gopie 2018).
D. Savings: D. Savings: Due to the reliability of an automated smart contract, the need for intermediaries is eliminated. All data is traceable and does not need to be verified by third parties because everything is in code (Gopie 2018).

E. Costs: The business processes require many operational expenses and resources like staff needs and tracking activities. Smart contracts can smoothly perform these processes and eliminate costs (Gopie 2018).
F. Transparency: Smart contracts provide full visibility and accessibility to all contents. Not being able to appeal after the smart contract is established increases the transparency between the parties.
G. Backup: All transactions that are performed are being recorded with essential details. The archive stays accessible every time, and data losses can be quickly returned.
H. Paper Free: The importance of nature is being noticed more by businesses from all over the world. Smart contracts allow the "go-green" movement because they are being operated on only virtual. It removes unnecessary paper usage (Chain Trade 2017). This situation overlaps with environmental policies.
I. Autonomy: The smart contracts are performing automatically on Blockchains. Thus, the manipulation risks posed by the third parties being involved in the process are being avoided.
J. Better Communication: Realising a project by smart contracts obliges setting all conditions detailed before the project starts. Thus, there is no place for mistakes and misunderstandings.

8.1.4.3 Types of Smart Contracts

"Smart contracts can be divided into two broad categories: Smart Legal Contracts and Code-Based. The Code-Based Smart Contracts, which are based on applications, can be further divided into three subtypes—DAO (Decentralised Autonomous Organisations), dApps (Distributed Applications), and IoT-combined contracts" (Grasso 2018).

A. Smart Legal Contracts: "A legally binding, digital agreement in which part or all of the agreement is intended to execute as algorithmic instructions" (Blycha et al. 2020).
B. Decentralised Autonomous Organisations (DAO): "An autonomous company or organisation that builds on a Blockchain in which governance and rules are encoded in the form of smart contracts" (IOSG VC 2020).
C. Distributed Applications (dApps): Decentralised applications; "like normal apps, and offer similar functions, but the key difference is they are run on a Peer-to-Peer network, such as a Blockchain" (Hussey and Chipolina 2020).
D. Smart Contracting Devices: Smart contracts that are combined with IoT devices (Grasso 2018).

8.1.4.4 Use-Cases of Smart Contracts

The usage area of smart contracts is spreading day by day. The applications of Blockchain with smart contracts are being seen everywhere of life and benefits to people: supply chains, eGovernment applications, gaming, gambling, banks, insurance, energy sector, IoT, creative industry, IT services, legal tech, mobility, accounting and auditing, digital identity (Sotnichek and Yatsenko 2018).

The British bank Barclays experiments with smart contracts to trade derivatives (CNBC 2017). OpenBazaar is a decentralised marketplace to buy and sell goods and services without a middleman (OpenBazaar 2019). "Houston, Texas-based Invizion thinks we can do a better job with our trash—by using a Blockchain to trace the waste stream from the home to the dump." (Cava 2020). "Smucker's is partnering with Farmer Connect, a start-up that utilises IBM's Blockchain to tackle traceability in the farm-to-fork journey" (Bourne 2020). "TrustVote is a decentralised e-voting system that is based on Blockchain and smart contracts. The goal of TrustVote is to facilitate managing electronic elections and to minimize humans' errors." (Soud et al. 2020). VANET is a secure message exchange platform for vehicles controlled by smart contracts (Kchaou et al. 2020).

Another application area of smart contracts is P2P energy trading. New promising business models in P2P energy trading emerged with the development of Blockchain technology and smart contracts. However, even though P2P energy trading businesses are promising, it is essential to indicate that these practises are not to replace the existing energy market but to supplement it (Küfeoğlu et al. 2019). Many companies have implemented a P2P business. WePower helps renewable energy producers to raise capital by connecting them with buyers, and they use the smart energy contract tokens along the process (WePower 2019). Some other P2P energy trading products are SunContract, Vandebron, PeerEnergyCloud, SonnenCommunity (Mandela 2019), (Zhang et al. 2018). These projects include the P2P energy trade with smart contracts in case there are any surplus or deficit in the energy produced for the house. The energy traded in the system is made entirely from renewable sources. P2P energy trading takes place in a safe, transparent, and systematic way through smart contracts.

8.1.5 United Nations Development Programme Sustainable Development Goals (SDG)

8.1.5.1 SDG 7 (Affordable and Clean Energy)

Ensuring access for everyone to accessible, reliable, renewable, and modern resources. While the main emphasis is on supplying electric power to the world's whole populations, there is also concern about significantly growing the share of renewable energy in the global energy mix (Silvestre et al. 2018).

8.1.5.2 SDG 9 (Industry, Innovation, and Infrastructure)

Develop resilient infrastructure, promote sustainable industrialisation, and inspire creativity. Given the development of electricity infrastructure, this objective is interpreted as follows: the development of quality, reliable, sustainable, and robust infrastructure, including regional and transborder infrastructure.

8.1.5.3 SDG 11 (Sustainable Cities and Communities)

Make cities inclusive, healthy, sustainable, and resilient. This latter objective includes ensuring access to secure, reliable, efficient, and sustainable transport systems for everyone by 2030 and enhancing road safety, in particular by expanding public transport, with a specific emphasis on the needs of disadvantaged people, women, children, people with disabilities and the elderly (Silvestre et al. 2018).

A growing number of people are moving from rural to urban areas. This trend persists, and 60% of the world's population is projected to live in urban areas by 2030. The critical reason for this vast influx is the smart services that a smart city provides to its inhabitants. By the year 2022, three hundred million smart homes will be built, according to a study. Blockchain-based systems will help immensely to remove threats and to promote a shared advantage for both smart home and smart city services (Moniruzzaman et al. 2020).

8.1.5.4 SDG 12 (Responsible Consumption and Production)

Achieving economic growth and sustainable development demand that we urgently reduce our ecological footprint by improving the way we manufacture and consume goods and services. Agriculture is the largest user of water worldwide, and irrigation now claims up to 70% of all freshwater for human use. To achieve this aim, the effective management of our common natural resources and the way we dispose of hazardous waste and pollutants are critical goals. It is equally important to encourage companies, businesses and customers to recycle and minimize waste, as well as to help developed countries to transition towards more sustainable consumption patterns by 2030 (UNDP 2020).

8.1.5.5 SDG 13 (Climate Action)

To achieve sustainable development, tackling climate change is a highly critical challenge. How to encourage global decarbonisation is the main issue. There are significant concerns such as how to decarbonise the power generation sector amid electrification and what the decarbonisation steps, energy sources, costs, and speed should be (Koyama 2019).

8.1.6 Aim of the Study

The aim of this chapter is to study Blockchain technology, in general, to pave the way for Peer-to-Peer (P2P) energy trading at the Open Digital Innovation Hub (ODIH). This study covers the following:

- Blockchain Architecture: Private Ethereum
- Blockchain-based energy trade mechanism
- A smart contract in which users interact.

When we build these structures, we will ensure that a sustainable platform can buy and sell energy on Blockchain without the involvement of third-parties and intermediaries. Our main goal is to keep efficiency at the highest level and turn ODIH into a structure that is self-sustaining in terms of economy and energy. Also, ODIH has established as a financial entity with peer to peer energy trading.

8.2 Methodology

In this part, we will explain how we can use a Blockchain-based P2P energy trading system in ODIH. To follow a step-by-step approach, we divided this part into three subsections as follows: Software, hardware, and regulations.

8.2.1 Software

P2P energy trading cannot be introduced without a software structure that allows Peer-to-Peer sharing of information and also allows device operators to track and manage the distribution network. Moreover, various trading rules established by the platform have also a major impact on the choices that peers make while trading with other peers.

The aim of our study is to implement a Blockchain developed for the energy sector that provides distributed market coordination for decentralised energy systems. By configuring the Ethereum platform to the micro-grid, sellers and buyers can make the desired contract using the smart contract without brokerage. Smart contracts based on Ethereum written in the Solidity programming language are used as a key element to facilitate market matching between individual electricity consumers and producers.

Distinctive high-level language for composing smart contracts incorporate Serpent (based on Python), Viper (based on Python), LLL (low-level Lisp), Solidity (based on JavaScript). The most unmistakably utilized language is Solidity since it is open source and straightforward (Karthik and Anand 2020). One of the critical highlights of Ethereum is its Turing-complete scripting language, which empowers the meaning of smart contracts. There are small applications that are executed at the top of the whole Blockchain organization. The Ethereum contract code is written in a low-level, stack-based bytecode language, for example, the Ethereum Virtual Machine (EVM) code. Opcodes are an intelligible portrayal of bytecodes. Smart contracts can be composed utilizing high-level languages that are aggregated into EVM bytecodes. Robustness is the most widely recognized language for writing smart contracts. There are two separate compilers accessible to gather Solidity Smart Contracts. One is written in JavaScript and called solc-js, while the other is written in C++ and is called solc (Bistarelli et al. 2020). How a smart contract is created is shown in Fig. 8.6.

Blockchain platforms are of two kinds: permissioned and public. Public Blockchains such as Bitcoin and Ethereum are public to everyone, i.e. anybody may enter the network and do any form of operation, including mining. Contrary to this, approved Blockchains, such as Hyperledger, Quorum, and Corda, regulate network access by specifying rules for authentication and authorisation.

Authentication rules define who is permitted to enter the network, while authorisation rules specify what user activities can be performed, such as Blockchain read/write permissions and miners. Furthermore, permissioned Blockchains can be split into two types: private and consortium Blockchain. The key difference between the two is that a single central miner governs a private Blockchain, while a set of pre-selected

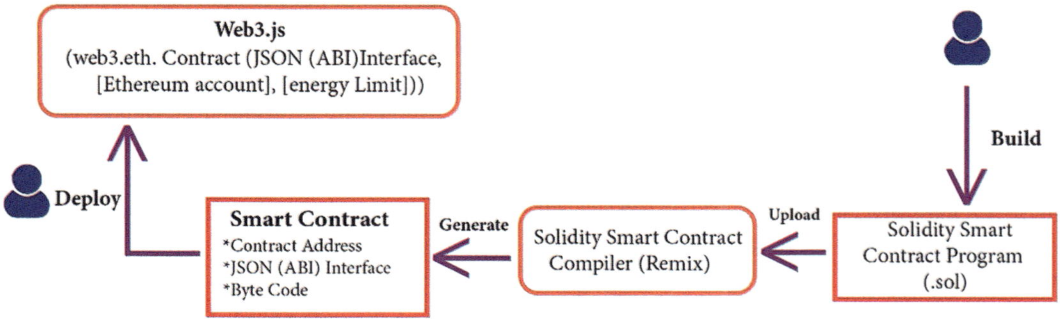

Fig. 8.6 Build smart contract

miners rule the consortium Blockchain (Abdella and Shuaib 2019).

Public Ethereum: Among prosumers and clients, the public Ethereum network permits anybody to look through the blocks, and anybody can trade and check. Yet additionally, those that are not explicitly included are hindering exchanges since they need to confirm and file exchanges. Moreover, examiners need to overhaul the brilliant meter consistently; however, there are various issues because of block check and generation speed. For this reason, public Ethereum isn't fruitful in direct exchanges with clients and examiners. PowerLedger can anyway be utilized to openly trade or store coins for use in open Blockchain contracts (Kim et al. 2018).

Private Ethereum: With a private Blockchain, individuals who are fit for creating power utilizing solar panels structure a Blockchain. Furthermore, the convention can be altered by the agreement of the members, with the goal that it tends to be updated consistently (Kim et al. 2018).

Ethereum, which was released in 2015, is a programmable public Blockchain named Ether with a native cryptocurrency. Although Ethereum 's structure is somewhat close to that of Bitcoin, a major difference is that the most recent state of each smart contract is often stored in Ethereum nodes, in addition to all other Ether transactions. The network needs to keep track of the current details of all of these applications for each Ethereum application, including the balance of each peer, all the smart contract code, and the location where it is all stored. Ether tokens appear in a wallet, much like bank account assets, and can be ported to another account (Tushar et al. 2020). Solidity smart contracts cannot be implemented directly by the Ethereum Virtual Machine (EVM) and are instead compiled into low-level machine instructions, i.e. bytecode. These instructions describe the complete language of Turing and are often expressed by the use of a readable format referred to as opcodes. Smart contracts are executed through Ethereum's distributed Blockchain-based computing platform (Bistarelli et al. 2020).

Solidity is a contract-oriented, high-level programming language with a JavaScript linguistic structure that is statically composed and bolsters legacy. It is intended to run on the Ethereum virtual machine (EVM) that is facilitated on the Ethereum nodes that are associated with the Blockchain. Contracts are worked as classes in object-oriented programming languages by utilizing Solidity to develop a contract. The Solidity code for smart contracts comprises factors and capacities that decipher and adjust them, as in traditional programming. A smart contract is created for P2P energy exchanging between the two companions locally of connected microgrid networks. The smart contract, i.e., created, contains an assortment of viewpoints, including mechanical and business perspectives. Specialized perspectives incorporate the type of supply (AC/DC), the voltage profile, the recurrence, and as far as possible for the boundaries as set out in the IEEE particulars, while business angles incorporate agreement terms and its separation understanding in any way. Essentials are needed prior to proceeding with the smart contract. This will incorporate the aim of the agreement, subtleties of people or substances, terms, and states of the agreement. Smart contracts can be written in different languages for various platforms, as depicted previously. Picking one of them depends on the need for the contract. Ethereum is the system picked for writing a smart contract since it is open source and offers a high level of encryption. Subsequent to choosing a Blockchain platform and programming language, the following move is to construct a smart contract. For example, smart contracts are produced in Solidity utilizing the predefined watchword "Contract."

```
Pragma solidity > = 0.4.22 < 0.6.0;
Contract trading {
// variables are defined;
//functions are defined;
}.
```

After the contract has been created, functions and variables are specified according to the specifications of the contract. Writing a smart contract can be done in Solidity with a method, considering all the requirements, obligations, specifics of the owner (producer/consumer), the technical and business aspects. These are used in the Solidity code using the functions and looping programming logic that performs the comparison and iteration tasks for the initiation of trading in accordance with the P2P trading algorithm (Karthik and Anand 2020).

A smart contract is sent on the Blockchain by methods for an exchange. The gas cap and the cost of gas are regularly identified with the execution of the smart contract. The former is the most extreme measure of gas that the client is eager to pay for the smart contract execution, while the last decides the pace of transformation of gas to the Ether cryptocurrency. Higher change rates would empower miners to pick contracts. Miners can lead a smart bytecode contract locally. Every instruction costs a specific measure of gas; the total sum charged for the contract is similar to the instructions that are given. A smart contract will, in any case, have a particular contract address when it is effectively associated with the Blockchain. Clients can go into a contract by creating an exchange containing the location of the contract, as far as possible, and the agreement technique to be named. The yield of the execution, the energy utilized, and other data is composed into a block. It is likely that the gas appointed to the exchange isn't sufficient to stop the gauge. For this situation, the execution is ended with an exemption, and the condition of the chain is gotten back to its underlying state (before the calculation is begun). Specifically, the execution of the contract should be repaid despite the fact that it has not been effectively finished. This prevents attacks on resource exhaustion (Bistarelli et al. 2020).

A smart contract can be called a tamper-proof program code that enables the safe execution of a predefined logic of a contract between parties who do not trust each other in a manner similar to the position of a trusted third party. Our proposed smart energy contract specifies various attributes in two structures which are *SellOrder* and *BuyOrder*. *SellOrder* consists of the address of the producer, 32 bytes of unsigned integer value *price*, 64 bytes of unsigned integer value *energy,* and *timestamp* value which is also in the type of 64 bytes of an unsigned integer. The other structure, *BuyOrder*, consists of an address type variable named *meterAdress* in addition to the *SellOrder* attributes address, price, energy, and timestamp. Methods of this contract are *sellEnergy* and *buyEnergy* whom can be callable by all prosumers;

- When the *sellEnergy* function is called, it first checks the seller's address and saves it in a variable. Another control is related to the amount of energy. If the energy amount is less than 1 kWh, the function cannot continue to work. Otherwise, it pushes the data it receives as a parameter to the *sellOrders* list in the contract. The *sellOrders* list keeps the generated energy offers created by the seller.
- When the *buyEnergy* function is called, the vendor address is checked first, and this address must be non-zero for the function to

work. Then it is checked whether the offer exists or not, and if the values in the *sellOrder* are correct, the data received as parameters are pushed to the *buyOrders* list as a *buyOrder* struct. Finally, the function checks if the consumer bought from the producer and stores it, and else it reverts all changes. Sample flowcharts of these functions can be seen in Fig. 8.7.

Ethereum is designed as a platform that allows developers to easily create Decentralised Applications (DApps) using Blockchain technology. Ethereum adds a full programming language, Solidity, which is Turing-complete (i.e. any type of smart contract can be created), for developers to set DApps rules by coding them as a smart contract. The smart contract is one form of identity that exists in the Ethereum network, and when other accounts call it, it will store the code in the main net and execute the code automatically. A decentralised application has been developed that can automatically execute smart contracts under the conditions specified in advance to provide fluent transactions (Xu et al. 2018). The following subsections give specifics of these contracts.

Algorithm 1

1: Call for input, requester
2: Check the status of contracts for P2P and P2G
3: Check market time
4: Verify the permission requestor and its status
5: If requester is seller call P2P.seller() else call P2P.buyer()
6: Check market time and call P2P.clearMarket()
7: If there is surplus/deficit energy call P2G.sell ()/P2G.buy ()
8: Get trading result
9: Get billing information (Khalid et al. 2020).

As shown in Algorithm 1, a principle smart contract is shaped to direct all energy exchanging measures to the nearby energy market. Market members draw in with this smart contract straightforwardly. It first tests the client's authenticity and empowers enrolled clients to participate in nearby exchanging activities. It

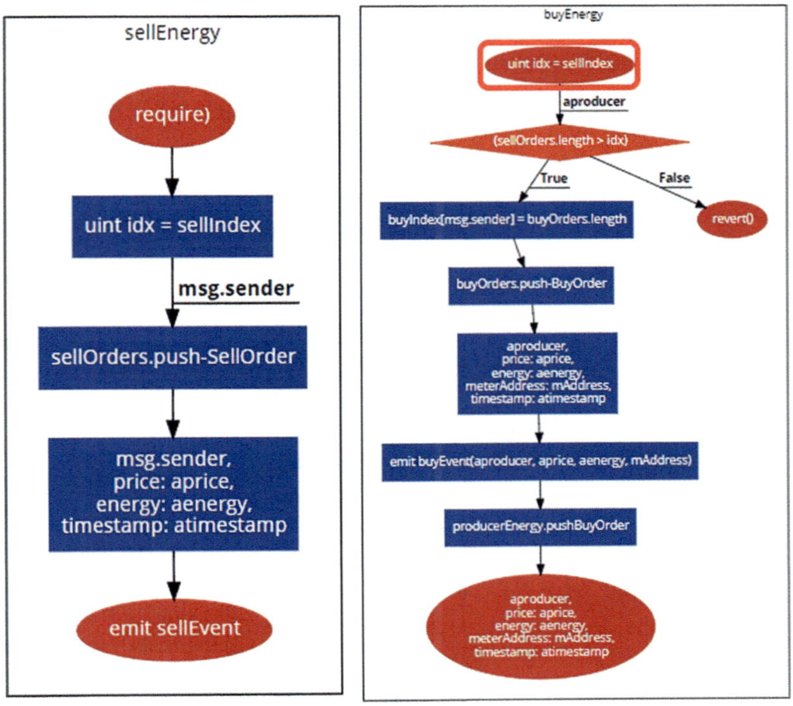

Fig. 8.7 Energy smart contract function flowcharts

tests the status of both P2P and P2G (peer-to-grid) smart contracts when a market member sends an energy excess or shortage request. Since the energy business is a closed auction market, it tests the time and calls the straightforward market highlight of the P2P smart contract after hitting the stamping freedom. Thus, all the offers are coordinated with purchasers from energy vendors, and reports are sent back to the principal contract (Khalid et al. 2020).

Algorithm 2

1: Input request, requester, sellers, buyers
2: function seller () {
3: Store seller in sellers and update price if it is lowest
4:}
5: function buyer () {
6: Store buyer in buyers
7:}
8: function clearMarket() {
9: Trade energy and call matchBid()
10:}
11: function matchBid(){
12: Exchange energy between buyers and sellers
13:} (Khalid et al. 2020)

The P2P smart contract (as demonstrated in Algorithm 2) gets the fundamental contributions from the main contract and the energy client or prosumer-related data. The vender highlight of this smart contract saves important information of the merchants that are utilized along these lines. It likewise audits the dealer's proposed price tag for power and differentiates it from the most minimal selling cost for power previously proposed. At the point when the offering opportunity arrives at an end, the main contract requires a reasonable market highlight. This component screens the tag of the buyer and exchanges energy at the very least distance between power consumers and prosumers to limit power misfortunes and make exchanging more successful (Khalid et al. 2020). For instance, if the dealer declares that the cost per unit of energy is \$0.25, and another vendor reports it as \$0.20. For this situation, the purchaser has two distinct options for buying power from the two of them, evidently from the subsequent dealer, because of a lower cost for every unit of energy. Exchange happens naturally in this manner due to smart contracts (Karthik and Anand 2020).

ODIH's participants in the market are represented with the consumer, and the prosumer is assigned an area tag accordingly. On the off chance that exchanging a similar zone isn't attainable (for example, prosumers of that territory don't have surplus energy), at that point, trading between adjoining regions is supported. Along these lines, this area is cleared, and the outcomes are gotten back to the principle smart contract. At the point when the labels of two members coordinate, the offer component is called to deal with demands. Suppose the consumer has more surplus energy than the matching prosumer. In that case, the consumer will be given energy and excluded from the buyer array, and the position of the consumer will be changed with the remaining surplus energy. On the other hand, if the consumer has fewer resources than the prosumer requires, the prosumer will be removed from the prosumer's energy transfer, and the consumer's power deficit will be updated by the new power deficit (Khalid et al. 2020).

8.2.1.1 Cost of Producing Electricity

In order to adapt quickly to the system for new technologies, they must offer various advantages. The cost advantage of decentralised P2P in energy trading is an important advantage for its rapid adaptation to ODIH.

A. Peak-to-Average Ratio (PAR)

Peak-to-Average Ratio is a significant factor in energy exchanging. It influences the proficiency and dependability of the main grid. The dependability of the main grid is fundamental to guarantee the unwavering quality and manageability of network activities and the board. An abatement in the PAR assists with keeping up the dependability of the utility, accordingly lessening the expense of power. It is mathematically shown in the (1) (Yahaya et al. 2020):

$$PAR = \frac{Maximum(p^{consumption})^2}{Average(p^{consumption})^2} \qquad (8.1)$$

B. Comparison of Production Cost Between Centralised and Decentralised(P2P) Structures

Decentralised P2P energy trading creates an important advantage in terms of cost when calculations are made over current values. In addition to our research, the evaluation we have produced using our own dataset is as follows: When solar energy is used to generate 1 MWh of electricity, an estimated 4.67 TL is required according to ODIH datasets. When we consider only the product without considering the installation and maintenance costs, the energy production unit cost of the lignite thermal power plant corresponds to 6.23 TL on average (Koç and Şenel 2013).

8.2.2 Hardware

The hardware section is divided into two general sections of the P2P energy trading for users. These are determined virtually and physically. In the first part, the answer to how virtual layers are in ODIH is sought. In the second part, how the physical layers are included in ODIH will be examined.

The market participants: The reason for P2P energy exchanging is to build the utilization of renewable energy or lessen reliance on the main grid, impacting the design of pricing plans and the market systems of the exchanging market. Additionally, the type of energy exchanged in the market should be resolved whether the energy is exchanged as electricity, heat, or a blend of both. It will be conceivable to isolate the important market members into two as a virtual layer and a physical layer.

The virtual layer basically provides participants with a protected connection to determine their power buying and selling parameters. It ensures that all members have an indistinguishable right of passage to a virtual platform, in which transfer of all types of data happens, purchase and sell orders are made, the ideal marketplace instrument is utilized to shape the buy and sell orders, lastly, monetary exchanges are accomplished upon effective coordinating of the requests.

On the other hand, the physical layer is largely a physical network that allows the transition of resources from vendors to consumers as soon as each party's financial settlements are completed over the virtual layer platform. This physical network may be the customary appropriated framework network provided and kept up through method of methods for the autonomous framework administrator or an extra, separate physical microgrid conveyance grid related to the conventional grid (Tushar et al. 2018). Figure 8.8 shows the layers of Blockchain stages in peer to peer energy trading.

8.2.2.1 Elements in the Virtual Layer

A. Information system: Most significant thing about a P2P energy network is a high-performing and secured data system. The data system can: permit all marketplace members to speak with one another for teaming up in power exchanging; combine the members inside a suitable commercial platform; supply the members indistinguishable get right of passage to the marketplace; monitor the marketplace activity, and wet guidelines on members' decisions to ensure local area security and dependability. Instances of such data systems consist of Blockchain-based smart contracts and consortium Blockchain (Zhang et al. 2018).

B. Market operation: the information system of a P2P network permits the marketplace operation along with marketplace allocation, value rules, and a plainly portrayed offering design. The fundamental objective of marketplace operation is to permit the members to encounter a productive energy purchasing and selling framework with the guide of utilizing coordinating the advance

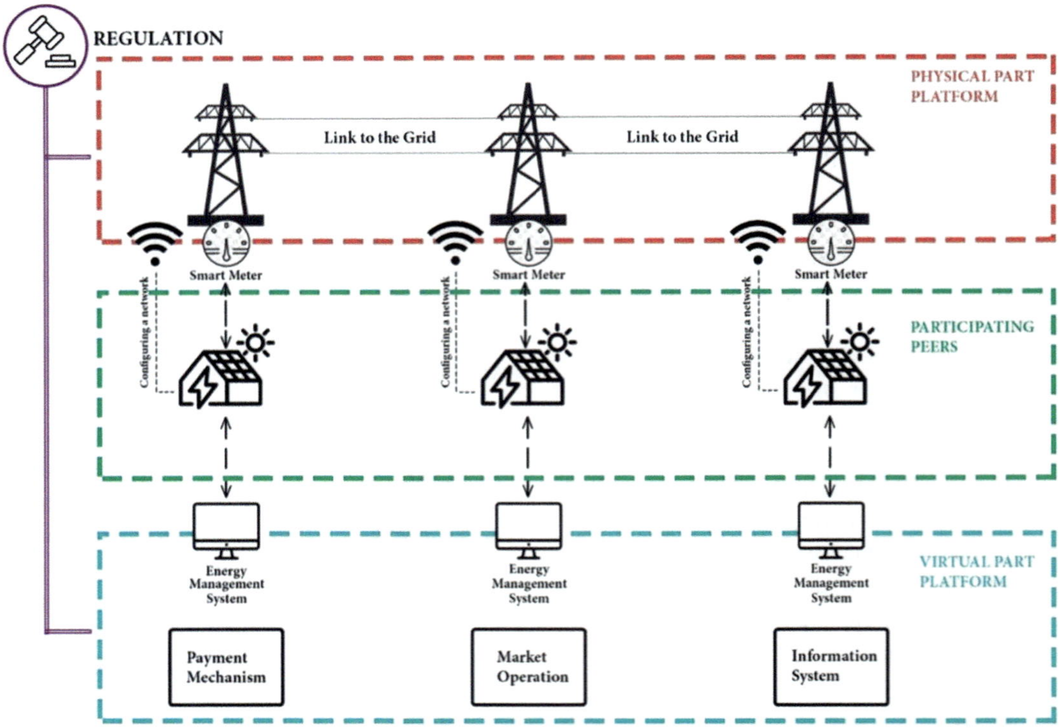

Fig. 8.8 Illustration of the physical and virtual layer platforms of a P2P electricity network (Tushar et al. 2020)

and buy orders in near-continuous granularity. In marketplace operations, the electricity production of each maker impacts the edges of a most and negligible assignment of power. Diverse marketplace-time horizons can likewise moreover exist inside the marketplace activity that should be equipped for creating adequate designation of energy at each degree of activity.

C. Pricing mechanism: Pricing systems are planned as components of marketplace operations and used to successfully adjust among the energy convey and request. Pricing systems utilized for P2P purchasing and selling have an essential distinction with that of the conventional energy markets, for example, in conventional energy markets, a critical piece of energy cost incorporates power overcharges and duties, notwithstanding, as sustainable power sources, for the most part, have low peripheral charges (Mengelkamp et al. 2018), prosumers will gain more income by setting their energy costs accordingly.

D. Energy management system: While partaking in P2P purchasing and selling through a particular offering component, the energy management system (EMS) of a prosumer makes sure about its conveyance of energy. With that in mind, an EMS has got passage to the ongoing organic market information of the prosumer through the transactive meter dependent on which it builds up the generation and admission profile of the prosumer and eventually chooses the offering technique to partake in the purchasing and selling for the benefit of the prosumer. For instance, The EMS of a level-headed prosumer may likewise moreover consistently purchase strength in the microgrid marketplace when the cost per unit of energy falls beneath its most rate threshold (Tushar et al. 2018).

8.2.2.2 Elements in the Physical Layer

A. Grid connection P2P exchanging might be completed for every framework connected and islanded microgrid frameworks. For adjusting the electricity demand and generation in a grid-connected system, it's far crucial to layout the association purposes of the main grid. By interfacing smart meters at those association points, it's far suitable to evaluate the general presentation of the P2P organization, for instance, in expressions of power and cost investment funds (Tushar et al. 2020). For islanded microgrids, then again, individuals ought to have adequate production potential to ensure the ideal degree of security and unwavering quality in providing power to purchasers.
B. Metering: Every prosumer must have a suitable metering infrastructure to have the ability to participate in P2P trading. Especially, every prosumer has to be prepared with a transactive meter, similar to a traditional electricity meter (Tushar et al. 2020). Transactive energy is able to determine whether or not to participate with the P2P marketplace primarily based totally at the demand and generation data in addition to the information to be had approximately marketplace conditions (price, overall demand, overall available generation, and network conditions). It can also communicate with other prosumers inside the community through any appropriate communication protocol. In our project, smart meters will be used as an element that can ensure accuracy during purchase and sale, provided that the production and consumption values on the system are monitored instantly. With relevant material, it will be possible to develop instant monitoring and decision-making options. Thanks to smart meters, we can follow our online account anytime, anywhere to see energy usage. Thus, we can manage our energy usage, and it can help to save money (Tushar et al. 2020).
C. Communication infrastructure: In P2P trading, the primary requirement of conversation is the invention of prosumers and data exchange in the community. A couple of P2P communication architectures exist inside the literature along with established, unstructured, and hybrid architectures (Jogunola et al. 2018). The selection of a communication structure needs to fulfil the overall performance necessities advocated by means of the IEEE 1547. three-2007 for the combination of DER that consist of latency, throughput, reliability, and safety (Jogunola et al. 2018). The information system helps prosumers inside a P2P energy sharing framework to choose energy boundaries by incorporating them to a reasonable market platform with equivalent admittance to every member, screen the market activity, and forcing limitations on prosumers' choices, whenever needed, for the organization security and unwavering quality purposes (Jogunola et al. 2018).

8.2.3 Regulations

There is a growing need for renewable energy all around the world. The reduction in unit costs required for the installation of renewable energy systems enabled these systems to spread. To meet this demand for renewable energy sources, various smart grids were needed (Arslan 2017). The most sustainable solution is to make long-term supply agreements between the consumer and the renewable energy producer. Still, these agreements will raise doubts about whether our electricity supplier can supply us entirely with renewable energy. Blockchain technology is a proven technology in terms of the security of trade data and network management.

Blockchain technology is critical in validating transactions by keeping on reliance and safety without any centralised governance. Blockchain has provided many solutions to several tasks and procedures by different use cases; thus, institutions have completed their digital transformation (Pradeepkumar et al. 2018). Through Blockchain technology, a Blockchain secure energy platform can be created, in which everyone from manufacturers to consumers, from system operators to regulatory organisations is a part (Arslan 2019).

It is a Peer-to-Peer energy platform where every element on the energy chain take part equally and transparently. This application can be realised by smart contracts, but there are legal deficiencies in this regard. The developments on the Blockchain are starting to show the potential of decentralised energy trading grids. Decentralised energy trading solutions have various regulatory issues on control and coordination. "For example, if the government wants to manage the power grids, it is difficult for them to have greater control and more decentralisation. In addition, when using the Blockchain, the price of energy trading should be regulated." (Ali et al. 2020).

All governments should prepare their legal infrastructure to provide to contribute and facilitate the development process of these systems. Some countries' parliaments have been inspecting smart contract reforms for several years (Skinner 2018). In the future smart contracts, which also include qualifications in accordance with contract law, are likely to be legally accepted as a conventional contractual relationship (Çelenk 2017), (BCTR 2019).

The creation of a thriving P2P market is only possible by creating good energy policies and regulations. Governments must create a structure for the market, set taxing and costing rules and how to integrate the P2P market with the existing traditional market structure. Moreover, with the correct implementation of the P2P energy market, ensuring the efficient use of renewable energy can reduce environmental damage.

Extensive P2P use and a large P2P energy market might end up loosening the control of the state over the energy markets. If the policymakers desire to retain their domination and control, this might mean hindering the advancement of P2P trade and related technologies. This is a significant challenge for Blockchain adoption (Ali et al. 2020).

8.2.3.1 Europe's P2P Trading Policies

Regulators ought to provide an equal working area for the platform businesses, usual public institutions, and merchandisers to take advantage of P2P energy trading ideally. EU Directive 2018/2001 of 11 December 2018—Clean Energy Package of legislation that has been published by the European Commission described important progress, P2P trading of renewable energy and its encouragement for the first time (IRENA 2020).

The European Green Deal is the EU's action plan to create a sustainable economy by overcoming environmental and climate challenges. With this action plan, Europe aims to be the first climate-neutral continent in the World by 2050. The European Commission proposed a "European Climate Law" to turn this political commitment and the goals set by the Green Deal into a legal obligation. It suggests a legally binding target of net-zero greenhouse gas emissions by 2050. To reach this target, actions towards the transition to clean energy are of great importance since the production and the use of energy account for more than 75% of the EU's greenhouse gas emissions. 40% of this energy is consumed by buildings (European Commission 2018). Given these data, some of the critical actions within the scope of the Green Deal are decarbonising the energy sector and renovating buildings to help people cut their energy bills and energy use (European Commision 2019).

Decarbonising the energy system is critical to the EU to reach its climate objectives. The European Commission determined the key principles to achieve decarbonisation as; prioritising energy efficiency and develop a power sector based largely on renewable sources having secure and affordable EU energy supply having a fully integrated, interconnected, and digitalised EU energy market (European Commision 2019).

EU Directive 2018/2001 of 11 December 2018 on the promotion of the use of energy from renewable sources is a directive that is part of the so-called "Clean Energy Package". In it, a framework for the promotion of renewable energy sources is established. It sets out rules for topics such as financial support in renewable electricity production and self-consumption for the Member States and their relationship with other countries.

Renewable sources accounted for 9.8% of the EU's energy consumption in 2010 and 17.5% in 2017. Under the Renewable Energy Directive, EU members have binding national targets for increasing the share of renewables in their total energy consumption by 2020. The 2020 target for the whole EU is 20% of the energy coming from renewables (European Commission 2020). The directive sets a binding Union target for the overall share of energy from renewable sources to be at least 32% by 2030 (European Commission 2018):

> Article 3, Binding overall Union target for 2030: Member States shall collectively ensure that the share of energy from renewable sources in the Union's gross final consumption of energy in 2030 is at least 32 %. The Commission shall assess that target with a view to submitting a legislative proposal by 2023 to increase it where there are further substantial costs reductions in the production of renewable energy, where needed to meet the Union's international commitments for decarbonisation, or where a significant decrease in energy consumption in the Union justifies such an increase.
>
> The Commission shall support the high ambition of Member States through an enabling framework comprising the enhanced use of Union funds, including additional funds to facilitate a just transition of carbon intensive regions towards increased shares of renewable energy, in particular financial instruments, especially for the following purposes:
>
> (a) reducing the cost of capital for renewable energy projects;
>
> (b) developing projects and programmes for integrating renewable sources into the energy system, for increasing flexibility of the energy system, for maintaining grid stability and for managing grid congestions;
>
> (c) developing transmission and distribution grid infrastructure, intelligent networks, storage facilities and intercom sections, with the objective of arriving at a 15% electricity interconnection target by 2030, in order to increase the technically feasible and economically affordable level of renewable energy in the electricity system; ... (European Union 2018).

> Article 11, Joint projects between Member States and third countries:
>
> One or more Member States may cooperate with one or more third countries on all types of joint projects with regard to the production of electricity from renewable sources. Such cooperation may involve private operators and shall take place in full respect of international law. (European Union 2018).

As the articles suggest, a clear target has been set, and the Commission bears a supportive role in promoting the renewable energy use of the member states. This supportive role is not only limited to EU borders. It is open for cooperation with third countries.

> Article 4, Support schemes for energy from renewable sources:
>
> In order to reach or exceed the Union target set in Article 3(1), and each Member State's contribution to that target set at a national level for the deployment of renewable energy, Member States may apply support schemes. Support schemes for electricity from renewable sources shall provide incentives for the integration of electricity from renewable sources in the electricity market in a market-based and market-responsive way, while avoiding unnecessary distortions of electricity markets as well as taking into account possible system integration costs and grid stability. Support schemes for electricity from renewable sources shall be designed to maximise the integration of electricity from renewable sources in the electricity market and to ensure that renewable energy producers are responding to market price signals and maximise their market revenues. Member States shall ensure that support for electricity from renewable sources is granted in an open, transparent, competitive, non-discriminatory and cost-effective manner. (European Union 2018).

The importance of renewables, self-consumers, prosumers, and the necessity for establishing a regulatory framework to empower prosumers to generate, consume, and sell electricity without disproportionate burdens is indicated. (European Union 2018):

> Renewables self-consumers should not face discriminatory or disproportionate burdens or costs

> and should not be subject to unjustified charges. Their contribution to the achievement of the climate and energy target and the costs and benefits that they bring about in the wider energy system should be taken into account. …
> With the growing importance of self-consumption of renewable electricity, there is a need for a definition of renewables self-consumers' and of 'jointly acting renewables self-consumers'. It is also necessary to establish a regulatory framework which would empower renewables self-consumers to generate, consume, store, and sell electricity without facing disproportionate burdens. …
> Empowering jointly acting renewables self-consumers also provides opportunities for renewable energy communities to advance energy efficiency at the household level and helps fight energy poverty through reduced consumption and lower supply tariffs. Member States should take appropriate advantage of that opportunity by, inter alia, assessing the possibility to enable participation by households that might otherwise not be able to participate, including vulnerable consumers and tenants. (European Union 2018).

The member states of the European Union have a policy position that supports the individual production and consumption practice mentioned in the relevant paragraph. In this way, it enables the appropriate people to take a role in this new application within certain limits.

> To that end, Member States should as a general principle not apply charges to electricity individually produced and consumed by renewables self-consumers within the same premises. However, in order to prevent that incentive from affecting the financial stability of support schemes for renewable energy, that incentive could be limited to small installations with an electrical capacity of 30 kW or less. In certain cases, Member States should be allowed to apply charges to renewables self-consumers for self-consumed electricity, where they make efficient use of their support schemes and apply non-discriminatory and effective access to their support schemes. Member States should also be able to apply for partial exemptions from charges, levies, or a combination thereof and support, up to the level needed to ensure the economic viability of such projects. (European Union 2018).

The conditions of production and consumption related to renewable energy are examined in the following paragraph, which includes the relevant regulations. At the same time, we see the appropriate cases for various rights.

> Article 21, Renewables self-consumers:
> Member States shall ensure that consumers are entitled to become renewables self-consumers, subject to this Article.
>
> 2. Member States shall ensure that renewables self-consumers, individually or through aggregators, are entitled:
>
> (a) to generate renewable energy, including for their own consumption, store and sell their excess production of renewable electricity, including through renewables power purchase agreements, electricity suppliers and peer-to- peer trading arrangements, without being subject:
>
> (b) in relation to the electricity that they consume from or feed into the grid, to discriminatory or disproportionate procedures and charges, and to network charges that are not cost-reflective;
>
> (c) in relation to their self-generated electricity from renewable sources remaining within their premises, to discriminatory or disproportionate procedures, and to any charges or fees; …
>
> (d) to receive remuneration, including, where applicable, through support schemes, for the self-generated renewable electricity that they feed into the grid, which reflects the market value of that electricity and which may take into account its long-term value to the grid, the environment and society.
>
> The renewables self-consumers installation may be owned by a third party or managed by a third party for installation, operation, including metering and maintenance, provided that the third party remains subject to the renewables self-consumers instructions. The third party itself shall not be considered to be a renewables self-consumer. (European Union 2018).

A key part of the P2P energy trade is the structuring of renewable energy generation and self-consumption. It is vital to establish a balanced structure by reviewing accessibility, economy and removing unfair barriers.

> 6. Member States shall put in place an enabling framework to promote and facilitate the development of renewables self-consumption based on an assessment of the existing unjustified barriers to,

and of the potential of, renewables self- consumption in their territories and energy networks. That enabling framework shall, inter alia:

(a) address accessibility of renewables self-consumption to all final customers, including those in low-income or vulnerable households;

(b) address unjustified barriers to the financing of projects in the market and measures to facilitate access to finance;

(c) address other unjustified regulatory barriers to renewables self-consumption, including for tenants;

(d) address incentives to building owners to create opportunities for renewables self-consumption, including for tenants;

(e) grant renewables self-consumers, for self-generated renewable electricity that they feed into the grid, non-discriminatory access to relevant existing support schemes as well as to all electricity market segments;

(f) ensure that renewables self-consumers contribute in an adequate and balanced way to the overall cost sharing of the system when electricity is fed into the grid. Member States shall include a summary of the policies and measures under the enabling framework and an assessment of their implementation respectively in their integrated national energy and climate plans and progress reports pursuant to Regulation (EU) 2018/1999. (European Union 2018).

Member states are guaranteed to provide a fair working environment for participation in renewable energy communities. Thus, sustainability will be achieved in the production and consumption of renewable energy. The energy produced by the generation units will be shared within the community, and opportunities to access energy markets will be created.

Article 22, Renewable energy communities:
Member States shall ensure that final customers, in particular household customers, are entitled to participate in a renewable energy community while maintaining their rights or obligations as final customers, and without being subject to unjustified or discriminatory conditions or procedures that would prevent their participation in a renewable energy community, provided that for private undertakings, their participation does not constitute their primary commercial or professional activity.

2. Member States shall ensure that renewable energy communities are entitled to:

(a) produce, consume, store and sell renewable energy, including through renewables power purchase agreements;

(b) share, within the renewable energy community, renewable energy that is produced by the production units owned by that renewable energy community, subject to the other requirements laid down in this Article and to maintaining the rights and obligations of the renewable energy community members as customers;

(c) access all suitable energy markets both directly or through aggregation in a non-discriminatory manner. (European Union 2018)

Frameworks will be established by the member states to support the development of renewable energy communities. Unfair legal barriers that prevent the development of these communities will be removed. Operators will act in partnership with communities. It will be transparent in all processes. After the cost calculations are made in a balanced way, they will be presented with transparency. No stakeholder of the system will be discriminated against each other. Accessibility to the system will be kept high, even for people with low income. Access to information will be provided easily.

4. Member States shall provide an enabling framework to promote and facilitate the development of renewable energy communities. That framework shall ensure, inter alia, that:

(a) unjustified regulatory and administrative barriers to renewable energy communities are removed;

(b) renewable energy communities that supply energy or provide aggregation or other commercial energy services are subject to the provisions relevant for such activities; 21.12.2018 L 328/121 Official Journal of the European Union EN

(c) the relevant distribution system operator cooperates with renewable energy communities to facilitate energy transfers within renewable energy communities;

(d) renewable energy communities are subject to fair, proportionate and transparent procedures, including registration and licensing procedures, and cost-reflective network charges, as well as relevant charges, levies and taxes, ensuring that they contribute, in an adequate, fair and balanced way, to the overall cost sharing of the system in line with a transparent cost-benefit analysis of distributed energy sources developed by the national competent authorities;

(e) renewable energy communities are not subject to discriminatory treatment with regard to their activities, rights and obligations as final customers, producers, suppliers, distribution system operators, or as other market participants;

(f) the participation in the renewable energy communities is accessible to all consumers, including those in low-income or vulnerable households;

(g) tools to facilitate access to finance and information are available; (European Union 2018).

The P2P energy trading definition was not previously included in the law. Including it by name will require it to be recognized by states. Also, in the definition of P2P energy trading, the guarantee that market conditions will work independently of any external factors and emphasize neutrality. This is important for the strengthening of P2P structures. With this directive peer to peer trading of renewable energy is defined for the first time by the European Commission as:

> … the sale of renewable energy between market participants by means of a contract with pre-determined conditions governing the automated execution and settlement of the transaction, either directly between market participants or indirectly through a certified third-party market participant, such as an aggregator. The right to conduct Peer-to-Peer trading shall be without prejudice to the rights and obligations of the parties involved as final customers, producers, suppliers or aggregators (European Union 2018).

8.2.3.2 Turkey's Energy Policies

A part of the country's energy strategy as stated by the Turkish Ministry of Foreign Affairs:

> Taking into account increasing energy demand and import dependency, prioritisation among energy supply security-related activities,
> Within the context of sustainable development, giving due consideration to environmental concerns all along the energy chain,
> Increasing efficiency and productivity, establishing transparent and competitive market conditions through reform and liberalisation,
> Expanding research and development activities on energy technologies.
>
> In this scope, the following goals are aimed:
> Increasing the ratio of indigenous and renewable energy sources in our energy mix,
> Increasing energy efficiency. (Ministry of Foreign Affairs 2020).

In addition, Turkey's being a founding member of the International Renewable Energy Agency (IRENA) on 26 January 2009 is evidence of the value given to renewable energy (Ministry of Foreign Affairs 2020). IRENA has important goals regarding P2P energy systems. Therefore, Turkey is also soon can be expected to provide such targets (IRENA 2019).

After the liberalisation of energy markets in the 80s, energy regulation bodies emerged throughout the world. Energy Markets Regulatory Authority (EPDK), the energy regulation authority in Turkey, had established control over the electricity markets but then added to regulation and monitoring of gas and oil markets as well (Güvenek 2009). EPDK, issuing licenses in the electricity market and determining tariffs, has the authority to make technical, legal, and financial evaluations (Koç and Gülşen 2018).

In 2019, a law, "Unlicensed Production Regulation" was introduced to meet the electricity needs of consumers from their generation facilities closest to the consumption point, bring small-scale production facilities to the country's economy in boosting security and reduce the amount of loss in the electricity network by ensuring the effective use of small-scale production resources without getting any production licenses (Gazete 2019). Thus, the use of roof-type solar panels has shown signs of moving towards such an electricity generation system in the future. The amount of installed power in our country in unlicensed electricity generation increases from year to year. In this context, the production made by photovoltaic solar energy (PV) comes first in resource maintenance. Highlighting the use of such systems shows that the system is not that far away from the transition to the P2P model. Despite all the advantages, there are two obstacles to the spread of P2P systems: energy market regulations and costs (Çadırcı and Tekdere 2019).

8.2.3.3 Regulatory Requirements to Apply P2P Trading and Promoting

Such a major transition in the energy system requires intensive effort from the state to the end-users. All elements must work hard to achieve a successful result. Governments should prepare the sector conditions for the new system by making the necessary regulations in the legal system and encourage the sector participants. There are some policies needed to meet regulatory requirements: Some technical efforts should be made to establish distributed power systems, integrate them into the existing network infrastructure, and use the infrastructure efficiently. The dissemination of the results and the benefits of the system should be presented by making simple test applications in the selected pilot areas. The developers of the system should be supported financially and morally so that better results should be provided (IRENA 2020).

Besides, some legal frameworks should be prepared for the implementation of P2P energy trading. Even though the system is decentralised, the state should stay away and facilitate it with laws. There are some regulatory requirements on the market and network sides to be able to apply a distributed energy system: Enabling energy trading between prosumers and consumers without any capacities; Taking preventions to ensure cybersecurity and privacy between peers; Ensuring that the responsibilities and roles of the stakeholders involved in the P2P system are clearly defined; Making sure that consumer rights are respected; Defining market rules about the P2P schemes; Determining ancillary services to be used in case of need; Determining network usage conditions and Ensuring the flexibility of the distribution system operator (IRENA 2020).

Municipalities may be involved in the process of establishing P2P energy exchange systems after the entire legal framework has been established by the country's legal mechanisms and the supervisory structure is established. Municipalities' involvement in the city-based implementation of distributed system projects can facilitate the spread of these systems (Acar et al. 2020).

Guaranteed procurement can be applied in the initial stages of the project implementation. In the model applied in Germany, the guarantee that all renewable energy produced within the scope of the guaranteed purchase tariff will be purchased has encouraged the users. Afterwards, although the guaranteed purchase rates decreased, energy sales continued with the increase in the number of users. Besides, users were encouraged to the system under a similar guaranteed purchase tariff applied in Australia. In a successful incentive model implemented in Denmark, it was ensured that users receive a bonus profit up to the source about how they got electricity (Acar et al. 2020).

Providing appropriate credit opportunities, insurances, and grants to cover the costs of those who want to integrate these types of energy systems into their buildings will also facilitate the spread of the systems.

8.3 Results

Blockchain is a distributed database with unique properties that traditional centralised information systems do not have access to. Immutability, accountability, and decentralisation are supported by Blockchain. It cannot be tampered with until the data is written into the Blockchain. The data are available for the participating parties and maintained in a globally distributed network, making it vulnerable to cyber-attacks.

P2P energy trading can lead to the development of distributed capital on a small scale and the creation of new markets. It is possible to make power production in urban regions to meet the end user's requirements, and it is possible to maximise the use of resources through the cooperative network between producers and consumers.

To make the P2P exchanging of power attainable, we should initially guarantee the organization's reasonability. All plans of action ought to have a structure that benefits both prosumers who sell renewable electricity and purchasers who purchase power from prosumers. Purchasers pay 25 cents/kWh to control generators, which is lower than the power value charged to buyer utilities and more prominent than the pay got by the Feed-in Tariff (FIT) for prosumers, to take a model from the German SonnenCommunity. This business model is feasible when the cost of providing renewable energy or power is lower than the current price of electricity. P2P trading of electricity will continue to grow in the future as the number of places where the electricity brokerage sector is enabled grows and the cost of renewable energy and storage equipment decreases (Xu et al. 2018).

Blockchain types have the function of solving gaps in the existing architecture. Table 8.1 contrasts the features of Blockchain solutions with conventional centralised and distributed architectures and highlights the stability, heterogeneity, and system transparency improvements of the Blockchain. Less central control and notable increases in productivity are also present. Depending on business strategy, payment methods, and demand responses, the discrepancies between the three Blockchain models are summarized.

The Blockchain is a network or device capable of keeping a stable circulated exchange records database. To create the Blockchain for

Table 8.1 Comparison of the energy trade models

Property	Solutions for blockchain			Solutions for non-blockchain	
	Type				
	Infrastructure-based P2P energy trading	Ad hoc P2P energy trading	Trading based on large scale energy storage	Decentralised	Centralised
Assets	Prosumers	Prosumers and grid stations	Prosumers, grid and energy storages	Prosumers, grid and energy storages	Prosumers, grid and energy storages
Single point of failure	No			Yes	
Anonymity of energy profile	Yes			No	
Blockchain	Smart contracts	Smart contracts	Smart contracts—ethereum	–	–
Decentralisation	Comparatively high	Comparatively high	Comparatively high	Comparatively low	Comparatively low
Verification of energy agreement	Consensus between all nodes via blockchain			By distributed consensus	By central authority
Trading of energy	Pure P2P	Hybrid P2P	Hybrid P2P	Hybrid P2P	-
Payment and Incentives	Energy coin or cryptocurrency	Energy coin or cryptocurrency	Energy coin or cryptocurrency	Visa/bank transactions	Visa/bank transactions
Approach to market	Limited to local area	Limited to local city	Global approach	Global approach	Global approach
Trust between trustless parties	Comparatively high	Comparatively high	Comparatively high	Comparatively low	Comparatively low
Answer to demand	Comparatively low	Comparatively low	Comparatively low	By central authority	
Central control	Minimal	Partial	Complete	Partial	Complete

archiving and validating worth trade, an enormous number of nodes are arranged. Under the Blockchain network, without incorporated exchange following and affirmation authority, any sort of significant worth could be moved straightforwardly from distributed. The up and coming age of IoT systems could consequently be planned with a computerized and safe exchange on top of Blockchain. With smooth exchanges and programmed coordination between devices, billions of connected devices could be checked by such frameworks. As all the partaking nodes take an interest in the exchange recording and affirmation, one single point of failure can be eliminated. With a stronger to-assault plot for different kinds of devices to run on, data exchanges will be more solid. Also, mechanized exchanges with completely independent framework firms, diminished expenses, and settlement time are given by the Blockchain-based IoT infrastructure (Xu et al. 2018). The system utilizes the smart contract on the Ethereum Blockchain to deal with access control strategy, data storage, and information stream management. With the smart contract, the utility meter could be followed and the guidelines can be set to control on and off-air conditioners and lights to save energy. An Ethereum private Blockchain is utilized to save all the exchanges of the smart home system. Proprietors in each house are given various degrees of power to control the openness of every proprietor to the system. In any case, this is only a proposed thought, and it isn't executed in a real testbed environment (Tonelli et al. 2018).

Blockchain use in P2P Energy trading is more reliable and more straightforward than traditional energy trading methods. In this research, we introduced an experimental energy exchange scheme to show the likelihood that Blockchain could be a successful solution, as indicated in previous studies. To introduce the energy exchange scheme, we selected a blockchain network called Ethereum. This study suggests using a private Blockchain local energy market in a small group with many sustainable energy systems. As stated in the introduction, users often trade using a basic exchange system to avoid the issue of auctions. A dynamic pricing mechanism and home energy management scheme have been put in motion to optimise economic gains at both community and personal levels. The simulation results show that the proposed solution decreases the cost of electricity and optimizes energy consumption, especially at peak hours. One of the key aims of this analysis is to suggest introducing an effective trading market for hybrid energy while reducing power prices and peak to average ratio (PAR). The outcomes show that our targets, cost, and PAR reduction are successfully achieved, according to the (1) The proposed model successfully decreases the PAR with formulas in (1), as seen in Fig. 8.9. This statistic indicates that the PAR was up to 3.8607 without P2P trading and was reduced to 3.6304. Because of these observations, it is clear that the proposed model of energy trade is favourable to the main grid as it clips the peak and lowers the PAR. On the other hand, for customers also, it is useful (Yahaya et al. 2020).

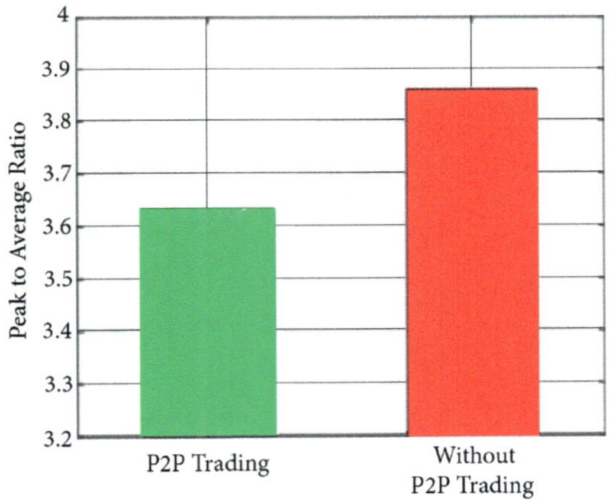

Fig. 8.9 Comparison of PAR for both approaches (Yahaya et al. 2020)

With Critical Peak Price (CPP), the energy costs are diminished by up to 44.73% when the booking algorithm is executed. It brought down the expense of power by up to 65.17% when utilizing the P2P market model and demurrage. Also, it brought energy costs by up down to 51.80% while actualizing the P2P evaluating the model and the peak-hour planning calculation. Also, with the RTP framework, the proposed approach that applies the P2P value model with the de-mining system and the booking calculation diminishes energy costs by up to 44.37% contrasted with 28.55% when the planning calculation is utilized and 35.09% when the P2P value model with de-mining component is actualized at top hours. What's more, the proposed approach accomplishes the greatest energy costs for each schedule opening comparative with utility rates and falls inside the doable district range. What's more, the proposed model adjusts nearby energy interest and family unit age without disregarding purchaser comfort. Furthermore, security weakness investigation is done, and the discoveries show that energy exchanging smart contract is steady and reliable against set-up weaknesses and assaults (Yahaya et al. 2020).

We made a comparison of operating costs without considering criteria such as installation cost and maintenance cost. It is estimated that the total cost will significantly reduce compared to the usage rate of Decentralised structures such as ODIH. For example, if we consider the annual cost of the electricity consumed in Turkey and we calculate the price of electricity consumed per household compared to the amount of electricity produced by the ODIH will emerge a graph as follows. Figure 8.10 shows the average electricity production cost of Turkey with estimated prosumer rates.

The Blockchain consortium is utilized to plan a hybrid P2P energy trade market where electricity producers and consumers transfer power with one another and the fundamental framework without confiding in any outsider. Two smart contracts are intended to implement the

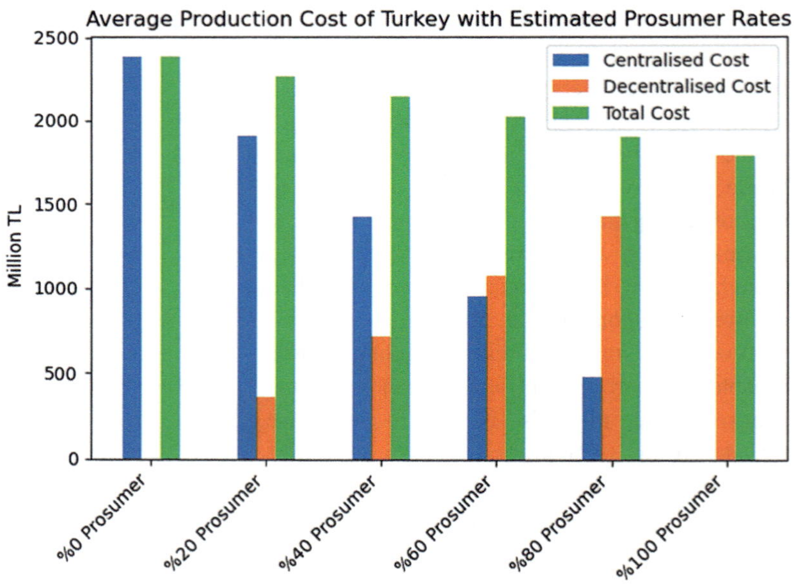

Fig. 8.10 Average electricity production cost of Turkey with estimated prosumer rates

neighbourhood energy market. The primary smart contract is answerable for the enrolment of the members and for the capacity of all essential information identifying with all exchanges. The P2P smart contract is liable for the administration of neighbourhood market exchanges. Recreation is done to confirm the presence of the proposed framework. The reproduction results demonstrate that the objectives, expenses and PAR decreases have been effectively accomplished. Later on, the proposed model will be contrasted and current firmly related models of energy the board to confirm its viability and importance. What's more, the effect of energy trade between contiguous locales will likewise be estimated (Khalid et al. 2020).

8.3.1 Opportunities for P2P Trading of Renewable Energy

- Allow prosumers to have more leverage over excess solar energy and to earn a potentially higher return compared to future fits.
- Connect local public prosecutors and customers in the neighbourhood.
- Theoretically give customers more competitive electricity relative to retail tariffs and clarity on the exact source of energy purchased.
- Current flat tariff systems do not promote P2P trading, although future market changes and the abolition of current guaranteed renewable energy buy-back rates could make trading more viable.
- Involve more households in sustainability problems and move to a cleaner energy mix.

Establishing the P2P energy market correctly will be possible with the correct observation of the energy to be produced and consumed in the system. The estimated electrical energy consumption and production by solar panels of ODIH from the data set can be examined by months in Table 8.2.

Comparison of energy usage amounts during the year is important in terms of balancing the market and creating necessary ancillary services. Failure to foresee fluctuations in production and to certain periods may cause problems in future. Figure 8.11 shows the comparison of estimated monthly energy consumption and production.

As can be observed in the graph, ODIH is self-sustainable in terms of electricity generation with solar energy for eight months a year and is in a position to sell the excess energy to the market.

As can be seen from the graph, ODIH is self-sustainable in terms of electricity generation with solar energy for eight months a year and is in a position to sell the excess energy released in the market.

Data from a bigger number of smart meters and more foundation data on the information would have been ideal for the current research. Future examination, including Blockchain-based Local Energy Markets (LEMs) ought to think about the conceivable expense of expectation mistakes. Besides, as far as anyone is concerned, no simulation of a Blockchain-based LEM with genuine utilization and yield information has been completed. Doing as such on a private Blockchain with an exchanging mechanism coded in a smart contract ought to be the following stage in evaluating conceivable specialized and applied weaknesses.

All in all, past exploration has indicated that Blockchain innovation and smart contracts joined with sustainable power creation can assume a significant part in handling the difficulties of environmental change (Kostmann and Hardle 2019).

Table 8.2 ODIH's energy usage

Months	Consumption (kWh)	Production (kWh)
January	1992.02	874.83
February	1644.62	1238.55
March	13,006.62	2593.95
April	975.62	3060.26
May	1019.886	2855.42
June	1250.886	3429.99
July	1354.886	3984.74
August	1213.886	2641.74
September	958.22	2154.01
October	992.62	1591.91
November	1317.02	950.29
December	1748.02	522.76

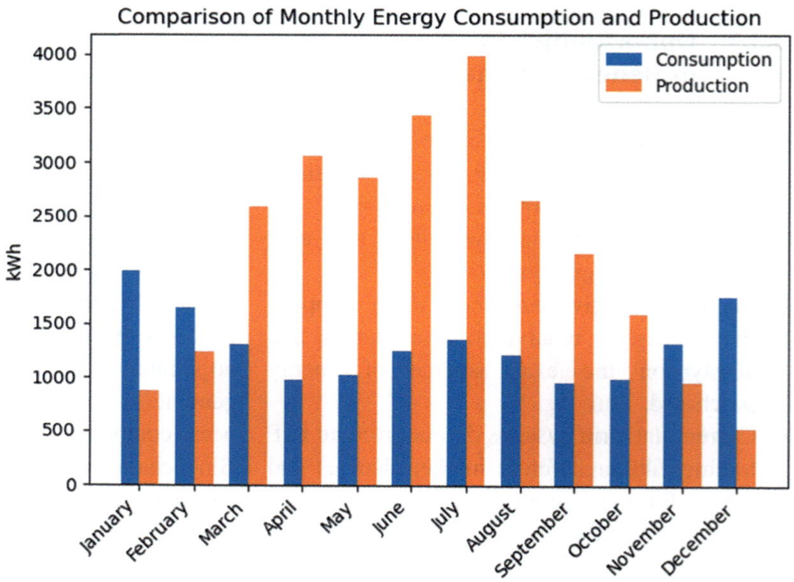

Fig. 8.11 Graph about ODIH's energy usage data

8.4 Discussion

We need to state that the Blockchain technology we use for Peer-to-Peer (P2P) energy trading in our project has a challenge in many aspects.

Most importantly, the advocates of the innovation should show that Blockchain can convey the adaptability, speed, and security needed for proposed use cases. Exploration endeavours on distributed consensus algorithms that are essential to accomplishing these objectives yet to continue. However, an answer that consolidates all the ideal highlights cannot be accomplished without huge trade-offs. Blockchain calculations are developed and secure. Then again, they are moderate and very energy-concentrated. Subsequently, Blockchain designers are progressively going to plans that are energy-proficient, quicker, and more adaptable.

Another significant challenge is that Blockchain frameworks, as of now, have high development costs. Blockchains can give huge expense reserve funds by bypassing merchants, anyway, for different use cases. They might not have the upper hand over arrangements effectively accessible in grounded markets. In the energy business, smart meters are at present being dispatched without huge registering capacities, so coordinating existing smart metering and grid infrastructure with dispersed records can involve critical expenses. Presently, the data in Blockchain frameworks can be moved at extremely low expenses. However, check and confirmation of information accompany high equipment and energy costs.

An extra challenge is that when a Blockchain framework is executed, any progressions to the overseeing conventions should be endorsed by the framework. In Blockchain environments, this has generally prompted clashes among designers and numerous framework forks. In the event that Blockchains are to a great extent embraced in energy systems, these issues can evoke frailty and discontinuity. In addition, Blockchain reception might be hampered by Bitcoin's relationship with reputation and criminal operations from its initial days, yet as Blockchains develop, this viewpoint may turn out to be less captivating over the long haul.

Although the trade in P2P energy could in the future give consumers and prosumers several advantages, the supply of renewable surplus energy is uncertain if the demand of the market is strong. The prosumers of the future may find it challenging to fulfil their own energy need during peak demand periods. On the other hand, state-of-the-art technology could shift from research laboratory to mass market and give rise to a growing number of prosumers who can store and exchange energy surpluses if appropriate. Overall, the profitability of Distributed Energy Resources (DER) trading may be adversely affected by the financial cost of updating and maintaining the centralised grid, unsupportive legislation, and cybersecurity and privacy issues. In addition, to avoid local outages, inadequate grid infrastructure can see regular ‘cut-out’ of exported electricity. As a result, many potential prosumers can withdraw from the central grid and become self-sufficient, leading to higher fixed grid costs for those who remain connected. On the other hand, new developments in grid management technologies may allow energy suppliers to incorporate DER into existing infrastructure effectively, securely, and reliably and encourage P2P trading and other DER monetisation markets. Furthermore, both consumers and prosumers should welcome P2P energy trading with enthusiasm as a trend towards diversification across the energy sector (Wu et al. 2019).

ODIH describes the energy system's contrasting futures for DER and speculates how P2P energy trading will work in those futures. Technological developments in solar PV, batteries, IoT, AI, and autonomous vehicles are already taking place to help consumers and the larger energy system unlock value from DER. The concern would be less about whether or not change will take place in the energy sector as we head towards 2050 but about the type and scale of the transformation. A key enabler for its future economic growth, especially in regional areas, will be the availability of affordable and reliable energy. A cohesive solution would entail future improvements to the energy system. Looking forward, it will be crucial to create partnerships, promote the exchange of information, and build long-term strategic planning and cooperation across governments, sectors, and communities. Strategic foresight resources can be used to promote constructive conversations among different stakeholders, and strategies and policies can be established to effectively manage future improvements for the modern energy system. ODIH represents a positive first step in informing decisions on future goals and investments in distributed energy systems and technologies.

8.4.1 Policy Recommendations

Developing technologies, studies in the field of Peer-to-Peer energy trading, and the needs of the market in this direction point to the necessity of legal regulations. A legal structure must be prepared to create any regularly functioning system with minimal problems. Fulfilling this condition is only possible with a good understanding of the system with its requirements, risks, duration, and possible consequences. Besides, studies abroad on this subject should be followed in detail, and appropriate parts of them should be taken as a sample.

This chapter has been prepared to provide suggestions and participate actively in the regular operation and use of the nascent P2P energy trading systems. There are some policy recommendations to ease the transition to P2P energy trading and promote people.

8.4.1.1 Defining and Legalising P2P Energy Trading

Although there are introductory legal frameworks for the transition to renewable energy and individual energy production, there is no official definition or regulation about P2P energy trading yet. Due to such uncertainties about P2P energy trading, attempts are avoided. It is a compulsory proviso to create a legal structure. Thus, question marks in this area will be eliminated, and more bold and concrete moves can be made. Financial and moral support should be determined to ensure the support to be given to test projects and to provide the necessary funds.

8.4.1.2 Supporting Pilot Studies, P2P System Developers

The lack of sufficient work in this area yet makes it even more necessary to conduct pilot studies such as ODIH. These pilot studies can only be done efficiently within appropriate test environments by providing the necessary facilities and comfort to the developers.

8.4.1.3 Setting an Efficient Smart Contract

Since smart contracts cannot be edited once they are published, the smart contract must be created correctly before the P2P network is being used. In addition, constantly switching between smart contracts to make changes will cause mistrust. Therefore, it is important to create a suitable smart contract and state legal adherence to this contract. This is a compulsory legal regulation, and it will increase the trust in the P2P network.

8.4.1.4 Determining the Responsibilities of System Participants

A free working environment should be provided for the distributed system operator, all peers, and all parties included in the system. Their independence must be legally guaranteed. Any pressure on any party will damage the decentralised structure of the system. It is a compulsory regulation to eliminate any oppressive elements and to prevent the transformation of state control into state administration.

8.4.1.5 Enabling Energy Trading without Any Capacities and Defining Market Rules

Different scenarios will arise where in-system electricity or grid electricity needs to be used in the process. The market rules of such different usage scenarios should be defined clearly and fairly. Under no circumstances should the user be harmed, and users must benefit from the system in a user-friendly way. Users should not be subject to any sanctions for using a P2P system where they can freely choose their energy supply method. P2P trading, which must be guaranteed by law to be done freely. Otherwise, it will cause problems in the later stages. A market whose rules are not determined by law can cause disputes and therefore crashes. In this way, security can be provided in the market.

8.4.1.6 Encouraging Sector Parts to Create P2P Systems and Individuals to Join Networks

Institutions and individuals may be timid in the first stage of transition to P2P systems and may avoid establishing or being involved in these networks for various reasons. There should be some short-term incentive programs and subventions to speed up and encourage transition. Although this is not a mandatory recommendation, the slow transition will affect the system negatively, and a narrow network will cause problems. Therefore, efforts should be made to accelerate the transition. Thus, the number of users of the system will increase, it will become more efficient, and the results will be better observed. There are some applications to encourage ones to applications:

A. Incentives, grants, and insurances: The application of P2P energy systems to buildings requires various costs. Due to a lack of the financial sources of small-scale prosumers, the transition to the application process may be slow. Users can be encouraged to transitioning by providing incentives with various allowance packages. In this direction, practices such as creating suitable credit opportunities, giving various grants, and securing the system by providing appropriate insurance opportunities can be implemented.
B. Guaranteed procurement: Guaranteed procurement programs can be created for a limited time to eliminate financial uncertainties that may occur in the minds of users. In this program, an incentive can be created by stating the minimum profit that the participants will make from the electricity they produce. The minimum profit rate will be covered by the state in case there is no demand in the system.
C. Municipalities support on the implementation of distributed projects: There may be some technical difficulties during the implementation of P2P energy systems to cities. At this point, it may be difficult for the state to directly intervene in every region. The involvement of the municipalities, or the local authorities, in the implementation of the systems, can facilitate these processes. The inclusion of the municipalities in the system can be described as technical consultancy.

8.4.1.7 Providing the Cyber-Security Between Peers and Ensuring Consumer Rights

Although Blockchain provides enough security for energy trading, other parts of the system can have security gaps. Despite some malicious peers and third parties trying to manipulate the system, necessary precautions should be taken, audits should be made, and deterrent punishments should be imposed. It is a compulsory regulation; thus, the security of the system will be ensured at every point.

8.4.1.8 Providing Energy Efficiency Use and Nature Protection

In a sustainable future, it is very important to ensure the continuity of the activities within the built environment with minimal damage to nature. A system whose by-products, such as energy-intensive Bitcoin mining, harming nature might lose its support in the long run. Therefore, the nature-friendliness level of the systems should be measured, and those who create risks should be intervened or discouraged.

8.5 Conclusion

Blockchain can be considered as a very successful technique that can simplify the metering and billing system for the Peer-to-Peer (P2P) energy trading industry. The promising features of Blockchain decentralised ledger technology indicate its feasibility to meet the prevailing challenges to current power structures, as reviewed ODIH. Blockchain provides revolutionary P2P energy trading platforms that allow participants to share excess energy and buy or sell carbon credits. This paradigm shift towards decentralised local energy exchange with P2P will significantly reduce transmission losses and delay costly network upgrades as well. The Blockchain distributed ledger, unlike centralised architectures, removes the intervention of third parties to maintain the system's integrity and protection. Blockchain uses automated technology for smart contracts that enhance cyber protection and optimises energy processes that can reduce transaction costs dramatically. This decentralised architecture which we used in ODIH, along with IoT integration such as advanced metering infrastructure, can give more competitive and productive energy markets that in turn add resilience to the power system network. Overall, Blockchain technology delivers phenomenal services in the energy society, as it facilitates recycling and decarbonisation with better renewable energy sources (RES) management. While the Blockchain appears to be a promising and revolutionary P2P energy trading application, it also presents certain problems that require some consideration. For example, the incorporation of IoT systems introduces smart devices that pose concerns such as privacy leakage and ever-expanding data storage and management volumes. Other problems related to P2P local energy markets include grid defection and the under-use of network assets. In addition, there are also research problems, such as policy and governance. However, Blockchain is a scalable medium and also a fast-moving area of research and development since there has been great success and growth in the energy sector in a short period of time. There are several research projects like ODIH that concentrate on creating and improving Blockchain technologies, shared networks, and ecosystems in energy trading. Other research activities, such as the possible integration of Artificial Intelligence, cloud computing, and Blockchain machine learning algorithms, can help unlock the full disruptive potential of this technology, making it suitable for market reliability and large-scale deployment. There are companies and collaborations that aim to increase such practices in order to increase productivity in the energy sector and improve the economy.

In ODIH, we presented our proposed business architecture design and implementation results for Smart Home System based on the Ethereum Blockchain. We implemented a smart contract using Solidity language to handle and record the transactions between the homeowners and Ethereum miners. ODIH shows that the integration of Blockchain technology in the hybrid P2P electricity market has a positive impact.

Implementations of this kind of P2P smart systems require a legal infrastructure to ensure the healthy and orderly operation of the system. Problems that may occur during the operation of the system can be minimised by doing appropriate legal planning. In the establishment of this legal framework, possible consequences should be predicted, and sample applications should be thoroughly examined and taken as a basis. Regulators have a great responsibility in terms of introducing the necessary laws and regulations.

In summary, ODIH will be a test-bed for promising results in distributed energy systems management for multi-sectoral demand and renewable generation through smart contract and Blockchain schemes.

8.5.1 Future Work

A wide range of transformations, tests, and investigations have been left for the future because of the absence of time and absence of information. Future work concerns a more profound investigation of specific systems, a new proposition to attempt various techniques, or basically interest.

There are some ideas during the development stage of our project which would be implemented in the future. Some of these ideas could not be implemented because of the whole project's situation. For instance, a Peer-to-Peer connection between houses or workplaces has not been implemented yet. Thus, we have not tested our product in real-life. It could be interesting if our project contains these functionalities:

8.5.1.1 Tokenisation

In the project, there are smart contracts working on the trade of Peer-to-Peer energy. These contracts use Ethereum as a payment method. There are some disadvantages of using the main network of Ethereum like; high transaction fees for low-cost operations, unstable exchange rates, and some issues in pricing. Otherwise, tokenisation leads to some advantages like; exchange rate prices can be directly proportional to the energy produced in the system, the value of individual electricity generation may increase due to the investment opportunity, etc.

8.5.1.2 Creating Own Blockchain Ledger and Network

In the current version of the project, our transactions are safe whenever the Ethereum network is safe, but this causes some problems which can be solved in the future, like having to wait for the block in which the transaction was recorded to be mined. It can be solved with a "Private Blockchain Network" which consists of consumers, producers, and prosumers in the energy trading system. Thus, it can be transformed into an easier to manage network. The sample features of this network can mainly be as follows, nodes with different options for consumer and prosumer can be created in the network and the decision mechanism of the processes can be shaped accordingly. On the network that our smart contracts will be connected to, the contract-guaranteed transaction can begin to be carried out quickly, and one or several different transactions can be sent to the Ethereum network later, and the transaction can be easily performed on that network.

8.5.1.3 Blockchain-Based Applications in Smart Cities

Smart cities of the future will harness the digital identity structure. In this way, energy trade will be more transparent and reliable. As the peer to peer energy trading develops, each consumer or institution participating in the system can have a blockchain-based digital identity. We can add the following items to our energy trading system in the future: Blockchain-based identity structure for system part, application interface gamification, carbon credit scoring, historical efficiency analyses and incentives.

References

Abdella JA, Shuaib K (2019) An architecture for blockchain based peer to peer energy trading. In: Sixth international conference on internet of things: systems, management and security, Spain

Acar A, Sarı AC, Taranto Y (2020) Binalarda çatı üstü güneş enerjisi potansiyeli—Türkiye'de çatı üstü güneş enerjisi sistemlerinin hayata geçmesi için finansman modelleri ve politikalar, s.l.: SHURA Enerji Dönüşüm Merkezi

Ali FS, Aloqaily M, Alfandi O, Özkasap Ö (2020) Cyberphysical blockchain-enabled peer-to-peer. Computer 53(9):56–65

Andoni M et al (2019) Blockchain technology in the energy sector: a systematic review of challenges and opportunities. Renewable and sustainable energy reviews, 100(SI: Women in Energy Special Edition: A Publisher's Selection of Highly Cited Articles by Female Scientists from Elsevier's Leading Energy journals), pp 143–174

Arslan C (2017) Enerji Sektöründe Kayıt Zinciri. s.l., Dünya Enerji Konseyi Türk Milli Komitesi

Arslan C (2019) Güvenli enerjinin anahtarı 'blockchain' uygulamaları [Interview] (18 12 2019).

BCTR (2019) Dünyada Blokzinciri Regülasyonları ve Uygulama Örnekleri Karşılaştırma Raporu, s.l.: BCTR.

Bistarelli S et al (2020) Ethereum smart contracts: analysis and statistics of their source code and opcodes. Internet of Things 12 March, 11 (100198):2542–6605

Blycha N, Horrigan B, Garside A, Garcia-Perrote G (2020) Pressure points: smart legal contracts—Shoring up supply chains in a time of crisis (Global). [Online] Available at: https://www.herbertsmithfreehills.com/latest-thinking/covid-19-australia-smart-legal-contracts-shoring-up-supply-chains-in-a-time-of. [Accessed 6 Sep 2020]

Bompolakis S (2020) Applications of the blockchain technology in the energy sector: the case of greek Islands'Microgrid, Athens: department of international and European studies MSc. In: Energy: strategy, law and economics

Bourne J (2020) Smucker's works with farmer connect for blockchain-based coffee transparency. [Online] Available at: https://blockchaintechnology-news.com/2020/07/smuckers-works-with-farmer-connect-for-blockchain-based-coffee-transparency/. [Accessed 5 Nov 2020]

Çadırcı BD, Tekdere M (2019) Lisanssız Elektrik Üretimi Kapsamında Güneş Enerjisinin Önemi. In: Enerji ve Çevre Ekonomisi. s.l.:Ekin Yayınevi, pp 63–87

Cava ND (2020) Invizion aims to use a blockchain track waste. [Online] Available at: https://decrypt.co/45612/invizion-wants-to-track-waste-on-the-blockchain. [Accessed 5 Nov 2020]

Çelenk BV (2017) Akıllı Sözleşmeler (Smart Contracts) Nedir?. [Online] Available at: https://startuphukuku.com/

akilli-sozlesmeler-smart-contracts-nedir/. [Accessed 9 Nov 2020]

Centrica (2018) Centrica and LO3 energy to deploy blockchain technology as part of local energy market trial in cornwall. [Online] Available at: https://www.centrica.com/news/centrica-and-lo3-energy-deploy-blockchain-technology-part-local-energy-market-trial-cornwall. [Accessed 5 Dec 2020]

Chain Trade (2017) 10 Advantages of using smart contracts. [Online] Available at: https://medium.com/@ChainTrade/10-advantages-of-using-smart-contracts-bc29c508691a. [Accessed 3 Nov 2020]

Christidis K, Devetsikiotis M (2016) Blockchains and smart contracts for the internet of things. IEEE Access 4:2292–2303

CNBC (2017) If power start-up drift van make it in New York, it may be lights out for traditional utilities. [Online] Available at: www.cnbc.com/2017/09/08/why-isnt-there-an-apple-or-amazon-of-utility-industry-its-coming.html. [Accessed 5 Dec 2020]

DEEP (2018) The 4Ds transforming the energy market. [Online] Available at: https://medium.com/@d33p/the-4ds-transforming-the-energy-market-1fb61fba385e. [Accessed 2 Nov 2020]

DevTeam.Space (2018) 10 use cases of smart contracts. [Online] Available at: https://www.devteam.space/blog/10-uses-for-smart-contracts/. [Accessed 2020 Nov 8]

European Commision (2018) A clean planet for all-A European strategic long-term vision for a prosperous, modern, competitive and climate neutral economy, s. l.: European Commision

European Commision (2019) Clean energy-The European green deal. [Online] Available at: https://ec.europa.eu/info/strategy/priorities-2019-2024/european-green-deal_en. [Accessed 12 Nov 2020]

European Commission (2020) 2020 climate & energy package. [Çevrimiçi] Available at: https://ec.europa.eu/clima/policies/strategies/2020_en [Erişildi: 10 Nov 2020]

European Union (2018) Directive (EU) 2018/2001 of the European parliament and of the council of 11 December 2018 on the promotion of the use of energy from renewable sources (recast). Official J Eur Union 82(328):82–209

Gates M (2017) Blockchain: ultimate guide to understanding blockchain, bitcoin, cryptocurrencies, smart contracts and the future of money. s.l.:CreateSpace Independent Publishing Platform

Gazete R (2019) Elektrik Piyasasında Lisanssız Elektrik Üretim Yönetmeliği. s.l.:Resmi Gazete

Gopie N (2018) What are smart contracts on blockchain?. [Online] Available at: https://www.ibm.com/blogs/blockchain/2018/07/what-are-smart-contracts-on-blockchain/. [Accessed 6 Nov 2020]

Grasso A (2018) Smart contracts: the business process enablers for blockchain. [Online] Available at: https://www.thedigitalenterprise.com/op-ed/smart-contracts-the-business-process-enablers-for-blockchain/. [Accessed 4 Nov 2020]

Güvenek B (2009) Devletin Regülasyonlar Yoluyla Piyasalara Müdahalesi ve Türkiye Enerji Piyasaları. SÜ İİBF Sosyal Ve Ekon Araştırmalar Dergisi 6(18):46–62

Hu J, Harmsen R, Crijns-Graus W (2017) Developing a method to account for avoided grid losses from decentralized generation: the EU case. Energy Procedia 141(Special Issue: Power and Energy Systems Engineering):604–610

Hussey M, Chipolina S (2020) What are Dapps?. [Online] Available at: https://decrypt.co/resources/dapps. [Accessed 6 Nov 2020]

Hydrogen (2019) 5 common blockchain applications in financial services. [Çevrimiçi] Available at: https://www.hydrogenplatform.com/blog/5-common-blockchain-applications-in-financial-services [Erişildi: 7 Dec 2020]

IEA (2017) Digitalisation and energy. IEA, Paris

Infinite Energy (2020) What is peer-to-peer energy trading?. [Online] Available at: https://www.infiniteenergy.com.au/peer-to-peer-energy-trading/. [Accessed 5 Dec 2020]

IOSG VC (2020) Overview of decentralized autonomous organization (DAO). [Online] Available at: https://medium.com/iosg-ventures/overview-of-decentralized-autonomous-organization-dao-f9ac47051d07. [Accessed 6 Nov 2020]

IRENA (2019) Innovation landscape for a renewable-power future, IRENA, Abu Dhabi

IRENA (2020) Peer-to-Peer electricity trading innovation landscape brief. IRENA, Abu Dhabi

Jogunola O, Ikpehai A, Anoh K, Adebisi B (2018) Comparative analysis of P2P architecture for energy trading and sharing. Energies 11(62):1–21

Kabessa (2017) Why PowerLedger deserves to be a top 10 coin?. [Online] Available at: https://steemit.com/crypto/@kabessa/why-powerledger-deserves-to-be-a-top-10-coin. [Accessed 4 Dec 2020]

Kabra N, Bhattacharya P, Tanwar S, Tyagi AS (2020) MudraChain: Blockchain-based framework for automated cheque clearance in financial institutions. Future Gener Comput Syst 102(SI: Blockchain-as-a-Service for Industrial Internet of Things and Big Data Applications):574–587

Karthik P, Anand R (2020) Energy trading in microgrids using BlockChain technology. In: 4th International conference on intelligent computing and control systems (ICICCS), Madurai

Kchaou A, Ayed S, Abassi R, El Fatmi G (2020) Smart contract-based access control for the vehicular networks. IEEE, Split, Hvar, Croatia, Croatia

Khalid R et al (2020) A blockchain-based decentralized energy management in a P2P trading system. In: ICC 2020—2020 IEEE international conference on communications (ICC). Dublin, Ireland

Kim S, Ryu S (2020) Analysis of blockchain smart contracts: techniques and insights. IEEE, Atlanta, GA, USA

Kim G, Park J, Ryou J (2018) A study on utilization of blockchain for electricity trading in Microgrid. In: 2018 IEEE international conference on big data and smart computing (BigComp), Shanghai

Koç ÖE, Gülşen MA (2018) Elektrik Enerjisi Piyasasında Regülasyon ve Bağımsız Düzenleyici Kurumlar: Türkiye Örneği. Sosyoekonomi 26(38):37–51

Koç E, Şenel MC (2013) Türkiye Enerji Potansiyeli ve Yatırım-Üretim Maliyet Analizi. Termodinamik Dergisi 245:72–84

Kostmann M, Hardle WK (2019) Forecasting in blockchain-based local energy markets. Energies 12 (14)(2718)

Koyama K (2019) "4d Challenges" in electrified society. In: The instute of energy economics. Japan

Küfeoğlu S, Liu G, Anaya K, Pollitt MG (2019) Digitalisation and new business models in energy sector, s.l.: University of Cambrige Energy Policy Research Group

Laurance T (2017) Blockchain for dummies (for dummies (computers)). 2 ed. s.l.: Wiley

Liu Y, Wu L, Li J (2019) Peer-to-peer (P2P) electricity trading in distribution systems of the future. Electricity J 32(4):2–6

Livingston DSVFM, FM (2018) Applying blockchain technology to electric power systems. In: Council on foreign relations: council on foreign relations. Washington DC

Lumenaza (2020) The software for the energy revolution. [Online] Available at: https://www.lumenaza.de/en/. [Accessed 5 Dec 2020]

Luu L et al (2016) Making smart contracts smarter. ACM, Vienna

Mandela WK (2019) Understanding P2P energy trading. [Online] Available at: https://medium.com/swlh/understanding-p2p-energy-trading-a477eb7b55e0. [Accessed 5 Nov 2020]

Mengelkamp E et al (2018) Designing microgrid energy markets: a case study: brooklyn microgrid. Appl Energy 210:870–880

Ministry of Foreign Affairs (2020) Turkey's energy profile and strategy. [Online] Available at: https://www.mfa.gov.tr/turkeys-energy-strategy.en.mfa. [Accessed 13 Nov 2020]

Mistry I, Tanwar S, Tyagi S, Kumar N (2020) Blockchain for 5G-enabled IoT for industrial automation: a systematic review, solutions, and challenges. In: Mechanical systems and signal processing, 135(SI: Special Issue on Industrial Internet of Things for Automotive Industry—New Directions, Challenges and Applications)

Moniruzzaman M, Khezr S, Yassine A, Benlamri R (2020) Blockchain for smart homes: review of current trends and research challenges. In: Department of electrical and computer engineering. Lakehead University, 955 Oliver Road, Thunder Bay ON P7B 5E1, Canada

Nejjari A (2018) Blockchain demystified: smart contracts. [Online] Available at: https://medium.com/the-archer/blockchain-demystified-smart-contracts-c23ab844ef4a . [Accessed 1 Nov 2020]

OpenBazaar (2019) How to sell digital goods for cryptocurrency on OpenBazaar. [Online] Available at: https://openbazaar.org/blog/how-to-sell-digital-goods-for-cryptocurrency-on-openbazaar/. [Accessed 8 Nov 2020]

Pradeepkumar DS, Singi K, Kaulgud V, Podder S (2018) Evaluating complexity and digitizability of regulations and contracts for a blockchain application design. ACM, Gothenburg, Sweden

PwC (2016) Blockchain—an opportunity for energy producers and consumers?, s.l.: PwC

Rosic A (2018) What is ethereum gas?. [Online] Available at: https://blockgeeks.com/guides/ethereum-gas/. [Accessed 8 Nov 2020]

Silvestre MLD, Favuzza S, Sanseverino ER, Zizzo G (2018) How decarbonization, digitalization and decentralization are changing key power infrastructures. Renew Sustain Energy Rev 93:483–498

Singh SK et al (2019) Smart contract-based pool hopping attack prevention for blockchain networks. Symmetry 11(7):941

Skinner C (2018) Research for reforms on blockchain smart contract laws begins in UK. [Online] Available at: https://www.financemagnates.com/cryptocurrency/regulation/research-for-reforms-on-blockchain-smart-contract-laws-begins-in-uk/. [Accessed 9 Nov 2020]

Sonnen (2019) What is the sonnen community?. [Online] Available at: https://sonnengroup.com/sonnencommunity/. [Accessed 5 Dec 2020]

Sotnichek M, Yatsenko M (2018) 5 security tips for writing smart ContractsIt was originally published on https://www.apriorit.com/. [Online] Available at: https://www.apriorit.com/dev-blog/581-security-tips-for-smart-contracts. [Accessed 8 Nov 2020]

Soud M, Helgason S, Hjálmtýsson G, Hamdaqa M (2020) TrustVote: on elections we trust with distributed ledgers and smart contracts. IEEE, Paris

Szabo N (1996) Smart contracts: building blocks for digital markets, s.l.: s.n

Szabo N (1997) The idea of smart contracts, s.l.: s.n

Tonelli R, Pinna A, Baralla G, Ibba S (2018) Ethereum smart contracts as blockchain-oriented microservices. Association for Computing Machinery, New York

Tushar W et al (2018) Transforming energy networks via peer-to-peer energy trading: the potential of game-theoretic approaches. IEEE Signal Process Mag 35 (4):90–111

Tushar W et al (2020) Peer-to-Peer trading in electricity networks: an overview. IEEE Trans Smart Grid 11 (4):3185–3200

UCL (2019) Transactive energy: knowledge sharing with Colombia and the UK. [Online] Available at: https://www.ucl.ac.uk/bartlett/sustainable/news/2019/oct/transactive-energy-knowledge-sharing-colombia-and-uk

UNDP (2020) GOAL 12: Responsible consumption and production. [Online] Available at: https://www.tr.undp.org/content/turkey/en/home/sustainable-development-goals/goal-12-responsible-consumption-and-production.html. [Accessed 17 Dec 2020]

UNFCCC (2020) United nations climate change. [Online] Available at: https://unfccc.int/climate-action/momentum-for-change/ict-solutions/solshare. [Accessed 5 Dec 2020]

Wang B, Alimohammadi N, Liaskos S (2019) Blockchain networks as adaptive systems. IEEE, Montreal
WEF (2017) The future of electricity: new technologies transforming the Grid Edge, s.l.World, Cologny
WePower (2019) WePower whitepaper, s.l.: WePower
Wu W et al (2019) The future of peer-to-peer trading of distributed renewable energy. CSIRO, Brisbane
Xu Q, He Z, Li Z, Xiao M (2018) Building an ethereum-based decentralized smart home system. In: 2018 IEEE 24th international conference on parallel and distributed systems (ICPADS). IEEE, Singapore
Yahaya AS et al. (2020) Blockchain baseed sustainable local energy trading considering home energy management and demmurage mechanism. Sustainability 12(3385)
Zhang C et al (2018) Peer-to-peer energy trading in a microgrid. Appl Energy, 26 May 220(SI: ICAE 2017):1–12

GPSR Compliance

The European Union's (EU) General Product Safety Regulation (GPSR) is a set of rules that requires consumer products to be safe and our obligations to ensure this.

If you have any concerns about our products, you can contact us on ProductSafety@springernature.com

In case Publisher is established outside the EU, the EU authorized representative is:

Springer Nature Customer Service Center GmbH
Europaplatz 3
69115 Heidelberg, Germany

Batch number: 10732801

Printed by Printforce, the Netherlands